高等院校信息技术规划教材

Android
移动应用开发实践教程

仲宝才 颜德彪 刘静 编著

清华大学出版社
北京

内容简介

本书系统地介绍Android应用开发的基本原理、四大组件、数据存储和网络应用，并提供相应的实践引导，主要包括Android基础入门、Activity与Intent、Android UI开发、数据存储与I/O服务与广播、Android多线程和网络编程等内容。

本书附有演示文件(PPT)、习题答案、学习建议等资源，并提供本书所有案例的源代码。

本书可以作为高等院校本、专科计算机相关专业的教材，也可以用作Android移动开发人员的参考书。

图书在版编目(CIP)数据

Android移动应用开发实践教程/仲宝才，颜德彪，刘静编著.—北京：清华大学出版社，2018
(2020.12重印)
(高等院校信息技术规划教材)
ISBN 978-7-302-50532-7

Ⅰ.①A…　Ⅱ.①仲…②颜…③刘…　Ⅲ.①移动终端—应用程序—程序设计—高等学校—教材
Ⅳ.①TN929.53

中国版本图书馆CIP数据核字(2018)第139399号

责任编辑：曾　珊
封面设计：常雪影
责任校对：徐俊伟
责任印制：丛怀宇

出版发行：清华大学出版社
网　　址：http://www.tup.com.cn，http://www.wqbook.com
地　　址：北京清华大学学研大厦A座　　**邮　　编**：100084
社 总 机：010-62770175　　**邮　　购**：010-83470235
投稿与读者服务：010-62776969，c-service@tup.tsinghua.edu.cn
质量反馈：010-62772015，zhiliang@tup.tsinghua.edu.cn
课件下载：http://www.tup.com.cn,010-83470236
印 装 者：三河市金元印装有限公司
经　　销：全国新华书店
开　　本：185mm×260mm　　**印　　张**：30.5　　**字　　数**：722千字
版　　次：2018年9月第1版　　**印　　次**：2020年12月第5次印刷
定　　价：79.00元

产品编号：071694-01

前言 Foreword

Android是Google公司于2007年推出的一款基于Linux自由及开放源代码的嵌入式操作系统，广泛应用于手机、平板电脑、穿戴设备等。自Android问世以来，Android应用开发相关书籍如雨后春笋般出现。近10年，我国市面上的Android教材主要有两类：一类是从欧美国家直接引进的英文原版教材或者中译本；另一类是由国内学者或者拥有丰富Android开发经验的企业工程师参考国外经典教材及Google官方API，结合自身的编程经验而编著的。这些教材纷繁多样，各有千秋，但是都或多或少存在一些问题，如过于详细，追求面面俱到；过于高深，需要读者具有较强的编程功底；内容过于陈旧，落后于知识的更新换代等，这些教材难以适合初学者。基于上述原因，在总结多年教学基础上，我们编撰了本书。

本书特点

1. 立足基础，由浅入深

本书内容立足Android应用开发涉及的基础知识，由浅入深地阐述Activity、Service、BroadcastReceiver和ContentProvider四大图形界面组件，Android网络编程，Android数据存储技术，Android多线程和Android高级编程相关知识。

2. 结构清晰，语言简练

本书面向Android应用开发初学者，内容为Android应用开发过程中的基础知识，共分为9章，每一章都围绕某一具体方面知识进行阐述，没有涉及复杂和高级的内容。全书结构清晰，语言简练。

3. 案例驱动

为了让读者更好地理解相关知识点，本书对每个知识点都有案例加以展示说明。

本书内容

全书共分9章。

第1章　Android入门基础，主要讲述Android系统起源、开发环境搭建、Android应用程序框架以及Android应用开发调试，通过学习本章，读者可以对Android及Android应用开发有初步的了解。

第2章　Activity与Intent，全面讲述Activity的使用和生命周期，对Intent的用法做了详细的阐述。

第3章　Android UI开发，主要讲解Android UI常用布局和常用控件的使用，对话框、菜单、导航栏、Adapter和AdapterView的使用。

第4章　Android数据存储技术，主要讲述Android中五种常用的数据存储方式。

第5章　服务与广播，主要讲述Service的用法、系统服务使用方法和广播接收器的使用。

第6章　Android多线程，主要讲解Android中的多线程以及线程之间的通信机制。

第7章　Android网络编程，主要讲解Android Http通信机制和网络数据解析机制。

第8章　Android高级编程，主要讲解Android多媒体和动画。

第9章　Android综合案例，主要以案例的形式讲述Android应用的开发过程和常用开源框架的使用。

第1、3、6、9章由颜德彪编撰，第2、4、5章由仲宝才编撰，第7、8章由刘静编撰，仲宝才负责全书的审阅和校订工作。

致谢

本书编撰过程中参考了Android官网和相关Android开发书籍，在此向相关作者表示诚挚的谢意。叶江霞、唐凯、赵辉阳、唐佳鑫、高国庆、杨攀、漆愚、胡飞等对于教程案例的编写提供了很大帮助，在此向他们表示谢意。

由于编者水平有限，书中难免存在不妥之处，敬请读者批评指正。

编　者

2018年3月

学习建议 Learning Tips

章　节	知 识 点	重　点	基 本 要 求	建议学时
第 1 章 Android 入门基础	(1) Android 发展历史； (2) Android 体系结构； (3) Android 开发环境搭建	Android 开发环境搭建	(1) 掌握 Android 开发环境搭建； (2) 了解 Android 发展历史； (3) 理解 Android 体系结构	2
第 2 章 Activity 与 Intent	(1) Activity 的创建与注册； (2) Activity 的启动方式； (3) Activity 之间的数据传递； (4) Activity 的生命周期； (5) Activity 的启动模式； (6) Intent 与 IntentFilter	(1) Activity 的创建与注册； (2) Activity 的启动方式； (3) Activity 之间数据传递； (4) Intent 与 IntentFilter	(1) 掌握 Activity 的创建与注册； (2) 掌握 Activity 的启动方式； (3) 掌握 Activity 之间的数据传递； (4) 理解 Activity 的生命周期； (5) 理解 Activity 的启动模式； (6) 掌握 Intent 的使用方式	8
第 3 章 Android UI 开发	(1) Android 常用布局的使用； (2) Android 常用控件的用法及常用的交互策略； (3) Adapter 的使用	(1) Android 常用布局的使用； (2) Android 常用控件的用法及常用的交互策略； (3) Adapter 的使用	(1) 掌握 Android 常用布局的使用； (2) 掌握 Android 常用控件的用法及常用的交互策略； (3) 掌握 Adapter 的使用	12

续表

章　　节	知　识　点	重　　点	基 本 要 求	建议学时
第4章 Android数据存储技术	(1) Android数据存储方式及其特点； (2) SharedPreferences存储数据方式； (3) File存储数据方式； (4) SQLite存储数据方式； (5) ContentProvider的创建与注册； (6) ContentProvider的使用； (7) ContentObserver的使用	(1) SharedPreferences存储数据方式； (2) File存储数据方式； (3) SQLite存储数据方式； (4) ContentProvider的使用	(1) 了解Android数据存储方式及其特点； (2) 掌握SharedPreferences存储数据方式； (3) 掌握File存储数据方式； (4) 掌握SQLite存储数据方式； (5) 掌握ContentProvider的创建与注册； (6) 掌握ContentProvider的使用； (7) 理解ContentObserver的工作原理	10
第5章 服务与广播	(1) Service的创建与注册； (2) Service的启动方式； (3) 常用系统Service的使用； (4) BroadcastReceiver的创建与注册； (5) 普通广播和有序广播的区别； (6) 监听系统广播	(1) Service的创建与注册； (2) Service的启动方式； (3) BroadcastReceiver的创建与注册； (4) 普通广播和有序广播的区别	(1) 掌握Service的创建与注册； (2) 掌握Service的启动； (3) 了解常用系统Service的使用； (4) 掌握BroadcastReceiver的创建与注册； (5) 掌握普通广播和有序广播的区别； (6) 了解如何监听系统广播	8
第6章 Android多线程	(1) Android多线程机制； (2) Handler线程通信机制； (3) AsyncTask	(1) Handler线程通信模型； (2) AsyncTask的使用	(1) 了解Android多线程机制； (2) 掌握Handler线程通信机制； (3) 掌握AsyncTask	4
第7章 Android网络编程	(1) 掌握Android Http的几种方法； (2) 掌握Android TCP Socket通信的基本原理； (3) 掌握XML文件和JSON数据解析的常规方法； (4) 了解WebView和WebService的实现原理	(1) Android Http网络通信； (2) JSON数据解析	(1) 掌握Android Http的几种方法； (2) 掌握Android TCP Socket通信原理； (3) 掌握JSON数据解析； (4) 了解Web Service的实现原理	8

续表

章　节	知 识 点	重　点	基 本 要 求	建议学时
第 8 章 Android 高级编程	(1) 熟悉 Android 多媒体编程； (2) 熟悉图像处理常用工具类； (3) 掌握 Android 动画编程	(1) MediaPlayer 的使用； (2) Bitmap 与 Bitmap-Factory 的使用； (3) Android逐帧、渐变和属性动画	(1) 熟悉 Android 多媒体编程； (2) 熟悉图像处理常用工具类； (3) 掌握 Android 动画编程	8
第 9 章 Android 综合案例	(1) 掌握 Android 应用的开发实现过程； (2) 了解 Android 快速开发框架的使用； (3) 了解 Web 后台服务器的配置与交互	(1) 掌握 Android 应用的开发过程； (2) 常用开源框架的使用	(1) 具备开发 Android 应用程序的能力； (2) 初步具备开发 Android 应用框架的能力	4

目录 Contents

第1章 chapter 1

Android 入门基础

主要内容：Android 系统介绍，Android 环境搭建，Android 应用程序结构，Android 程序调试方法

建议课时：2 课时

知识目标：(1) 了解 Android 系统特性及体系结构；

(2) 掌握 Android 应用开发环境的搭建；

(3) 理解 Android 应用程序结构；

(4) 掌握 Android 应用程序的常规调试方法。

能力目标：(1) 具备搭建 Android 应用开发环境的能力；

(2) 初步具备开发 Android 应用程序的能力。

目前常用的 Android 开发环境有以下三种：

(1) JDK＋Eclipse＋Android SDK＋ADT 插件；

(2) JDK＋ADT Bundle 包；

(3) JDK＋Android Studio。

前两种工具以 Eclipse 为集成开发环境（Integrated Development Environment，IDE)，在早期的 Android 应用开发中采用较多，现在更多地采用 Google 官方推荐的 Android Studio 作为 Android 应用开发的工具。本书中的案例均采用 Android Studio 作为开发环境。

Android 开发环境需要完整 JDK，本书采用 JDK8，读者可以在 Oracle 官网（http://www. oracle. com/technetwork/java/javase/downloads/jdk8-downloads-2133151. html）下载 JDK。图 1-1 为 JDK 8u111 的下载页面，请根据自己计算机的操作系统，选择合适的 JDK 版本，这里选择的是 Windows x64 版本。

下载运行 jdk-8u111-windows-x64. exe 文件，根据安装向导逐步完成安装。本书将其安装在 C:\Program Files\Java\jdk1. 8. 0_111 文件夹中。

接下来配置 Java 环境变量。右击“我的电脑”，在快捷菜单中选择“属性”命令，打开“属性”对话框，选择“高级系统设置”选项卡，点击“环境变量”按钮。在打开的对话框中点击“新建系统变量”按钮，弹出“新建系统变量”对话框，在“变量名”文本框输入 JAVA_HOME，在“变量值”文本框输入 JDK 的安装路径，点击“确定”按钮，如图 1-2 所示。

在“系统变量”选项区域中查看 Path 变量，在 Path 变量的内容起始位置添加“.；%

Java SE Development Kit 8u111

You must accept the Oracle Binary Code License Agreement for Java SE to download this software.

○ Accept License Agreement ◉ Decline License Agreement

Product / File Description	File Size	Download
Linux ARM 32 Hard Float ABI	77.78 MB	jdk-8u111-linux-arm32-vfp-hflt.tar.gz
Linux ARM 64 Hard Float ABI	74.73 MB	jdk-8u111-linux-arm64-vfp-hflt.tar.gz
Linux x86	160.35 MB	jdk-8u111-linux-i586.rpm
Linux x86	175.04 MB	jdk-8u111-linux-i586.tar.gz
Linux x64	158.35 MB	jdk-8u111-linux-x64.rpm
Linux x64	173.04 MB	jdk-8u111-linux-x64.tar.gz
Mac OS X	227.39 MB	jdk-8u111-macosx-x64.dmg
Solaris SPARC 64-bit	131.92 MB	jdk-8u111-solaris-sparcv9.tar.Z
Solaris SPARC 64-bit	93.02 MB	jdk-8u111-solaris-sparcv9.tar.gz
Solaris x64	140.38 MB	jdk-8u111-solaris-x64.tar.Z
Solaris x64	96.82 MB	jdk-8u111-solaris-x64.tar.gz
Windows x86	189.22 MB	jdk-8u111-windows-i586.exe
Windows x64	194.64 MB	jdk-8u111-windows-x64.exe

图 1-1 JDK 8u111 的下载页面

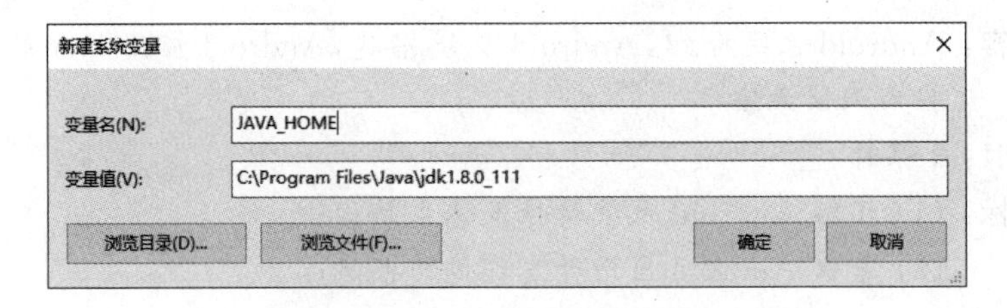

图 1-2 Java 环境变量配置图

JAVA_HOME%\bin;%JAVA_HOME%\jre\bin;”。新建变量 CLASSPATH，设置变量值为“.;%JAVA_HOME%\lib\dt.jar;%JAVA_HOME%\lib\tools.jar;”。

1.1 Android 系统介绍

Android 是 Google 公司基于 Linux 平台开发的手机及平板电脑的操作系统。自推出以来，备受关注，并成为移动平台最受欢迎的操作系统之一。

1.1.1 Android 平台特性

Android 平台具有如下特性：

(1) 应用程序框架支持组件的重用与替换。这样我们就可以把系统中不喜欢的应用程序删除，安装喜欢的应用程序。

(2) Dalvik 虚拟机专门为移动设备进行了优化。Android 应用程序将由 Java 编写、编译的类文件通过 DX 工具转换成一种后缀名为.dex 的文件来执行。Dalvik 虚拟机是基于寄存器的，相对于 Java 虚拟机速度要快很多。从 Android 5.0 开始，Dalvik 虚拟机已由 ART 虚拟机替代。

(3) 内部集成浏览器基于开源的 WebKit 引擎。有了内置的浏览器，意味着 WAP 应用的时代即将结束，真正的移动互联网时代已经来临，手机就是一台“小电脑”，可以在网络信息海洋中随意遨游。

(4) 优化的图形库。包括 2D 和 3D 图形库，3D 图形库基于 OpenGL ES 1.0，强大的图形库给游戏开发带来福音。

(5) 多媒体支持。包括常见的音频、视频和静态图像文件格式，如MPEG4、H.264、MP3、AAC、AMR、JGP、PNG、GIF。

(7) 提供对GSM电话(依赖于硬件)、蓝牙(Bluetooth)、EDGE、3G、4G、Wi-Fi(依赖于硬件)的支持。

(8) 提供对照相机、GPS、指南针和加速度计(依赖于硬件)等的支持。

(9) 丰富的开发环境，包括SDK、大量的类库、设备模拟器、调试工具、内存及性能分析图表插件。

1.1.2 Android系统版本

Android在正式发行之前拥有两个以著名的机器人命名的内部测试版本，分别是铁臂阿童木(Astro，Android 1.0)和发条机器人(Bender，Android 1.1)。由于涉及版权问题，Android 1.5发布时，Google将Android系统命名规则变更为用甜点作为系统版本的代号。随着Android系统版本的更新，作为版本代号的甜点按照英文字母顺序依次为：纸杯蛋糕(Cupcake Android 1.5)，甜甜圈(Donut, Android 1.6)，松饼(Eclair，Android 2.0/2.1)，冻酸奶(Froyo, Android 2.2)，姜饼(Gingerbread, Android 2.3)，蜂巢(Honeycomb, Android 3.0)，冰淇淋三明治(Ice Cream Sandwich，Android 4.0)，果冻豆(Jelly Bean，Android 4.1和Android 4.2)，奇巧巧克力(KitKat，Android 4.4)，棒棒糖(Lollipop，Android 5.0)，棉花糖(Marshmallow，Android 6.0)，牛轧糖(Nougat，Android 7.0)，奥利奥(Oreo，Android8.0)，见表1-1。

表1-1 Android系统版本

Android版本号	API级别	版本主要特性
Android 1.0Astro(铁臂阿童木)	API Level 1	网页浏览器，照相机支持； 支持E-mail传输，Google相关应用； 多媒体播放器，通知、声音识别器； 支持Wi-Fi和蓝牙等
Android 1.1Bender(发条机器人)	API Level 2	用户搜索企业和其他服务时，下方会显示出其他用户搜索时对该搜索信息的评价和留言； 加强了电话功能，改进了免提功能； 支持对邮件附件的保存和预览功能； 增加了长按任意界面弹出多选框的功能
Android 1.5Cupcake(纸杯蛋糕)	API Level 3	拍摄/播放影片，并支持上传到Youtube； 支持立体声蓝牙耳机，同时改善自动配对性能； 最新的采用WebKit技术的浏览器，支持复制/粘贴和页面中搜索； 提供屏幕虚拟键盘； 主屏幕增加音乐播放器和相框widgets； 应用程序自动随着手机旋转； 相机启动速度加快，拍摄图片可以直接上传到Picasa； 来电照片显示

续表

Android 版本号	API 级别	版本主要特性
Android 1.6Donut(甜甜圈)	API Level 4	重新设计的 Android Market 手势； 支持 CDMA 网络； 文字转语音系统(Text-to-Speech)； 快速搜索框； 查看应用程序耗电； 支持虚拟私人网络(VPN)； 支持更多的屏幕分辨率； 支持 OpenCore2 媒体引擎； 新增面向视觉或听觉困难人群的易用性插件
Android 2.0/2.0.1/2.1Eclair(松饼)	API Level 5 API Level 6 API Level 7	优化硬件速度； Car Home 程序； 支持更多的屏幕分辨率； 改良的用户界面，新的浏览器的用户接口，支持 HTML5； 新的联系人名单； 更好的白色/黑色背景比率； 改进的 Google Maps 3.1.2； 支持 Microsoft Exchange； 支持内置相机闪光灯； 支持数码变焦； 改进的虚拟键盘； 支持蓝牙 2.1； 支持动态桌面的设计
Android 2.2/2.2.1Froyo(冻酸奶)	API Level 8	整体性能大幅度提升； 3G 网络共享功能； Flash 的支持； App2sd 功能； 全新的软件商店； 更多的 Web 应用 API 接口的开发
Android 2.3Gingerbread(姜饼)	Android 2.3-2.3.2 API Level 9 Android 2.3.3-2.3.7 API Level 10	增加了新的垃圾回收和优化处理事件； 原生代码可直接存取输入和感应器事件、EGL/OpenGL ES、OpenSL ES； 新的管理窗口和生命周期的框架； 支持 VP8 和 WebM 视频格式，提供 AAC 和 AMR 宽频编码，提供了新的音频效果器； 支持前置摄像头、SIP/VOIP 和 NFC(近场通信)

续表

Android 版本号	API 级别	版本主要特性
Android 3.0/3.1/3.2Honeycomb（蜂巢）	API Level 11 API Level 12 API Level 13	全新设计的 UI 增强网页浏览功能； 允许用户随意访问自己的文件管理器； 经过优化的 Gmail 电子邮箱； 全面支持 Google Maps； 对 widget 支持的变化，能更加容易地定制屏幕 widget 插件； 任务管理器可滚动，支持 USB 输入设备（键盘、鼠标等）； 支持 Google TV； 支持 XBox 360 无线手柄； 支持 7 英寸设备； 引入了应用显示缩放功能
Android 4.0 Ice Cream Sandwich（冰淇淋三明治）	Android 4.0-4.0.2 API Level 14 Android 4.0.3-4.0.4 API Level 15	全新的 Chrome Lite 浏览器，有离线阅读、16 标签页、隐身浏览模式等； 截图功能； 更强大的图片编辑功能； 自带照片应用堪比 Instagram，可以加滤镜、加相框，进行 360°全景拍摄，照片还能根据地点来排序； Gmail 加入手势、离线搜索功能，UI 更强大； 新功能 People：以联系人照片为核心，界面偏重滑动而非点击，集成了 Twitter、Linkedin、Google＋等通信工具
Android 4.1/4.2./4.3Jelly Bean（果冻豆）	API Level 16 API Level 17 API Level 18	特效动画的帧速提高至 60fps，增加了三倍缓冲； 增强通知栏； 全新搜索；搜索将会带来全新的 UI、智能语音搜索和 Google Now 三项新功能； 桌面插件自动调整大小； 加强无障碍操作； 语言和输入法扩展；新的输入类型和功能； 新的连接类型； Photo Sphere 全景拍照功能； 键盘手势输入功能； 改进锁屏功能，包括锁屏状态下支持桌面挂件和直接打开照相功能等； 可扩展通知，允许用户直接打开应用； Gmail 邮件可缩放显示； Daydream 屏幕保护程序

续表

Android 版本号	API 级别	版本主要特性
Android 4.4KitKat(奇巧巧克力)	API Level 19 API Level 20	支持语音打开 Google Now; 在阅读电子书、玩游戏、看电影时支持全屏模式(Immersive Mode); 新的电话通信功能; 旧有的 SMS 应用程序集成至新版本的 Hangouts 应用程序; Emoji Keyboard 集成至 Google 本地键盘; 支持 Google Cloud Print 服务,让用户可以利用户内或办公室中连接至 Cloud Print 的打印机,打印出文件; 支持第三方 Office 应用程序直接打开及存储 Google Drive 内的文件,实时同步更新文件; 支持低电耗音乐播放; 全新的原生计步器; 全新的 NFC 付费集成; 全新的非 Java 虚拟机运行环境 ART(Android Runtime); 支持 Message Access Profile(MAP); 支持 Chromecast 及新的 Chrome 功能; 支持隐闭字幕
Android 5.0/5.1Lollipop(棒棒糖)	Android 5.0-5.0.2 API Level 21 API Level 22	使用一种新的 Material Design 设计风格; 改良的通知界面及新增优先模式; 预载省电及充电预测功能; 新增自动内容加密功能; 新增多人设备分享功能,可在其他设备登录自己账号,并获取用户的联系人、日历等 Google 云数据; 强化网络及传输连接性,包括 Wi-Fi、蓝牙及 NFC; 强化多媒体功能,例如支持 RAW 格式拍摄
Android6.0Marshmallow(棉花糖)	API Level 23	新系统的整体设计风格依然保持扁平化的 MeterialDesign 风格; 新增运行时权限概念,开发者需要手动请求系统授予权限; 新增瞌睡模式和待机模式; 选择文本时,会在文本附近弹出悬浮框,悬浮框中会有类似"剪切""复制""粘贴"的选项; 对软件体验与运行性能进行了大幅度的优化

续表

Android 版本号	API 级别	版本主要特性
Android7.0/7.1Nougat(牛轧糖)	API Level 24	支持分屏多任务； 全新下拉快捷开关页； 通知消息快捷回复； 通知消息归拢； 夜间模式； 流量保护模式； 全新设置样式； 改进的 Doze 休眠机制； 系统级电话黑名单功能； 菜单键快速应用切换
Android8.0Oleo(奥利奥)	API Level 26	借助通知渠道对通知进行精细化管理； 支持 Activity 中的画中画模式； 固定快捷方式和小部件——Pinning shortcuts； 限制后台 APP 活动； 可调整的图标

1.1.3 Android 体系架构

Android 系统采用分层的架构，如图 1-3 所示。从架构图看，Android 分为四个层，从高层到低层分别是应用程序层、应用程序框架层、系统运行库层和 Linux 内核层。

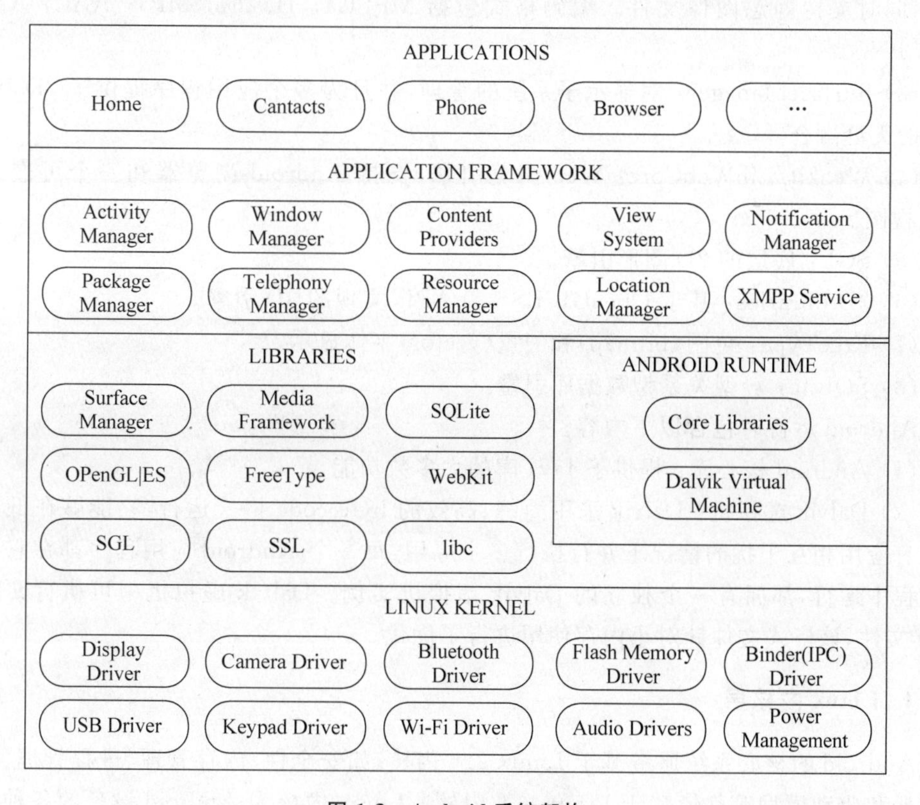

图 1-3 Android 系统架构

1. 应用程序层

Google最开始时就在Android系统中捆绑了一些核心应用程序,同Android系统一起发布,如Email客户端、SMS短消息程序、日历、地图、浏览器、联系人管理程序等。应用是用Java语言编写的运行在虚拟机上的程序,开发者可以利用Java语言设计和编写属于自己的应用程序。

2. 应用程序框架层

应用程序框架层向开发人员提供构建应用程序时用到的各种API。Android自带的核心应用就是使用这些API完成的,开发者可以利用这些API开发自己的应用程序。

3. 系统运行库层

Android包含一些C/C++库,这些库被Android系统中不同的组件使用。它们通过Android应用程序框架为开发者提供服务。核心库包含以下内容。

(1) 系统C库(Libc):从BSD继承来的标准C系统函数库Libc,专门为基于embedded Linux的设备定制。

(2) 媒体库:基于PacketVideo OpenCORE,支持多种常用的音频、视频格式回放和录制,同时支持静态图像文件。编码格式包括MPEG4, H.264, MP3, AAC, AMR, JPG, PNG。

(3) SurfaceManager:对显示子系统的管理,并且为多个应用程序提供了2D和3D图层的无缝融合。

(4) Webkit/LibWebCore:Web浏览引擎,支持Android浏览器和一个可嵌入的Web视图。

(5) SGL:底层的2D图形引擎。

(6) 3D libraries:基于OpenGL ES 1.0 APIs实现的3D引擎。

(7) FreeType:位图(bitmap)和矢量(vector)字体显示。

(8) SQLite:轻型关系型数据库引擎。

Android运行时包含以下内容。

(1) Android核心库:提供了Java库的大多数功能。

(2) Dalvik虚拟机:Dalvik采用简练、高效的byte code格式运行,它能够在低资耗和没有应用相互干扰的情况下并行执行多个应用,每一个Android应用程序都在它自己的进程中运行,都拥有一个独立的Dalvik虚拟机实例。Dalvik虚拟机中可执行文件为.dex文件,该格式文件针对小内存使用进行了优化。

4. Linux内核层

Android的核心系统服务基于Linux 2.6内核,如安全性、内存管理、进程管理、网络协议栈和驱动模型等都依赖于Linux 2.6内核。Linux内核为Android设备的各种硬件提供驱动程序，例如显示驱动、照相机驱动、音频驱动、蓝牙驱动、Wi-Fi 驱动、Binder IPC 驱动和 Power Management 等。

1.1.4 Android 四大组件

Android 开发四大组件分别是：①Activity（活动），用于表现功能；②Service（服务），后台运行服务，不提供界面呈现；③BroadcastReceiver（广播接收器），用于接收广播；④ContentProvider（内容提供器），支持在多个应用中存储和读取数据，相当于数据库。

1. Activity

Activity（活动）是 Android 中最基本和最常用的组件，是一个负责与用户进行交互的组件，Activity 中所有操作都与用户密切相关。一般认为一个 Activity 就是一个单独的屏幕，上面可以显示一些控件与用户进行交互，监听并处理用户的操作事件并做出响应。关于 Activity 更加深入详细的内容将在第 2 章讲解。

2. Service

Service（服务）是一个没有用户界面，可以在后台运行执行耗时操作的应用组件。其他应用组件能够启动 Service，并且当用户切换到其他的应用场景时，Service 将持续在后台运行。Service 在很多应用场景中被使用，比如播放音乐、检测 SD 卡上文件的变化和记录地理信息位置的改变等。关于 Service 更加深入详细的内容将在第 5 章讲解。

3. BroadcastReceiver

在 Android 中，Broadcast（广播）是一种广泛运用的在应用程序之间传输信息的机制。BroadcastReceiver（广播接收器）是对发送出来的 Broadcast 进行过滤接收并响应的一类组件。使用 BroadcastReceiver 可以方便实现全局监听，完成不同组件之间的通信。例如来电话了、来短信了、手机没电了等系统发送的消息，都是以广播的形式通知应用程序。关于 BroadcastReceiver 更加深入详细的内容将在第 5 章讲解。

4. ContentProvider

ContentProvider（内容提供器）是 Android 提供的第三方应用数据的访问方案。在 Android 中，对数据的保护是很严密的，除了放在 SD 卡中的数据，一个应用所持有的数据库、文件等内容，都不允许其他应用直接访问。Andorid 当然不会真的把每个应用都做成一座孤岛，它为所有应用都准备了一扇窗，这就是 ContentProvider。Android 平台提供了 ContentProvider，使一个应用程序可以把指定的数据集提供给其他应用程序，其他应用程序通过 ContentResolver 从该内容提供器中获取或存入数据。ContentProvider 支持多个应用程序的数据共享，是跨应用共享数据的唯一方法。关于 ContentProvider 更加深入详细的内容将在第 4 章讲解。

1.2 Android开发环境搭建

Android Studio 是一个基于 IntelliJ IDEA 的 Android 集成开发工具，它用于Android应用程序的开发和调试。我们可以在 Google 的 Android 中国开发者社区(https://developer. android. google. cn/studio/index. html)官网，下载所需要的 Android Studio 版本。这里以 Android Studio 2. 2 正式版本为例。

运行 android-studio-bundle-145. 3276617-windows. exe，进入 Android Studio 安装向导，如图 1-4 所示。

图 1-4 Android Studio 安装向导(一)

点击 Next 按钮后，跳转至图 1-5 所示的 Choose Components(组件选择)对话框，可在其中选择需要安装的组件。这里选择安装 Android SDK 和 Android Virtual Device。

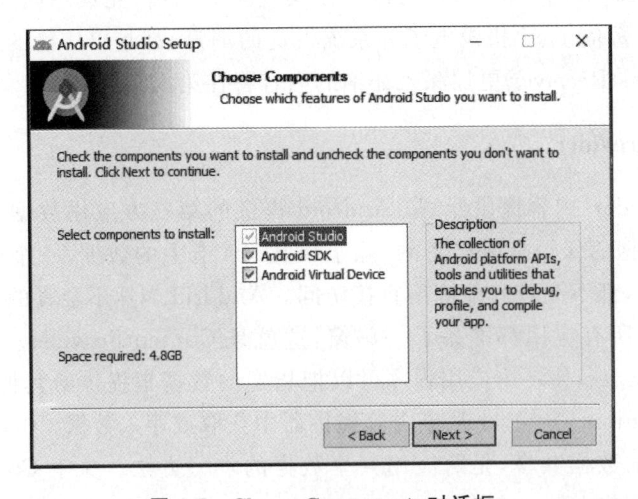

图 1-5 Choose Components 对话框

点击 Next 按钮，接受 License，选择 I Agree，设置 Android Studio 的安装路径和 Android SDK 的安装路径，如图 1-6 所示。

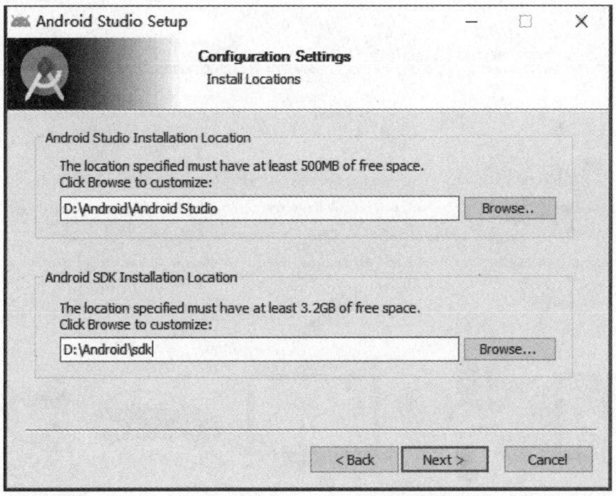

图 1-6 Configuration Settings 对话框

点击 Next 按钮，再点击 Install 按钮，完成 Android Studio 初步安装，如图 1-7 所示，勾选 Start Android Studio，点击 Finish 按钮，启动 Android Studio，如图 1-8 所示。

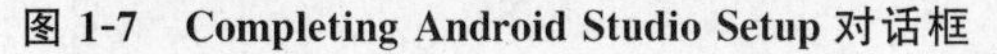

图 1-7 Completing Android Studio Setup 对话框

图 1-8 Android Studio 启动

启动 Android Studio 之后，再次进入 Android Studio 安装向导，如图 1-9 所示。

点击 Next 按钮，设置将要安装的 Android Studio 类型，如图 1-10 所示。初学者可以选择 Standard 模式默认安装，也可以选择 Custom 模式自定义安装。这里选择 Custom 模式，以强化大家对 Android Studio 工具的认识。

自定义安装中，需要设置 Android Studio 的 UI 主题风格，如 IntelliJ、Darcula，如图 1-11 所示。Darcula 主题号称为程序员专用主题背景，这里我们也选择 Darcula 主题。

下一步是 SDK 组件的设置，如图 1-12 所示。在这里选择需要的组件，以及 Android SDK 的位置。接下来就是 Android 模拟器的设置，如图 1-13 所示。

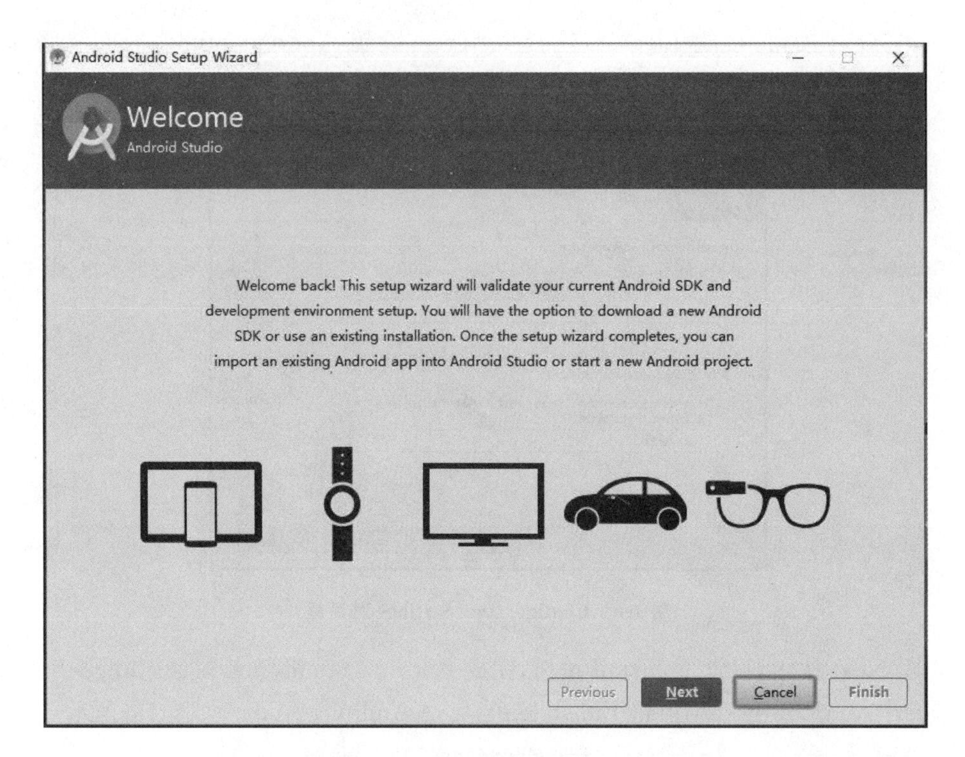

图 1-9 Android Studio 安装向导(二)

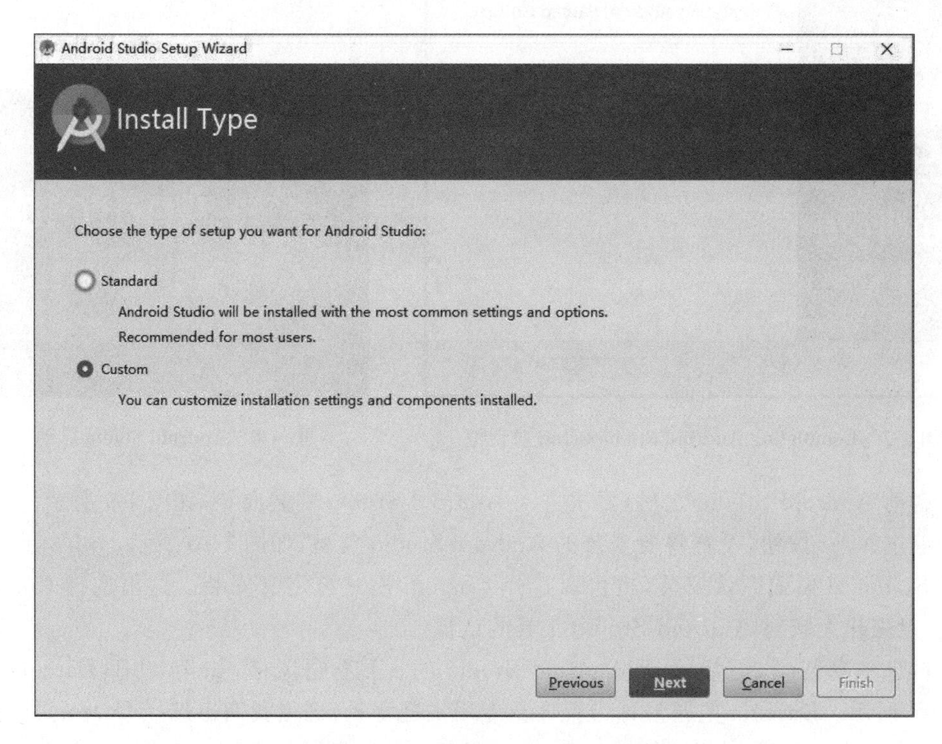

图 1-10 Install Type 对话框

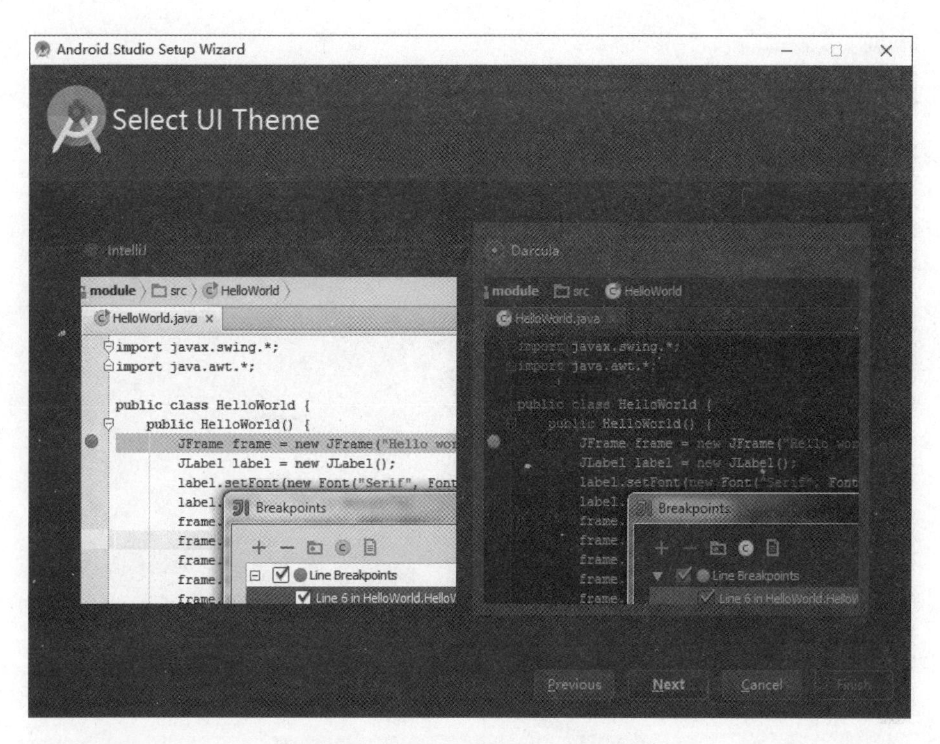

图 1-11　Select UI Theme 对话框

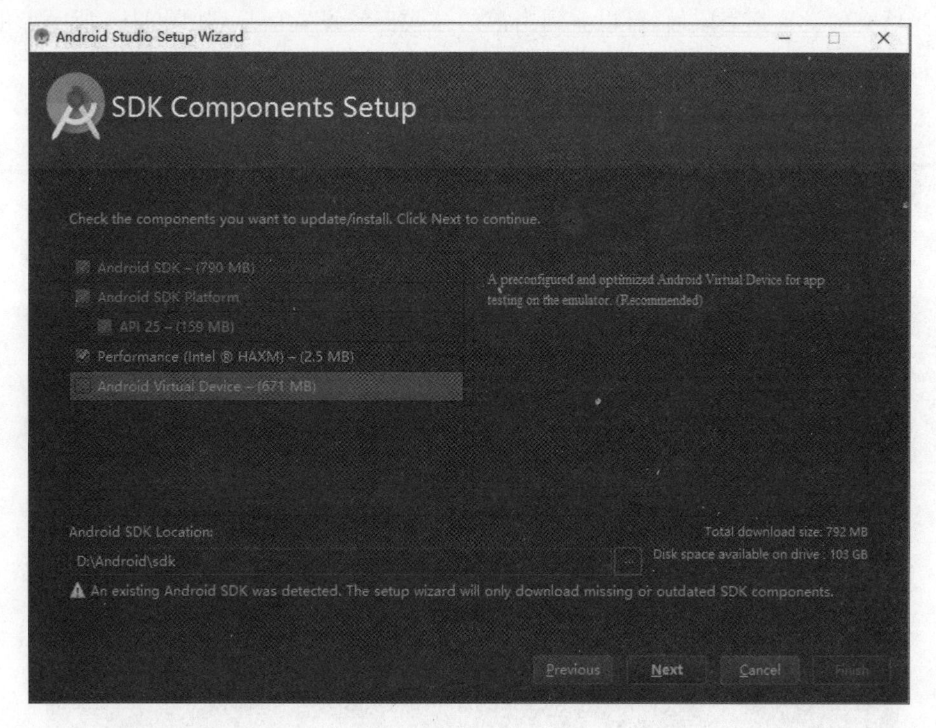

图 1-12　SDK 组件设置

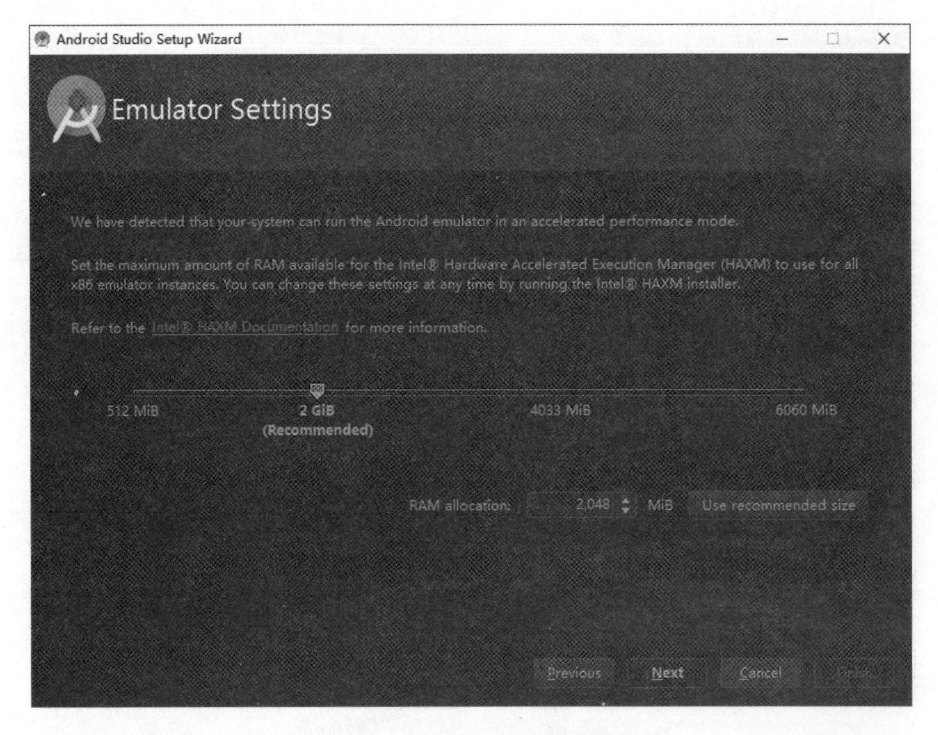

图 1-13 Android 模拟器设置

当所有的设置都完成后，可以看到当前的一些安装设置信息，如图 1-14 所示。确认这些信息后，点击 Finish 按钮，系统将下载相应的组件，如图 1-15 和图 1-16 所示。

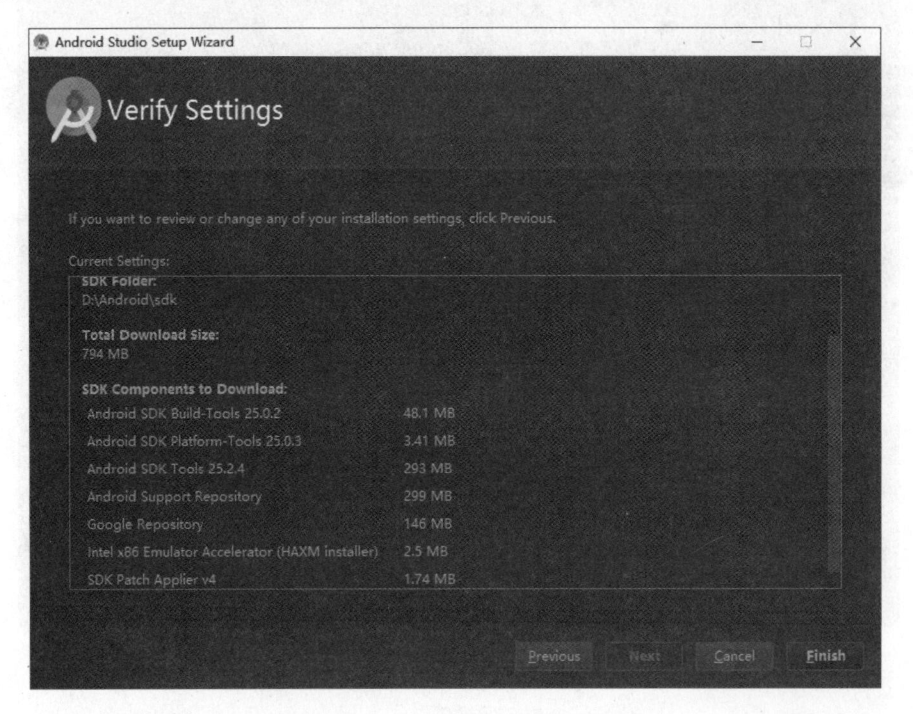

图 1-14 Verify Settings

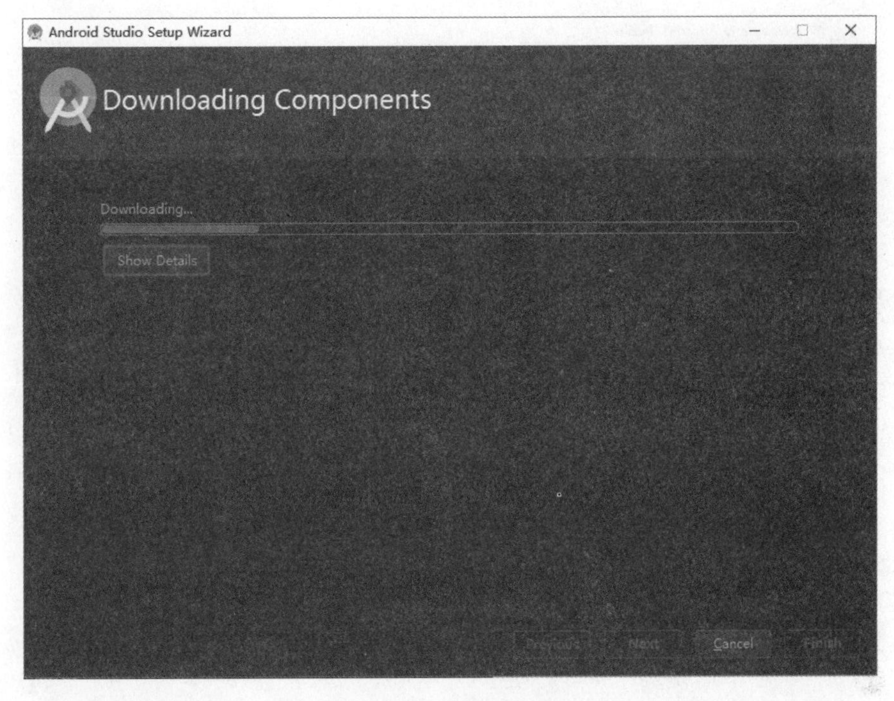

图 1-15　下载组件

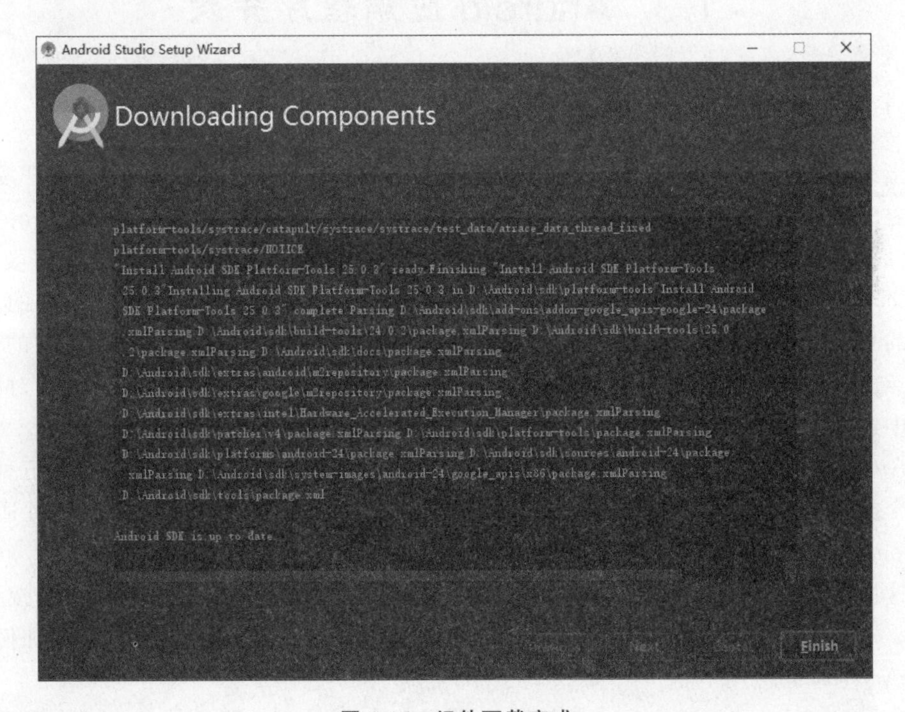

图 1-16　组件下载完成

当所有组件下载完成，将看到 Android Studio 的欢迎页面，如图 1-17 所示，至此，就完成了 Android Studio 的安装。

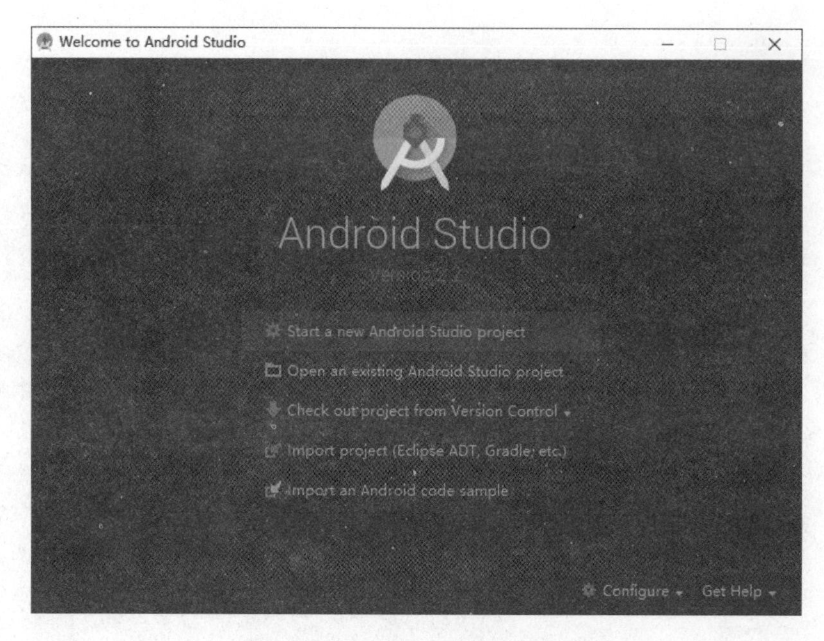

图 1-17 Android Studio 欢迎页面

1.3 Android 应用程序开发

1.3.1 第一个 Android 程序

当第一次启动 Android Studio 时，首先停留在图 1-17 所示的欢迎界面，供开发者选择需要的操作，如果要创建一个新的工程，可点击 Start a new Android Studio project，进入图 1-18 所示的 Create New Project（创建新工程）的对话框。开发者需要输入 Application name 和 Company Domain，其中 Application name 将作为应用程序的名字，反转的 Company Domain 将默认为包名，Project location 表明工程文件在磁盘中的位置。

点击 Next 按钮，进入图 1-19 所示 Target Android Devices（目标设备选择）对话框，在其中选择应用类别和 API 级别。

Android 应用类别有 Phone and Tablet、Wear、TV、Android Auto 和 Glass 五种。针对每个类别，需要指定应用支持的 SDK 最低版本，Minimum SDK 设置越低，可适用的 Android 版本越多。如果想要了解更多的信息，可点击 Help me choose。

这里选择开发一个手机应用，并设置 Minimum SDK 为 API 15，点击 Next 按钮，弹出图 1-20 所示的 Add an Activity to Mobile 对话框，在该对话框中可以选择添加何种类别的 Activity 到应用中。

这里选择 Empty Activity，点击 Next 按钮，进入图 1-21 所示的 Activity 配置对话框。在图 1-21 中，输入 Activity 名称和对应的 Layout 名称。

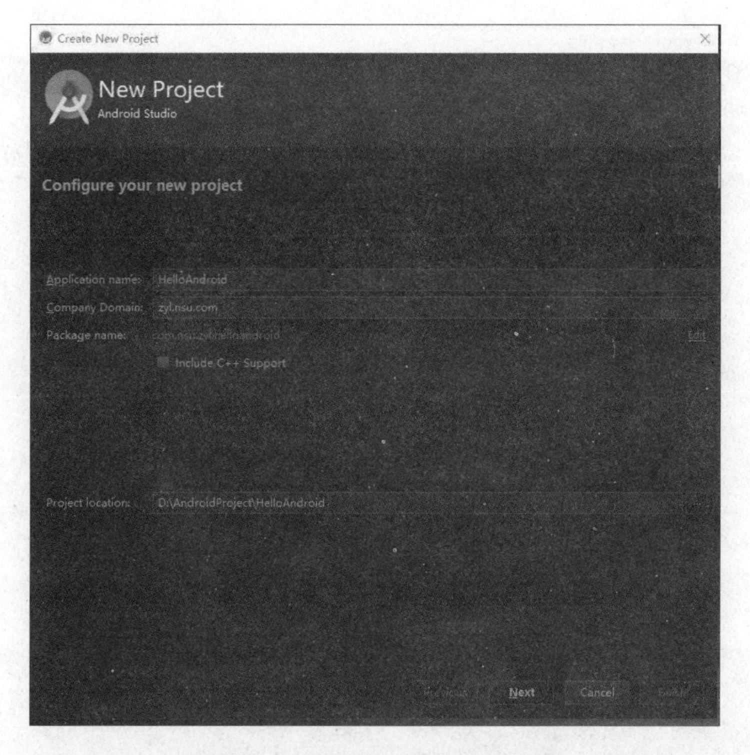

图 1-18 Create New Project 对话框

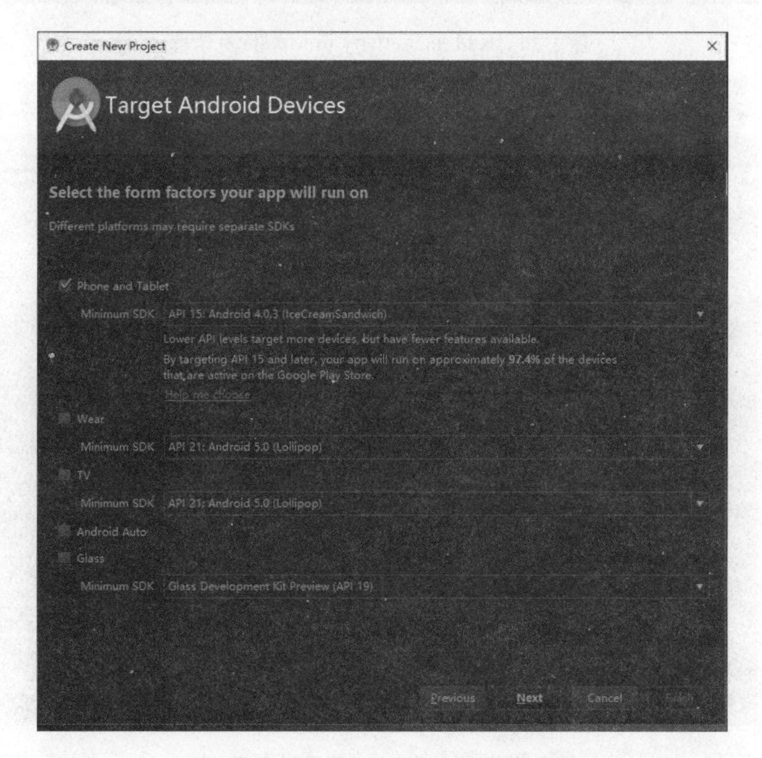

图 1-19 Target Android Devices 对话框

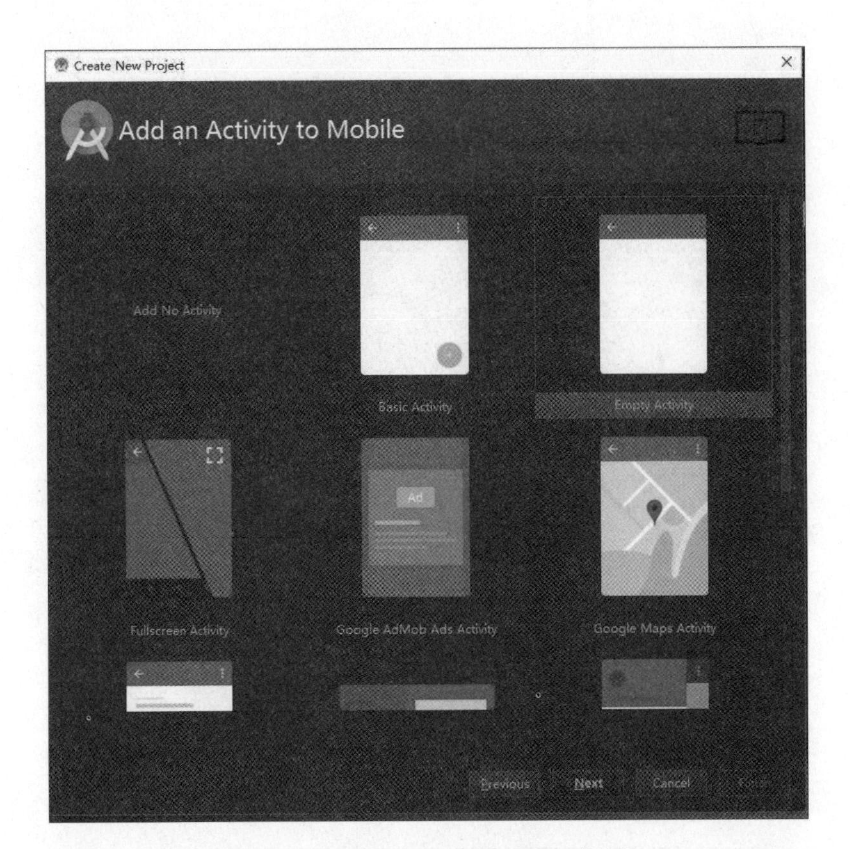

图 1-20　Add an Activity to Mobile 对话框

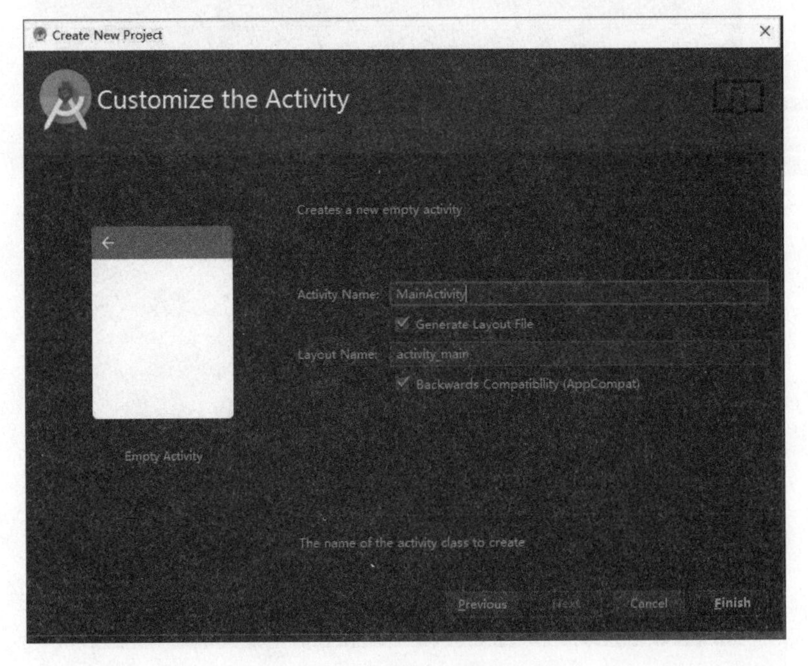

图 1-21　Activity 配置对话框

点击 Finish 按钮，Android Studio 会创建一个默认的工程目录结构的 Android 工程——HelloAndroid，Android Studio 工具页面如图 1-22 所示。

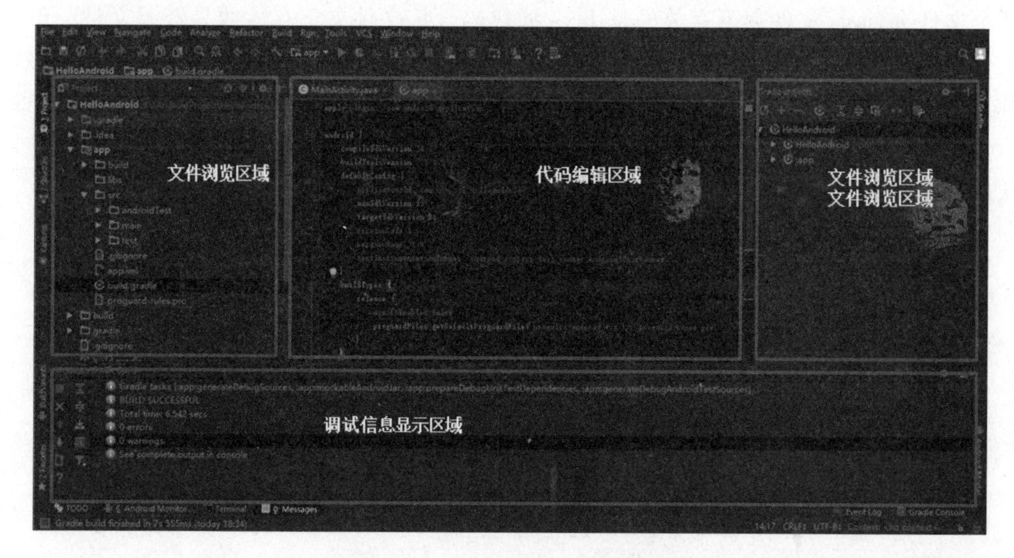

图 1-22　Android Studio 工具页面

如果根据前面的设置能看到这个界面，且没有提示错误信息，说明 Android Studio 配置正确，就可以进行 Android 应用开发了。

1.3.2　Android 程序结构

创建 Android 应用程序后，就可以浏览当前项目工程，熟悉应用程序的结构。Android Studio 提供了多种浏览方式，如 Project、Android、Packages 等。

当选择 Project 浏览方式时，Android 工程将以资源管理器的树形结构展开(如图 1-23 所示)，当你希望看看磁盘上存在的关于这个项目的文件夹和目录时，就可以使用 Project 浏览方式。

在 Project 浏览方式下，目前只需要关注以下几个文件和文件夹。

(1) app 文件夹：这是工程产生后，Android Studio 自动创建的 module 所在文件夹，应用程序的源代码和资源文件就放在这个 module 当中。

(2) build：编译后的文件存放的位置，最终生成的 apk 文件就在这个目录中。

(3) libs：添加的 *.jar 或 *.so 等文件存放的位置。

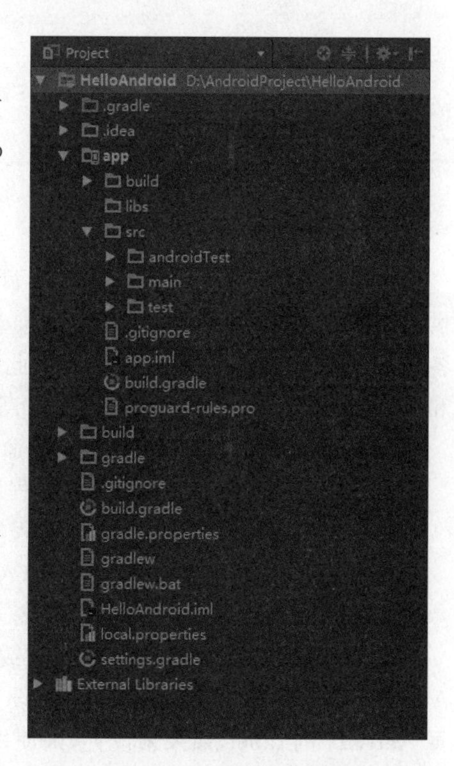

图 1-23　Project 工程视图

（4）src文件夹里有3个子文件夹：androidTest、main和test。androidTest和test从名字就可以看出其中存放的是和测试相关的内容。main文件夹下又分为java和res两个文件夹，java文件夹下存放的是Java源代码，res文件夹下存放的是资源文件。main文件夹下面的AndroidManifest.xml文件是当前Android应用的配置文件，包括程序名称、图标、访问权限等整体属性信息；另外，程序中定义的组件（Activity、Service、ContentProvider和BroadcastReceiver）需要在AndroidMainfest.xml文件中注册后才能使用。

HelloAndroid项目对应的AndroidManifest.xml文件内容如文件清单1-1所示。

文件清单1-1　AndroidManifest.xml

```
<?xml version="1.0" encoding="utf-8"?>
<manifest xmlns:android="http://schemas.android.com/apk/res/android"
    package="com.nsu.zyl.helloandroid">

    <application
        android:allowBackup="true"
        android:icon="@mipmap/ic_launcher"
        android:label="@string/app_name"
        android:supportsRtl="true"
        android:theme="@style/AppTheme">
        <activity android:name=".MainActivity">
            <intent-filter>
                <action android:name="android.intent.action.MAIN" />
                <category android:name="android.intent.category.LAUNCHER" />
            </intent-filter>
        </activity>

    </application>
</manifest>
```

AndroidManifest.xml是整个项目的配置文件，描述了当前应用的package中提供的各种组件（Activity、Service、BroadcastReceiver、ContentProvider等）的实现类以及各种能被处理的数据和启动位置；此外，还能声明程序中需要的permissions（权限）和instrumentation（安装控制和测试）等。

xmlns:android：定义android的命名空间。

package：指定本应用内Java主程序的包名。

application：声明了每一个应用程序的组件及其属性。

android:allowBackup：将程序加入系统的备份和恢复架构中。

android:icon：表示App的图标。

android:label：表示App的名字或者Activity页面的标题。同时，在Application和Activity中设置label时，Activity中设置的label优先级更高。App的名字取决于启动页Activity的label，如果未设置，则使用Application里设置的label。

android:theme：android的主题。

android:name：表示当前 activity 的名字。

intent-filter：包含 action、data 和 category 三种子标签。

action：只有 android:name 属性，常见的是 android. intent. action. MAIN，表示此 activity 是作为应用程序的入口。

category：android:name 属性，常见的是 android. intent. category. LAUNCHER，决定应用程序是否显示在程序列表里。

build. gradle 文件：是 module 编译时的配置文件，其内容大致如文件清单 1-2 所示。

文件清单 1-2 build. gradle

```
apply plugin: 'com.android.application'

android {
    compileSdkVersion 24
    buildToolsVersion "25.0.2"
    defaultConfig {
        applicationId "com.nsu.zyl.helloandroid"
        minSdkVersion 15
        targetSdkVersion 24
        versionCode 1
        versionName "1.0"
        testInstrumentationRunner "android.support.test.runner.AndroidJUnitRunner"
    }
    buildTypes {
        release {
           minifyEnabled false
           proguardFiles getDefaultProguardFile('proguard-android.txt'),
'proguard-rules.pro'
        }
    }
}

dependencies {
    compile fileTree(dir: 'libs', include: ['* .jar'])

    androidTestCompile('com.android.support.test.espresso:espresso-core:2.2.2', {
        exclude group: 'com.android.support', module: 'support-annotations'
    })
    compile 'com.android.support:appcompat-v7:24.2.1'
    testCompile 'junit:junit:4.12'
}
```

minSdkVersion 表示这个应用允许安装的最低 API Level，例如这里写的 15，说明这个应用只能安装到 15 或 15 版本以上的安卓设备上(即 Android 4.0 及以上版本)，其他的安卓设备(例如 Android 3.X)都不行。targetSdkVersion 表示这个应用的理想运行系统版本。例如，targetSdkVersion 24 指明这个应用使用的是 Level 为 24 的 SDK，说明该应用运行的理想系统版本是 Android 7.0，当在理想系统上运行该应用时，会省略软件的兼容性判断，提高程序的运行效率。versionName 表示版本号，可以通过查看该应用在设备上的详细信息获取。

“Android”浏览方式下工程视图如图 1-24 所示。目录结构有了一些变化，但核心内容基本相同，大家可以在后面的学习中逐步体验。

图 1-24　Android 工程视图

manifests：AndroidManifest. xml 是 APP 的配置信息，内容同上。

java：主要为源代码和测试代码。

res：主要是资源目录，存储所有的项目资源。

drawable：存储图片和 xml 文件，* dpi 表示存储分辨率的图片，用于适配不同的屏幕。

- mdpi：320×480；
- hdpi：480×800、480×854；
- xhdpi：至少 960×720；
- xxhdpi：1280×720。

layout：存储布局文件。

mipmap：存储原生图片资源，缩放时有性能优化。

values：存储 App 引用的一些值。

- colors. xml：存储了一些 color 样式；
- dimens. xml：存储了一些公用的 dp 值；
- strings. xml：存储了引用的 string 值；
- styles. xml：存储了 App 需要用到的一些样式。

Gradle Scripts：build. gradle 为项目的 gradle 配置文件，内容同文件清单 1-2。

程序源文件 MainActivity. java 见文件清单 1-3。

文件清单 1-3　MainActivity. java

```
package com.nsu.zyl.helloandroid;
import android.support.v7.app.AppCompatActivity;
import android.os.Bundle;

public class MainActivity extends AppCompatActivity {

    @Override
    protected void onCreate(Bundle savedInstanceState) {
        super.onCreate(savedInstanceState);
        setContentView(R.layout.activity_main);
    }
}
```

表面上看 MainActivity 继承了 AppCompatActivity，但本质上继承的还是 Activity，具体

内容将在第2章详细讲解。重写onCreate()方法，其中，super.onCreate方法是调用父类的onCreate方法，然后setContentView方法就是为当前的activity引入了名为activity_main的布局，这样第一个应用程序就完成了。

1.3.3 Android模拟器

在1.3.2节我们创建了一个Android应用程序，该程序需要Android手机才可以运行。为了看到程序的运行效果，可以使用Android真机来演示，也可以使用Android模拟器来演示。关于模拟器的选择，提供以下两种方案。

1. Android自带的模拟器

打开Android Studio，在工具栏中找到AVD Manager的图标，点击打开Android Virtual Device Manager窗口，如图1-25所示。

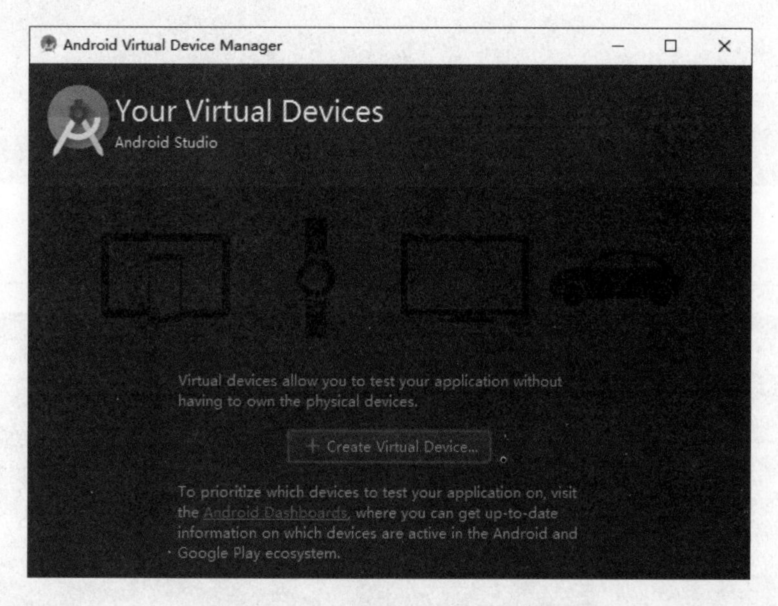

图1-25 Android Virtual Device Manager窗口

点击Create Virtual Device按钮，进入模拟器分辨率选择对话框，如图1-26所示。

(1) 选择目标设备：手机、平板、手表、电视。

(2) 选择建议的设备尺寸，例如当前选择的是Nexus 4，4.7寸，分辨率为768×1280。

选定后，点击Next按钮，进入模拟器详细数据设置对话框，如图1-27所示。

选择相应的Android版本，点击Next按钮，进入下一个对话框，如图1-28所示。

最终为模拟器设置对话框，如果想切换分辨率和api版本，点击相应的Change按钮就可以更改。点击Finish按钮，回到Android Virtual Device Manager窗口，可以查看当前已创建好的模拟器，如图1-29所示。

需要注意的是，由于模拟器无SIM卡，无Wi-Fi网络，硬件资源受限，在Android应

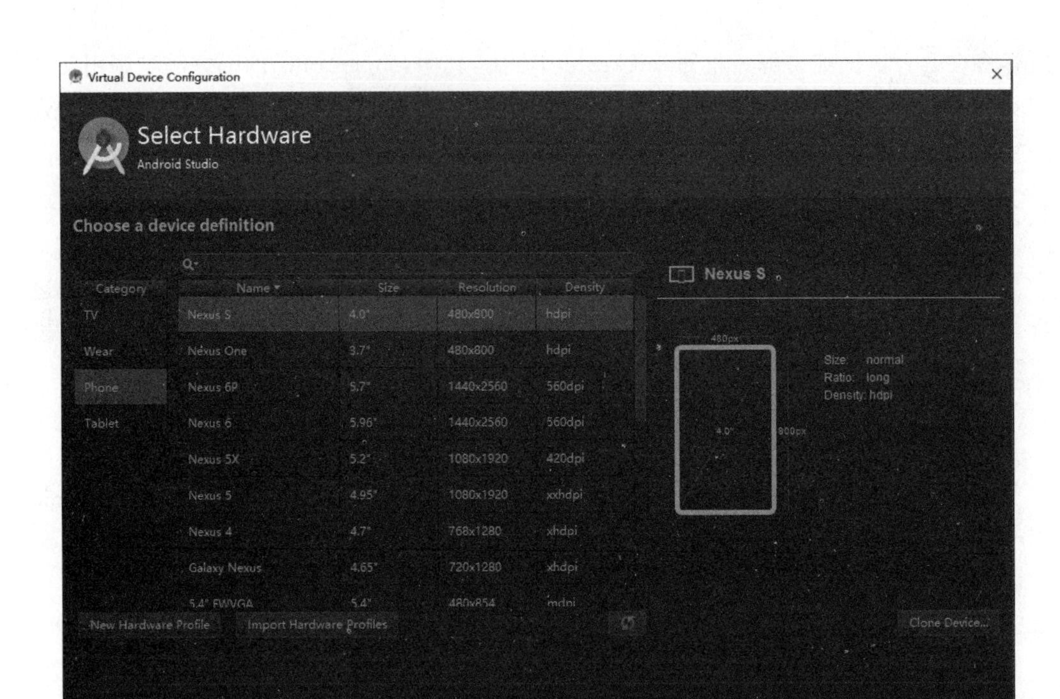

图 1-26 AVD 模拟器分辨率选择对话框

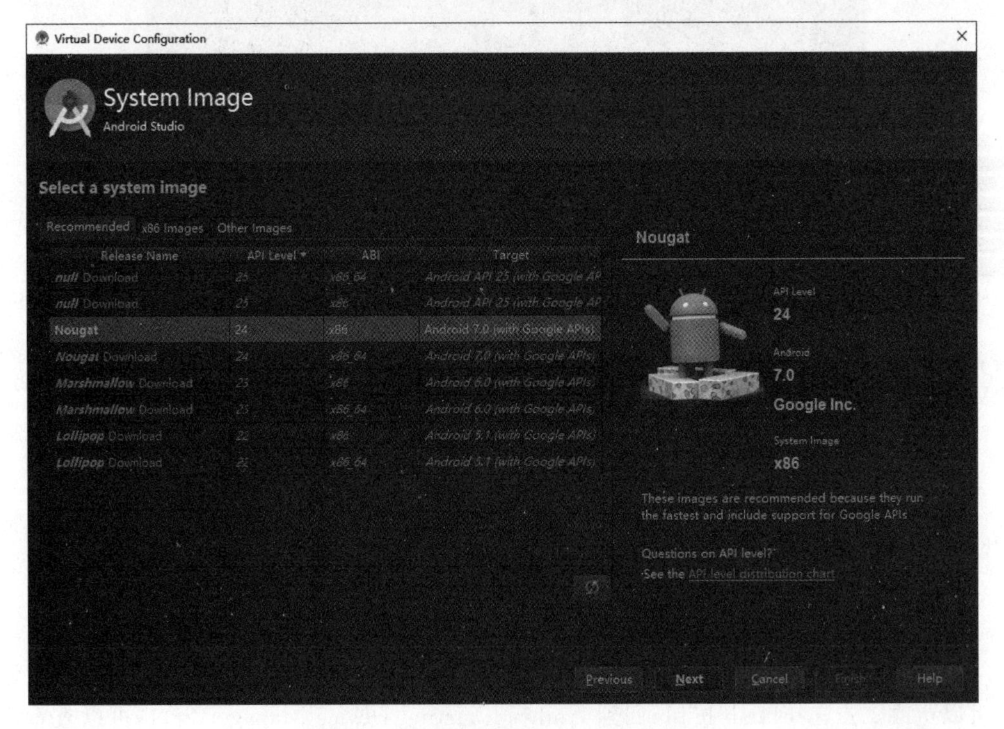

图 1-27 AVD 模拟器 API 选择对话框

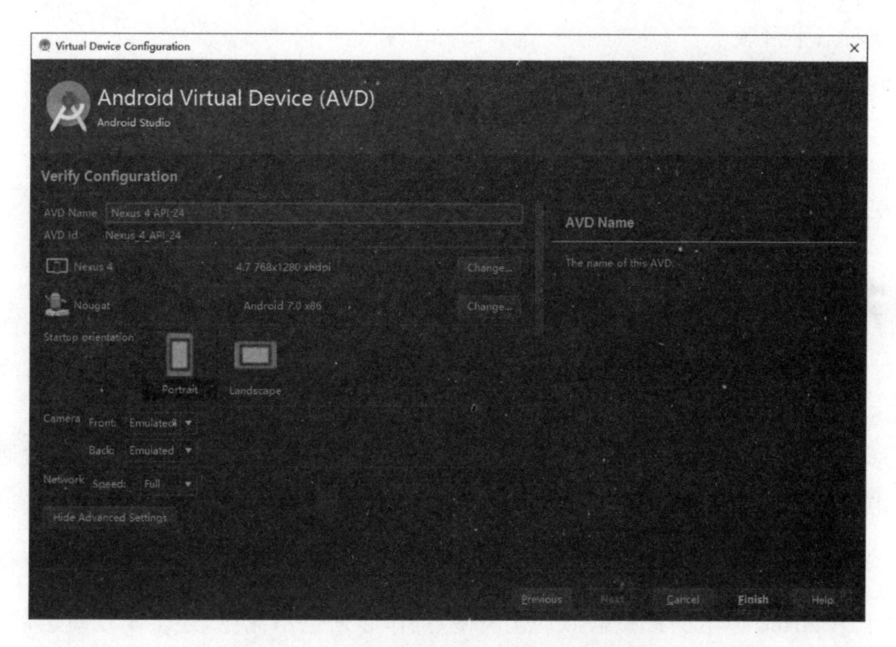

图 1-28　AVD 模拟器详细参数设置对话框

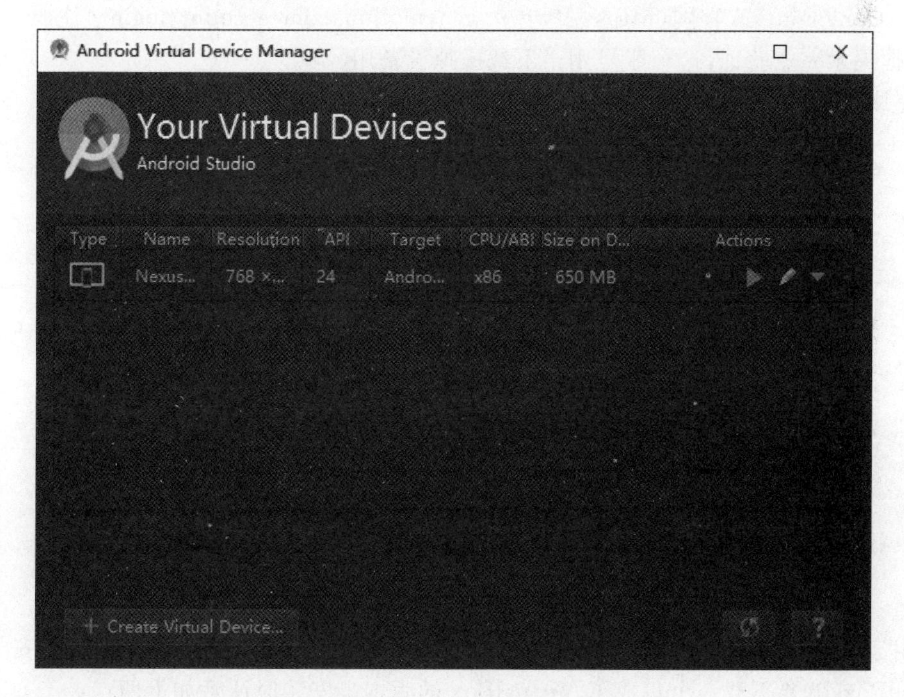

图 1-29　Android Virtual Devices 窗口

用开发中，推荐在真机上调试程序。

Android 模拟器常用配置信息说明如下：

Name：模拟器名字。

Device：屏幕分辨率。

Target：平台版本。

Keyboard：使用硬件键盘。

BackCamera：后摄像头。

Memory Options：内存选项。

RAM：手机内存。

VM Heap：堆内存。

Internal Storage：ROM，存放安装到模拟器上的 APP。

SD Card：SD 卡大小。

Snaphot：快照功能，加快 AVD 的启动，但会影响程序调试，导致对代码的修改不能立即反应在 AVD 上。

Use Host GPU：使用宿主机的 GPU 加速，一般在调试 3D 游戏时开启，不过通常在真机上做 3D 游戏的调试。

2. 第三方模拟器

由于 Android 原生的模拟器启动比较慢，操作起来也不流畅，还会出现莫名的问题，所以可以使用第三方 Genymotion 模拟器来运行 Android 应用程序。

在 Genymontion 官网(https://www.genymotion.com/account/login/)注册一个账号，根据提示完成验证，这个账号用于下载虚拟设备用。

完成注册后，选择相应的版本下载安装，如图 1-30 所示。

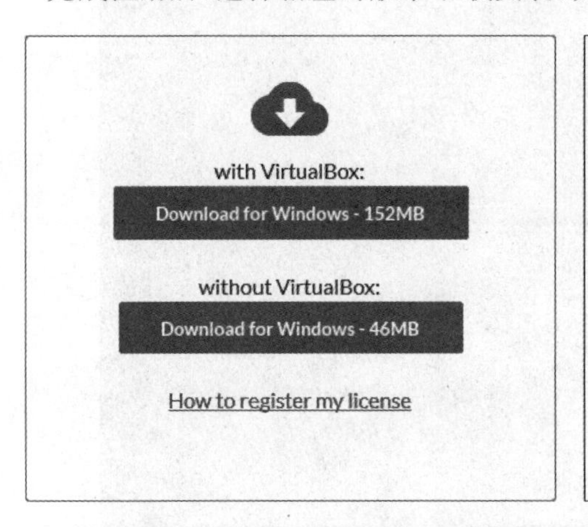

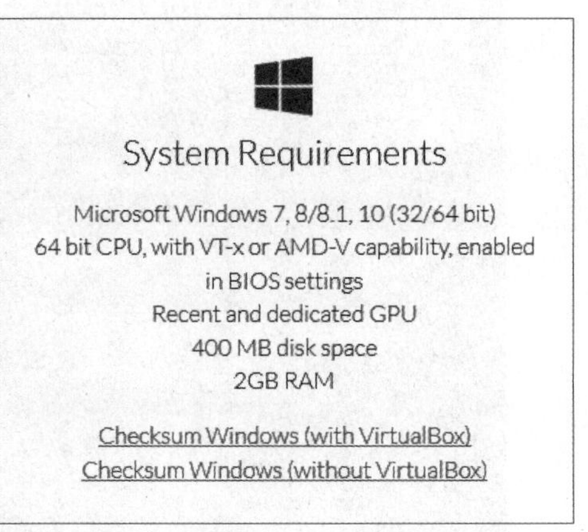

图 1-30 Genymotion 版本下载

由于 Genymotion 模拟器运行需要 VirtualBox 支持，如果之前没有安装过 VirtualBox，就要下载一个集成了 VirtualBox 的版本。安装时注意如下两点：

(1) 两者须安装在同一个盘上。

(2) 不要安装在中文目录。

安装成功后，启动 genymotion.exe，输入先前注册的账号和密码，点击 Add 按钮下载虚拟设备，下载完成如图 1-31 所示。

接下来需要在 Android Studio 上安装 Genymotion 插件，有以下两种方法。

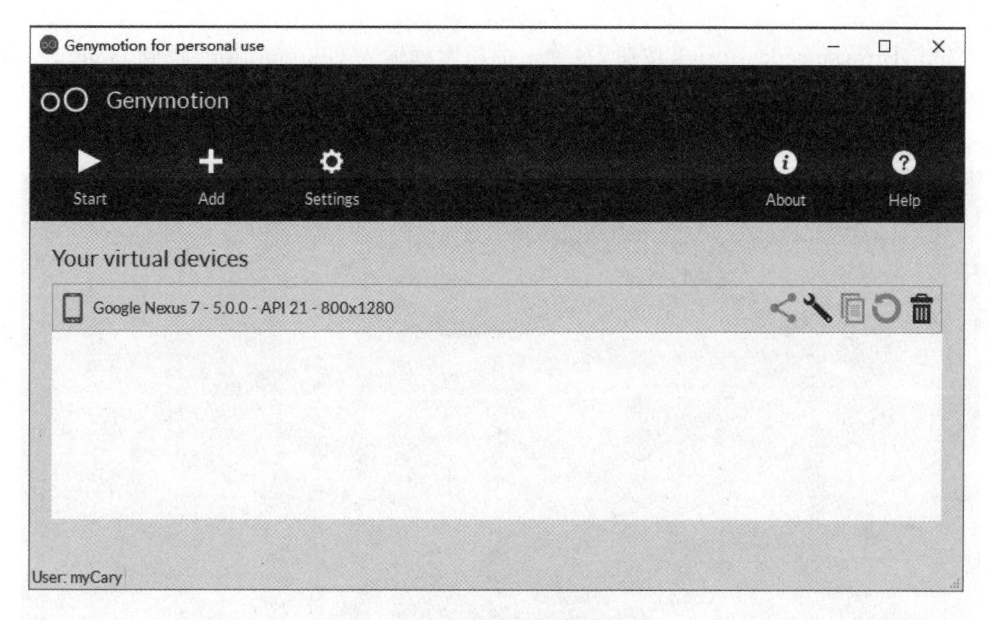

图 1-31　Genymotion 虚拟设备

方法一：打开 Android Studio，打开 File 菜单，点击 Setting 按钮，选择 IDE Settings 对话框，在下拉菜单中选择 Plugins，如图 1-32 所示。

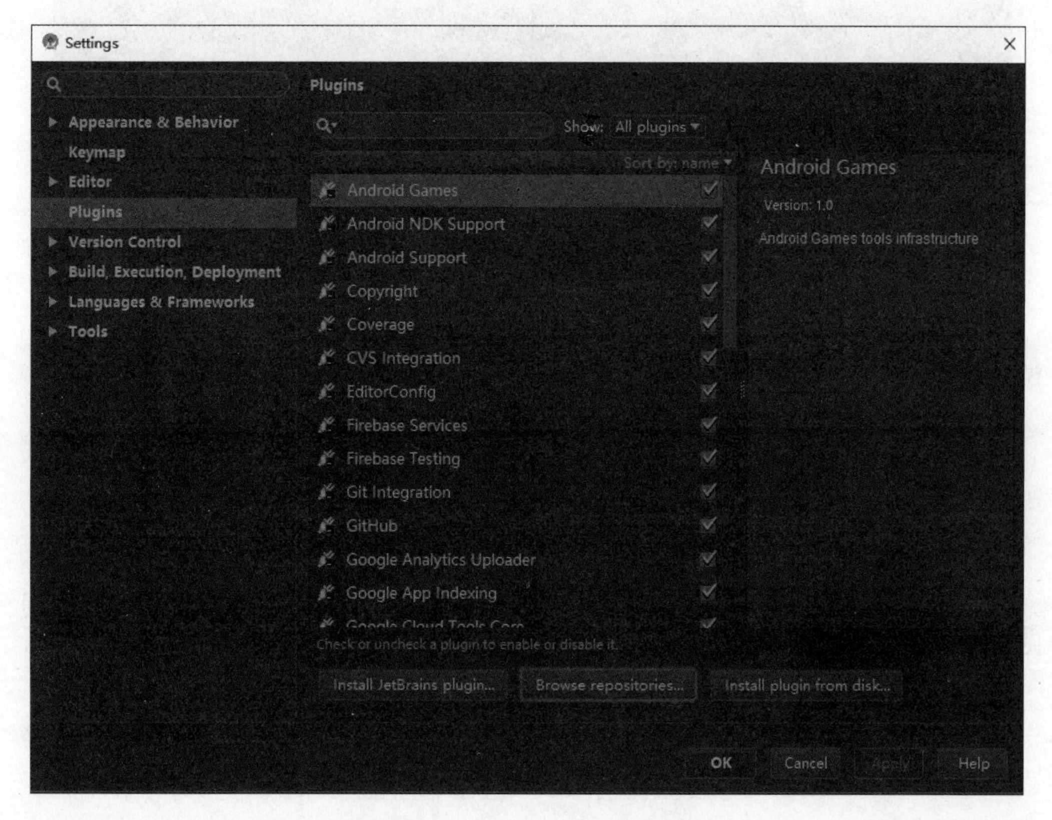

图 1-32　Android Studio 插件搜索

右击 Browse repositories 按钮，在弹出的搜索框输入 Genymotion，找到 Genymotion 插件，如图 1-33 所示。点击 Install 按钮，完成 Genymotion 插件的安装。

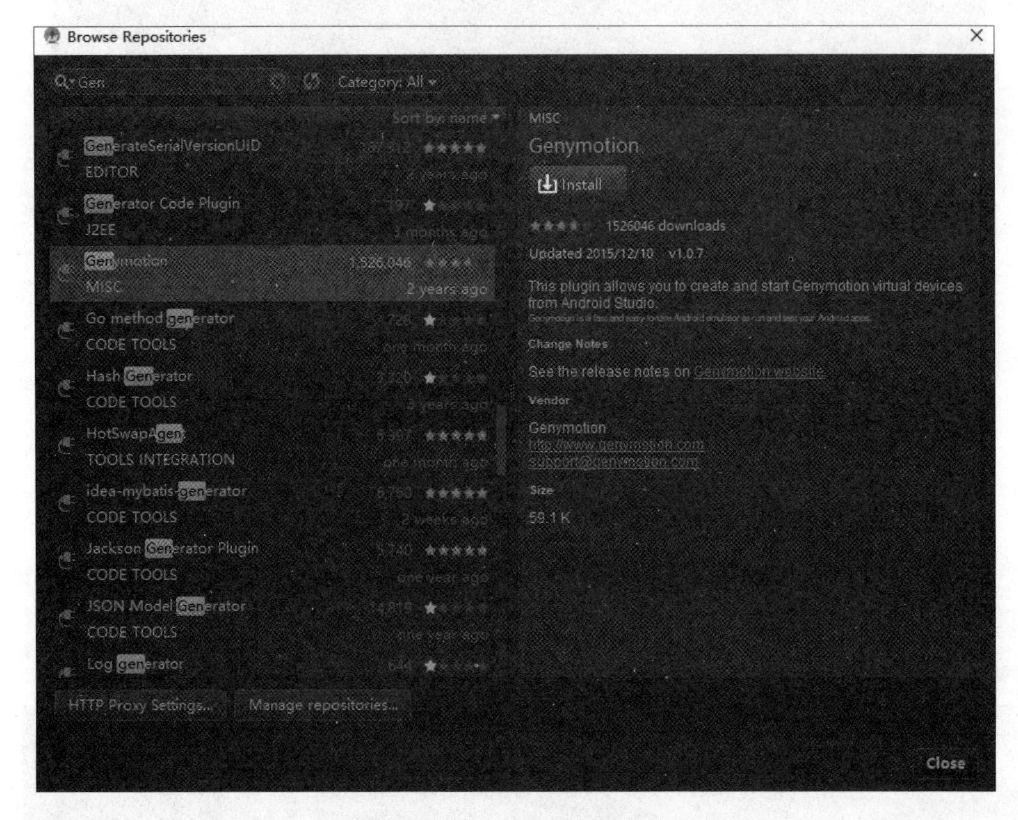

图 1-33 Genymotion 插件安装

方法二：在 Genymotion 官网下载安装包的页面上找到 IDEA Plugins，下载，重复方法一中的步骤。

Genymotion 插件安装完成后，重启 Android Studio，工具栏上会多出一个 Genymotion 模拟器图标，如图 1-34 所示。

图 1-34 Android Studio 工具栏

点击 Genymotion 模拟器图标，第一次需要配置指向 Genymotion 的安装路径，如图 1-35 所示。这里将 Genymotion 安装在 D:\Genymobile\Genymotion 目录下。

第一次配置后，点击图标直接进入 Genymotion 设备管理对话框，如图 1-36 所示。

在 Genymotion 设备管理对话框中，选择一个模拟器，点击 Start 按钮，可快速启动该 Genymotion 模拟器，如图 1-37 所示。

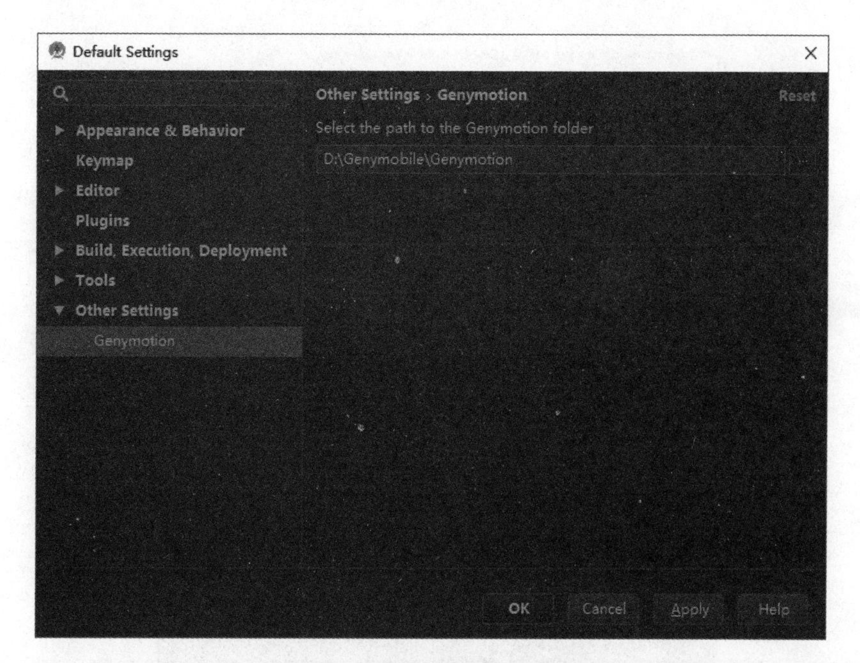

图 1-35　Genymotion 设置对话框

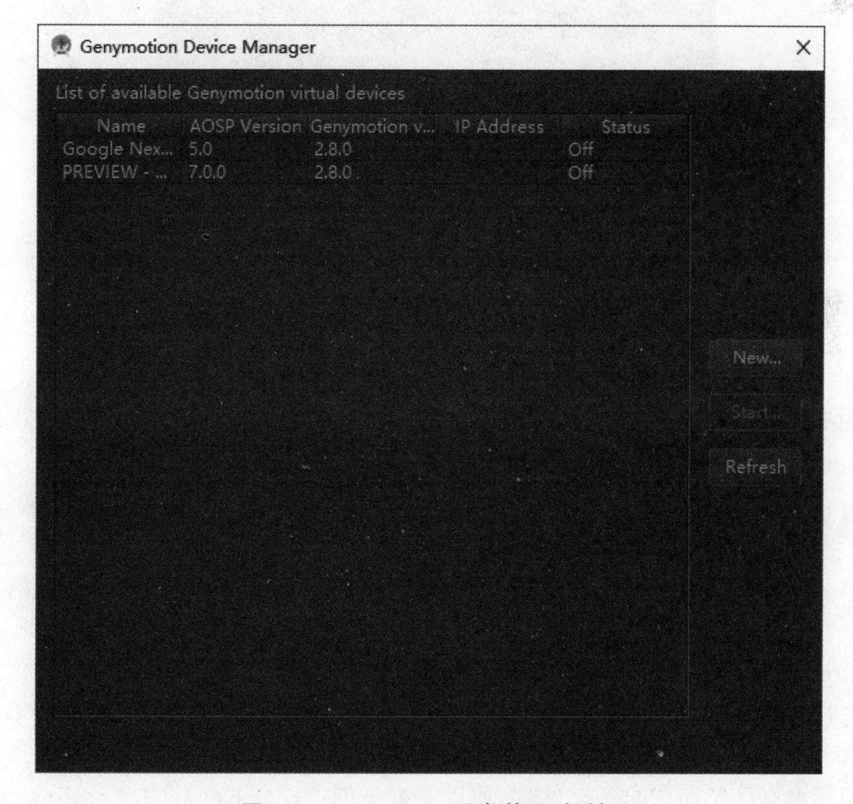

图 1-36　Genymotion 设备管理对话框

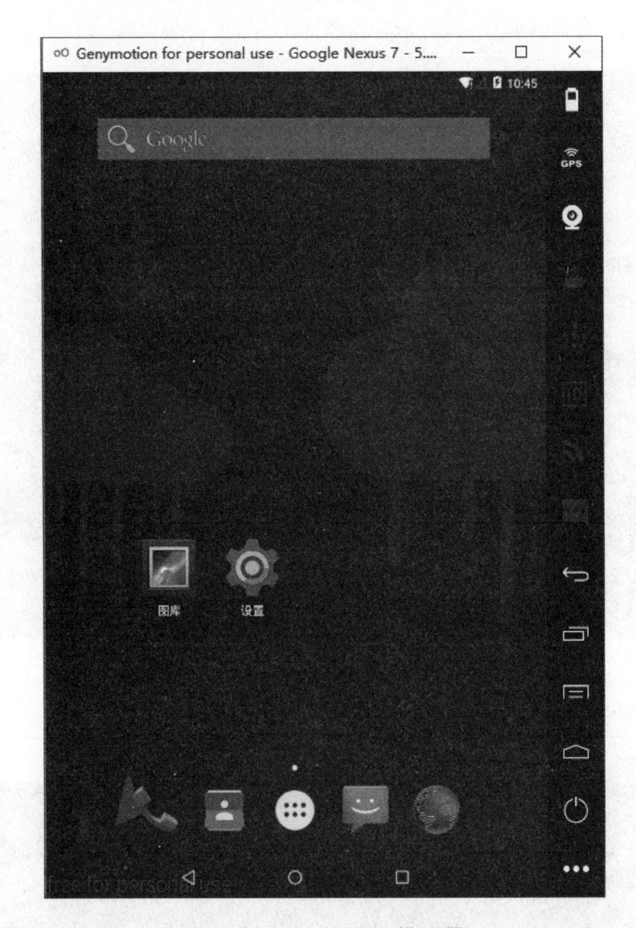

图 1-37 Genymotion 模拟器

1.3.4 Android 应用程序的打包与发布

Android 应用程序的打包分为非签名打包和签名打包两种，签名打包的 apk 才能正式在 Android 应用市场发布。

在 adt 调试时会自动生成 apk，我们称之为非签名打包，仅供调试使用。

Eclipse 下，该 apk 位于 workspace/工程名/bin/目录下：workspace/工程名/bin/xxx.apk。

而 Android Studio 将其放在了 module 中，具体位置：android studio 工程的存储路径下 app/build/outputs/apk，其中 app 是 module，在对应的 module 下即可找到。

Android APP 都需要一个证书对应用进行数字签名。

首先打开 Android Studio，选择 Build→Generate Signed APK 命令，进入 Generate Signed APK（生成签名 APK）的向导对话框，如图 1-38 所示。

我们还没有已经存在的 Key 文件，所以需要先创建一个 Key，点击 Create new 按钮，可根据自己的需要填写相关项，如图 1-39 所示。

点击 OK 按钮，完成 Key 的创建，返回 Generate Signed APK 对话框，自动填写刚创

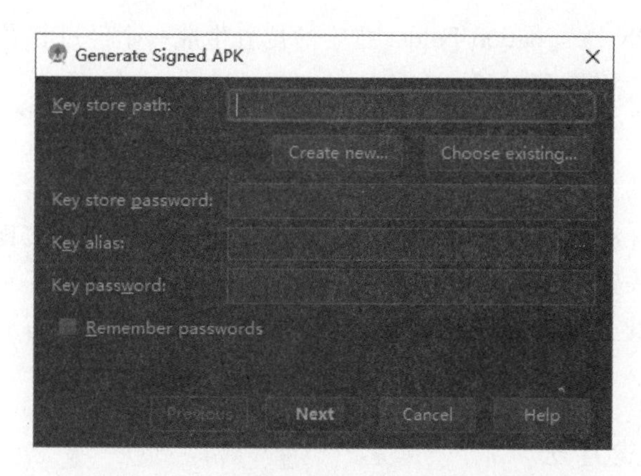

图 1-38 Generate Signed APK 对话框

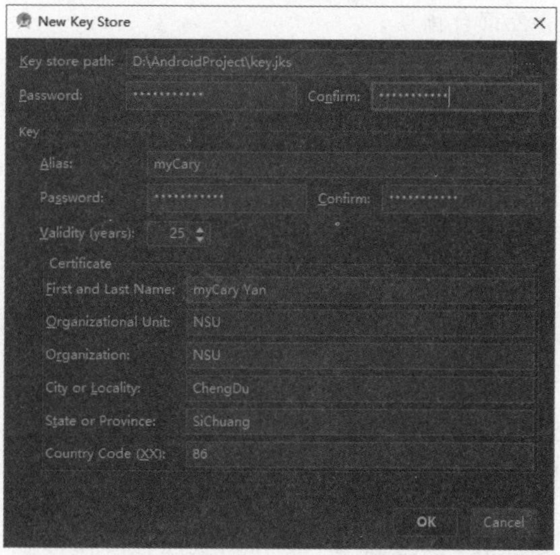

图 1-39 New Key Store 对话框

建的 Key 的 alias 和密码,如图 1-40 所示。

点击 Next 按钮,再点击 Finish 按钮,如图 1-41 所示。

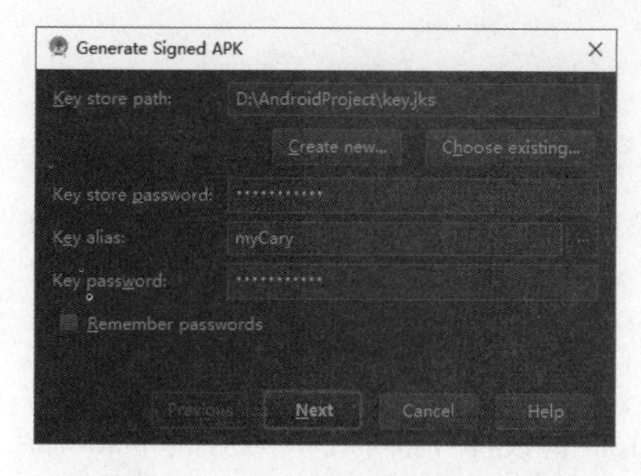

图 1-40 Generate Signed APK 对话框

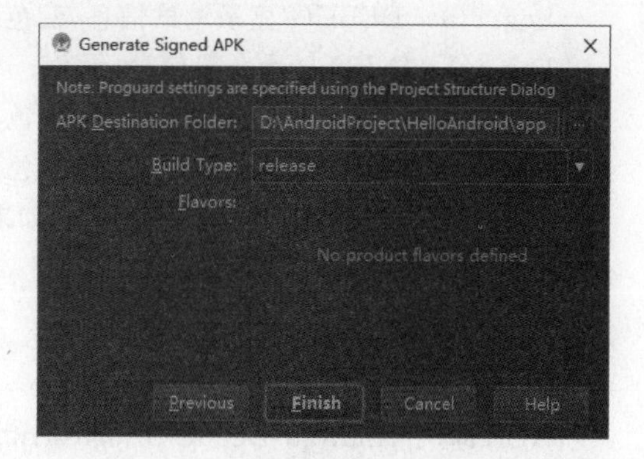

图 1-41 生成签名 APK 文件

完成后会在 APK Destination Folder 指定的位置生成一个经过签名打包的 apk 文件。

1.4 Android 程序调试方法

Android Studio 提供了部署应用到设备上运行的方法和调试工具。将应用程序通过 Android Studio 部署到设备上有两个方式：run app 和 debug app。debug app 模式下，程序运行起来可以直接进入断点调试模式，对代码进行静态调试；而 run app 模式只能通过 attach 的方式进入断点调试模式。

1.4.1 静态调试方法

所谓静态调试就是冻结应用程序运行的状态，仿佛时间停止了一般，然后逐一观察此时程序的各个参数是否符合预期。

首先，在希望代码暂停运行的地方打断点——在代码前点击一下，出现一个红色的圆点，如果想取消，再点击一次即可。

然后，用 debug run 的方式部署程序。

当程序运行到这段代码的这个位置时，程序将停止下来，切换到 Debug 窗口。这时，就可以观察各个参数了。

1.4.2 LogCat 的使用

对于那些和时间相关的程序(不能让程序暂停，等你慢慢观察)，就不能使用静态调试方法了，须采用动态调试、添加 Log 的方式。

Log 的中文名字称作日志，在编程界表示程序运行过程中打印出的信息。根据 Log 就可知道现在程序运行到什么地方了，Log 还可以携带程序中某些变量的信息输出，让我们更精准地知道程序当前运行的状态。在代码中添加一段函数，就能通过特别的工具输出这些 Log。

Logcat 区域的日志共分为 5 级：

Log. v() VERBOSE 显示全部信息，黑色；

Log. d() DEBUG 显示调试信息，蓝色；

Log. i() INFO 显示一般信息，绿色；

Log. w() WARN 显示警告信息，橙色；

Log. e() ERROR 显示错误信息，红色。

颜色只是方便查看，并无其他意义。如果日志信息过多，也可以对日志过滤，以方便查看。

1.4.3 ADB 常用命令及使用

ADB 即为 Android Debug Bridge(Android 调试桥)，指存在于 SDK 的 platform-tools 目录中的 adb. exe 工具。熟练使用 ADB 命令将会大大提升开发效率。ADB 的命

令有很多，这里总结在 Android 应用开发中常用到的一些 ADB 命令。

adb version：查看 adb 版本；

adb device：查询已连接的 Android 设备与模拟器；

adb install < apk 文件路径>：安装一个 apk；

adb uninstall < package >：直接卸载 Android 应用程序；

adb shell pm list packages：列出手机装的所有 APP 的包名；追加-s，列出系统应用的所有包名；追加-3，列出除了系统应用的第三方应用包名。

当 Android 设备连接出错，或者断开一个连接后，需要重启 ADB 服务，可采用如下两个命令：

adb kill-server：停止 adb server；

adb start-server：启动 adb server。

1.4.4 DDMS 的使用

DDMS(Dalvik Debug Monitor Service)为 IDE、emulator 及真正的 Android 设备架起了一座桥梁。开发人员可以通过 DDMS 看到目标机器上运行的进程/线程状态，可以查看进程的 heap 信息、Logcat 信息，可以查看进程分配内存情况，可以向目标机发送短信以及打电话，可以向 Android 开发发送地理位置信息。

在 Android Studio 中，点击 Tools 菜单，选择 Android→Android Device Monitor 命令，在弹出的对话框中就可以看到 DDMS，如图 1-42 所示。

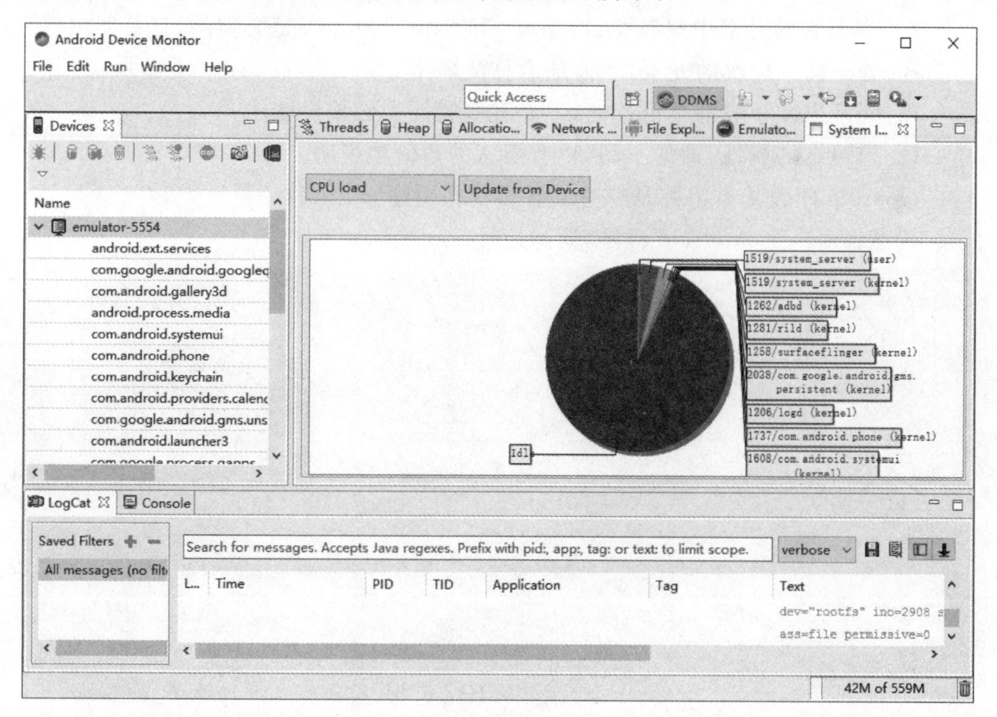

图 1-42 DDMS 视图

本章小结

本章学习 Android 应用开发的基础，主要讲解 Android 入门基础知识。首先对 Android 系统的历史、平台特性、版本、体系结构、四大组件进行介绍，了解这些知识有助于读者对 Android 系统有初步的理解。在本章的后几节详细介绍了 Android 开发环境的搭建，Android 项目的创建与运行，Android 程序结构，以及 Android 程序常用调试方法。相信读者在学习本章之后，能够掌握在 Android Studio 环境下开发 Android 应用项目的基本步骤，并能开发调试基本的 Android 应用程序。

习　题

1. Android 设备上安装的 QQ 聊天工具，属于 Android 体系结构的哪一层？（　　）
   A. 应用程序层　　B. 应用程序框架层
   C. 核心类库　　D. Linux 内核
2. 应用程序层是一个核心应用程序的集合，下面属于该层的是（　　）。
   A. 活动管理器　　B. 短信程序
   C. 音频驱动　　D. Dalvik 虚拟机
3. 关于 AndroidManifest.xml 文件，以下描述错误的选项是（　　）。
   A. 在所有的元素中只有<manifest>和<application>是必需的，且只能出现一次
   B. 处于同一层次的元素，不能随意打乱顺序
   C. 元素属性一般都是可选的，但是有些属性是必须设置的
   D. 对可选的属性，即使不写，也有默认的数值项说明
4. Android 四大基本组件是哪些？各有什么用途？
5. 任意编写一个 Android 程序，调试运行。

第2章 chapter 2

Activity 与 Intent

主要内容：Activity 的创建与注册，Activity 启动方式，Activity 生命周期，Intent 与 IntentFilter

建议课时：8 课时

知识目标：(1) 掌握 Activity 的创建与注册；

(2) 掌握 Activity 的启动方式；

(3) 熟悉 Activity 的生命周期；

(4) 掌握 Intent 与 IntentFilter 的使用。

能力目标：(1) 具备 Activity 开发基本能力；

(2) 具备借助 Intent 启动 Activity 的能力。

为了更好地通过示例讲解 Activity 与 Intent 的相关知识点，需创建 Android 项目 Chapter02Application，在该项目中完成本章的示例代码。

首先打开 Android Studio，弹出图 2-1 所示欢迎界面。

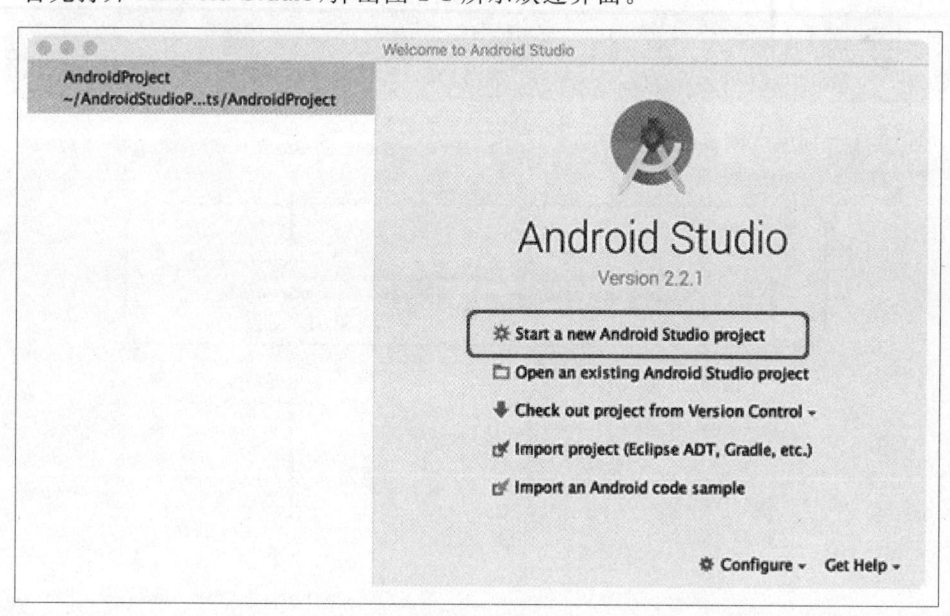

图 2-1　Android Studio 欢迎界面

在图 2-1 界面点击 Start a new Android Studio project，弹出图 2-2 所示 project 配置对话框。

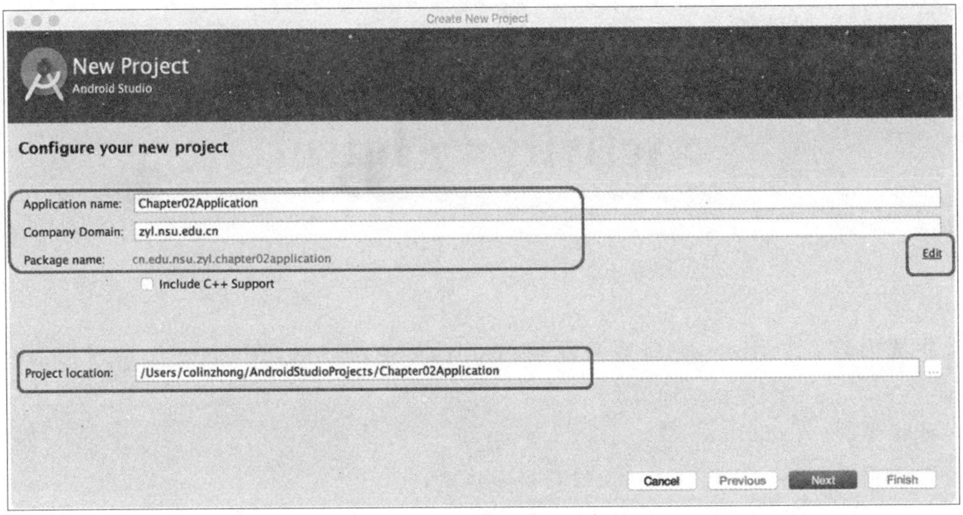

图 2-2 project 配置对话框

在对话框中，输入项目的 Application name 和 Company Domain 信息，其中 Application name 是应用程序名，Company Domain 是公司域名，Android Studio 会自动根据输入的公司域名和应用程序名，将公司域名倒序＋应用名作为当前应用的 Package name。如果开发者对生成的 Package name 不满意，可以点击页面中的 Edit 按钮，修改当前应用的 Package name。编辑完成后，点击 Next 按钮进入图 2-3 所示对话框。

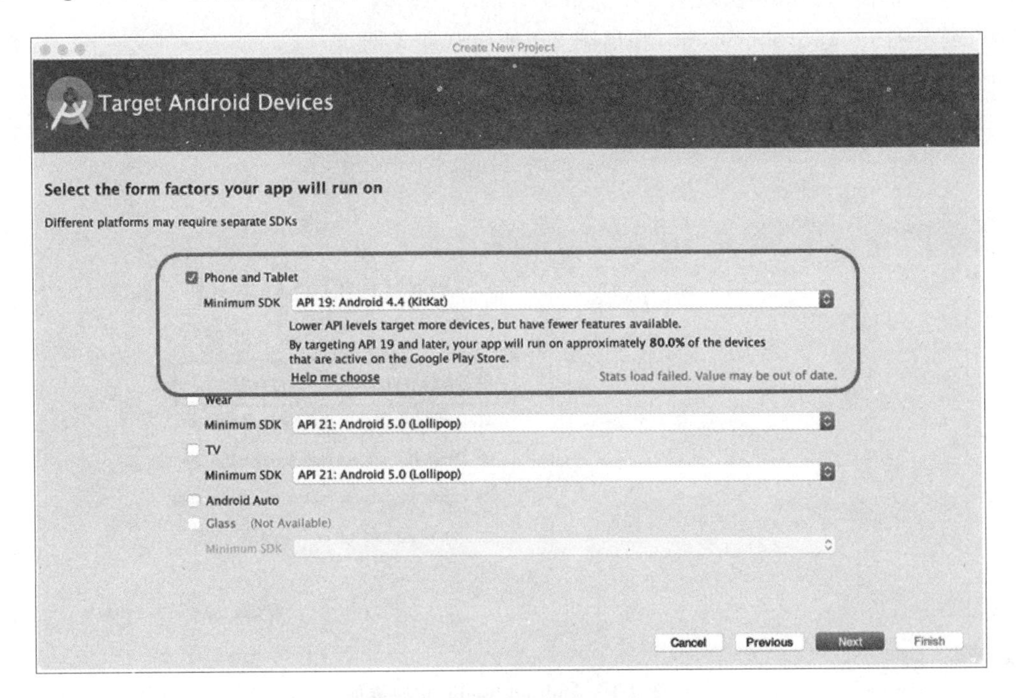

图 2-3 设置目标设备对话框

在该对话框中，选择开发的应用运行的目标设备以及目标设备支持的SDK最低版本。目前支持Phone and Tablet(手机和平板)、Wear(穿戴设备)、TV(电视)和Android Auto(Android车载软件)。本节围绕“手机和平板”开发，因此选择Phone and Table，然后点击Next按钮，进入图2-4所示对话框。

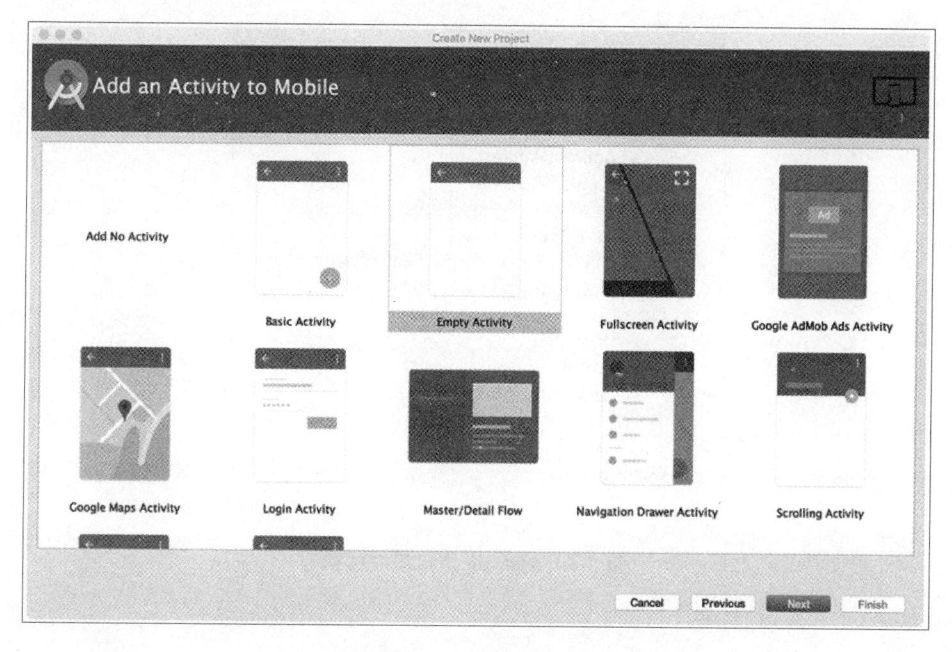

图2-4　创建Activity

Android Studio提供了许多Activity模板，可以根据所创建Activity的实际用途，选择不同的Activity模板。这里选择Empty Activity，然后点击Next按钮，弹出图2-5所示对话框。

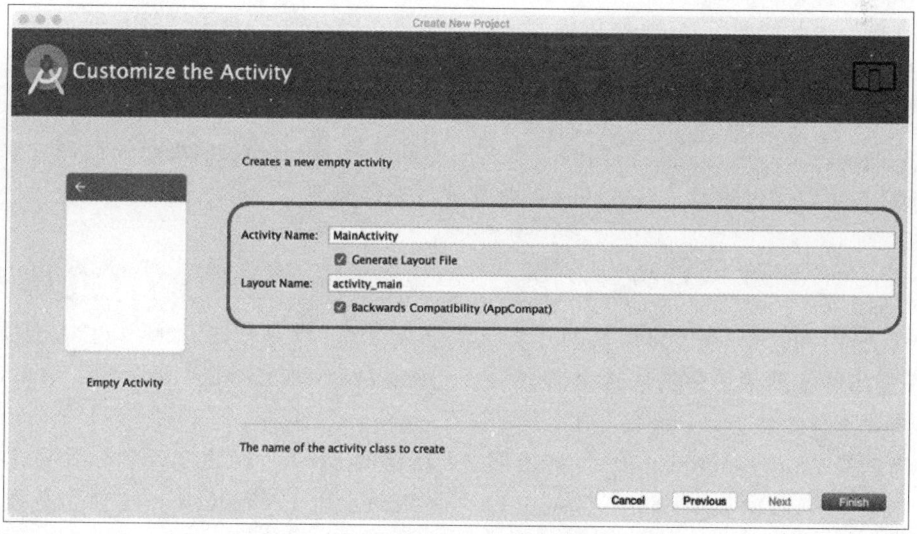

图2-5　Activity配置对话框

输入 Activity Name 和 Layout Name，其中 Activity Name 是创建的 Activity 名字，Layout Name 是 Activity 对应的布局文件名。点击 Finish 按钮，完成 Android Project 的创建流程。

新创建的 Android 项目程序结构如图 2-6 所示。

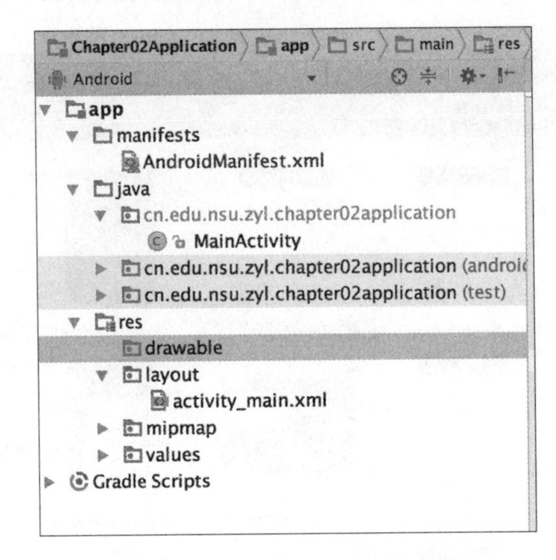

图 2-6 Android 应用程序结构

2.1 Activity 的使用

2.1.1 Activity 简介

Activity 是 Android 程序中最基本的组件，主要用于显示界面以及处理用户在界面上的操作。通常情况下，一个应用程序会包含若干个 Activity，每个 Activity 负责一个界面的展现。

Activity 类中定义了大量回调方法。当 Activity 部署到 Android 应用中后，随着应用程序的运行，Activity 会在不同的状态之间不断切换，Activity 中特定的回调方法将会自动调用，可以通过重写这些方法来对业务逻辑进行处理。

2.1.2 Activity 的创建

在 Android Studio 中可以借助开发向导为 Android 项目新建 Activity。右击项目包名，在弹出的快捷菜单中依次选择 New→Activity 命令，弹出图 2-7 所示的子菜单，根据实际需要选择新建 Activity 的类型。

当选择 Empty Activity 项后，弹出图 2-8 所示的 Configure Activity 对话框，在对应的文本框中分别输入 Activity Name、Layout Name 和 Package name，它们分别对应 Activity 名、布局文件名和包名。

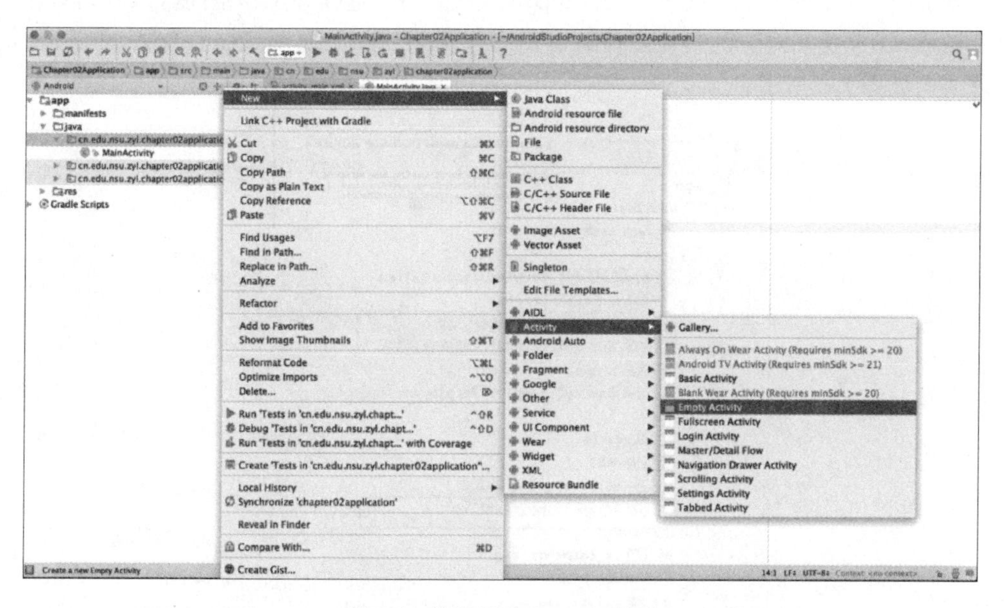

图 2-7 选择新建 Activity 类型

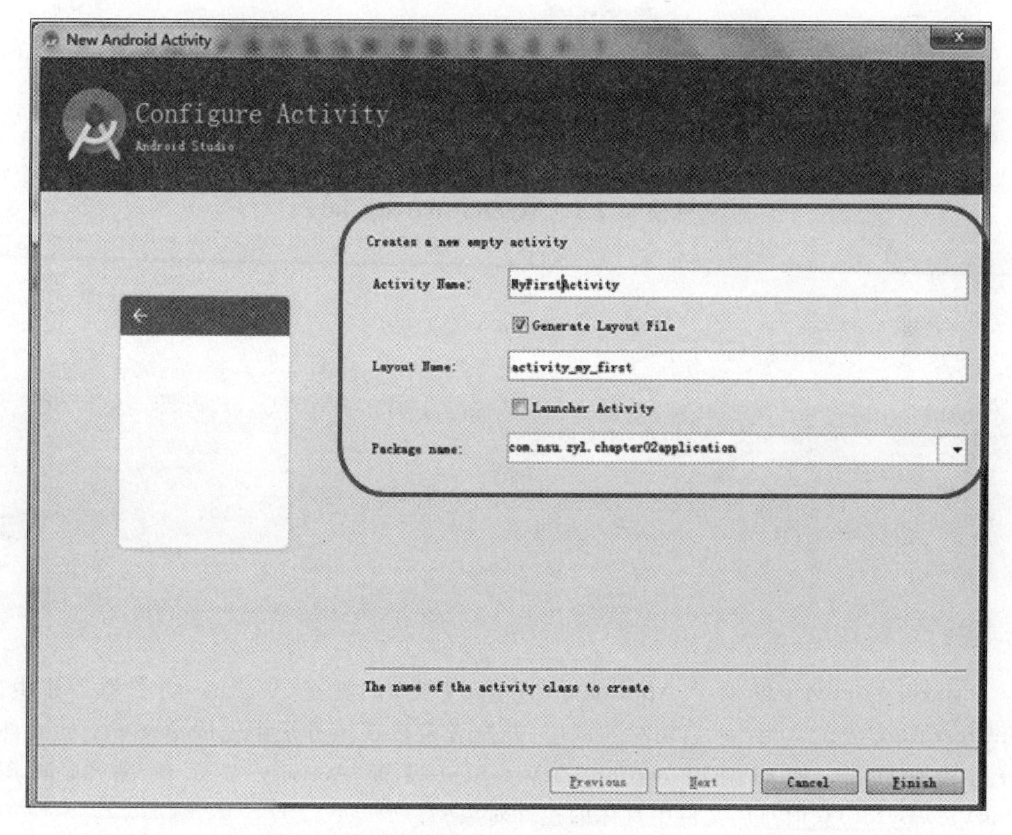

图 2-8 Configure Activity 对话框

点击图 2-8 中的 Finish 按钮，完成 Activity 创建工作，这样在工作空间可以看到新建的 MyFirstActivity 类和对应的布局文件 activity_my_first. xml。现在 Chapter02Application 的目录结构如图 2-9 所示。

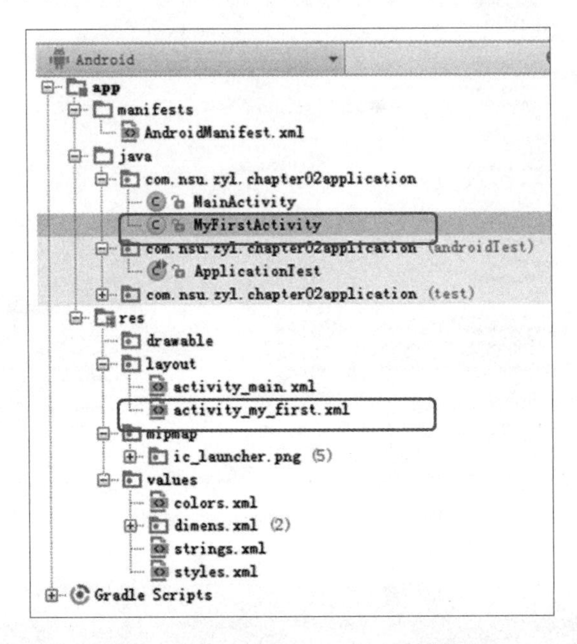

图 2-9　Chapter02Application 目录结构

生成的 MyFirstActivity 类如文件清单 2-1 所示。

文件清单 2-1　MyFirstActivity. java

```
package com.nsu.zyl.chapter02application;

import android.support.v7.app.AppCompatActivity;
import android.os.Bundle;

public class MainActivity extends AppCompatActivity {

    @Override
    protected void onCreate(Bundle savedInstanceState) {
        super.onCreate(savedInstanceState);
        setContentView(R.layout.activity_main);
    }
}
```

MyFirstActivity 继承了 AppCompatActivity 类，该类是 Activity 的子类。其中，onCreate()方法在 Activity 启动时被调用，开发者可以在该方法中完成 Activity 的大部分初始化工作，例如调用 setContentView() 设置 Activity 显示的视图，调用 findViewById()获取布局文件中具体的控件对象。

自动生成的 MyFirstActivity 对应的布局文件 activity_my_first. xml 如文件清单 2-2 所示。

文件清单 2-2 activity_my_first.xml

```
<?xml version="1.0" encoding="utf-8"?>
<RelativeLayout xmlns:android="http://schemas.android.com/apk/res/android"
    xmlns:tools="http://schemas.android.com/tools"
    android:layout_width="match_parent"
    android:layout_height="match_parent"
    android:paddingBottom="@dimen/activity_vertical_margin"
    android:paddingLeft="@dimen/activity_horizontal_margin"
    android:paddingRight="@dimen/activity_horizontal_margin"
    android:paddingTop="@dimen/activity_vertical_margin"

    tools:context="com.nsu.zyl.chapter02application.MyFirstActivity">

</RelativeLayout>
```

<RelativeLayout>表示当前布局采用相对布局方式，相对布局是 Activity 默认的布局方式。双击 MyFirstActivity 对应的布局文件 activity_my_first.xml，在 Android Studio 编辑区域显示图 2-10 所示界面，开发者可以通过点击 Design 或 Text 实现图形化和文本两种编辑方式的切换。

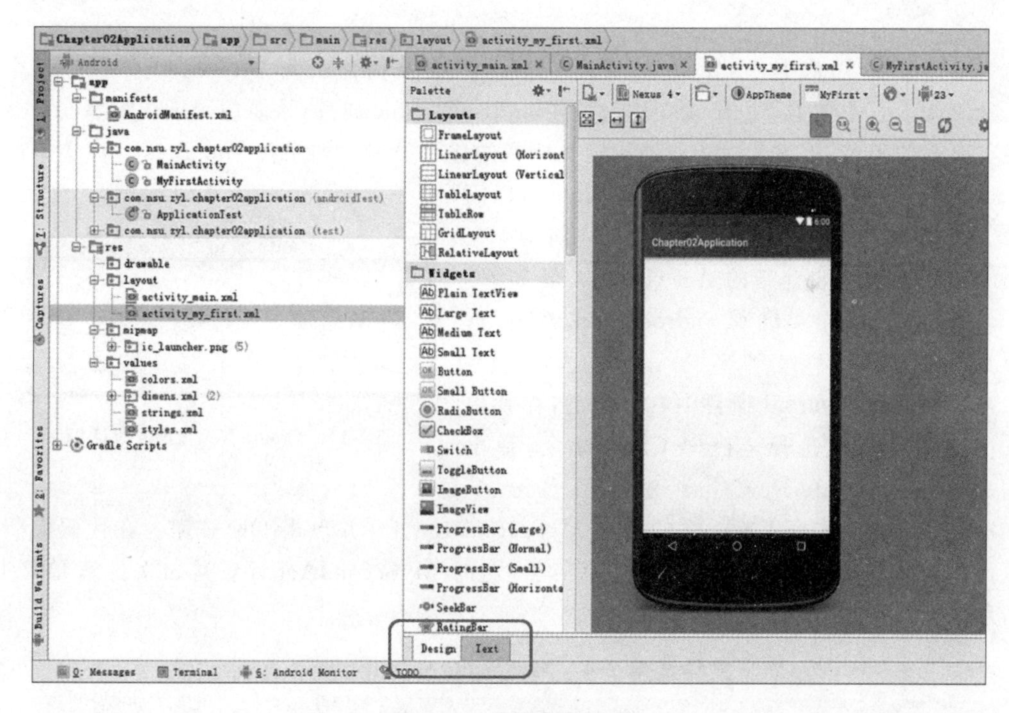

图 2-10 编辑布局文件

定义好的 Activity 如果要在应用中使用，需要在文件清单 2-3 中注册。对于由 Android Studio 创建的 Activity，IDE 会在创建 Activity 的同时，完成该 Activity 在 AndroidManifest.xml 中的注册。

文件清单 2-3　AndroidManifest.xml

```
<?xml version="1.0" encoding="utf-8"?>
<manifest xmlns:android="http://schemas.android.com/apk/res/android"
    package="com.nsu.zyl.chapter02application">

    <application
        android:allowBackup="true"
        android:icon="@mipmap/ic_launcher"
        android:label="@string/app_name"
        android:supportsRtl="true"
        android:theme="@style/AppTheme">
       <activity android:name=".MainActivity">
         <intent-filter>
           <action android:name="android.intent.action.MAIN" />
           <category android:name="android.intent.category.LAUNCHER" />
         </intent-filter>
        </activity>
        <activity android:name=".MyFirstActivity"></activity>
    </application>

</manifest>
```

上述黑体代码是新增的 MyFirstActivity 注册信息。每个 Activity 对应一个<activity>元素，在该<activity>中设置了 android:name 属性，该属性用于指定 Acitivity 的实现类的类名，该类名需要明确指定类所在的包和类名。如果在<manifest>标签中的 package 属性指定了应用程序包，则 name 可以简写为"类名"。

如果不借助 Android Studio 的向导，在 Android 应用中，创建新的 Activity 的步骤如下：

(1) 定义一个类继承 android.app.Activity 或其子类。

在前面的 Android 项目中右击包名，在弹出的快捷菜单依次选择 New→Class 命令，弹出图 2-11 所示 Create New Class 对话框。

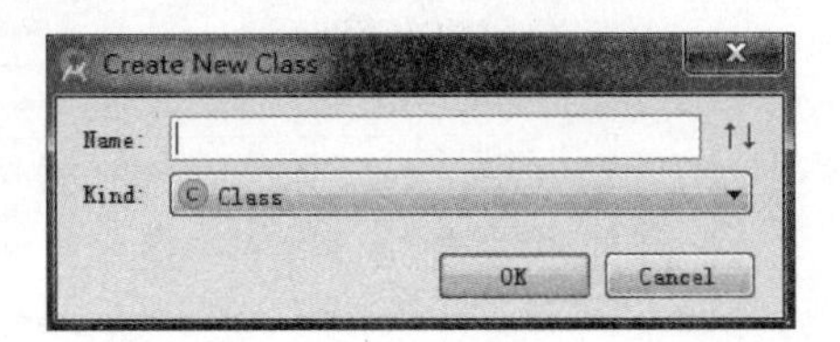

图 2-11　Create New Class 对话框

在 Name 文本框中输入类的名称 MySecondActivity 并点击 OK 按钮。在代码编辑页面让新建的类继承 Activity 类作为父类，这时 MySecondActivity 类如文件清单 2-4 所示。

文件清单 2-4　MySecondActivity.java

```
import android.app.Activity;

public class MySecondActivity extends Activity {

}
```

(2) 在 res/layout 目录下创建 Activity 对应的 xml 布局文件。

右击 layout 文件夹名,在弹出的快捷菜单依次选择 New→XML→Layout XML File 命令,弹出图 2-12 所示对话框,在 Layout File Name 文本框输入布局文件名 activity_my_second,在 Root Tag 文本框输入布局文件根元素标签。

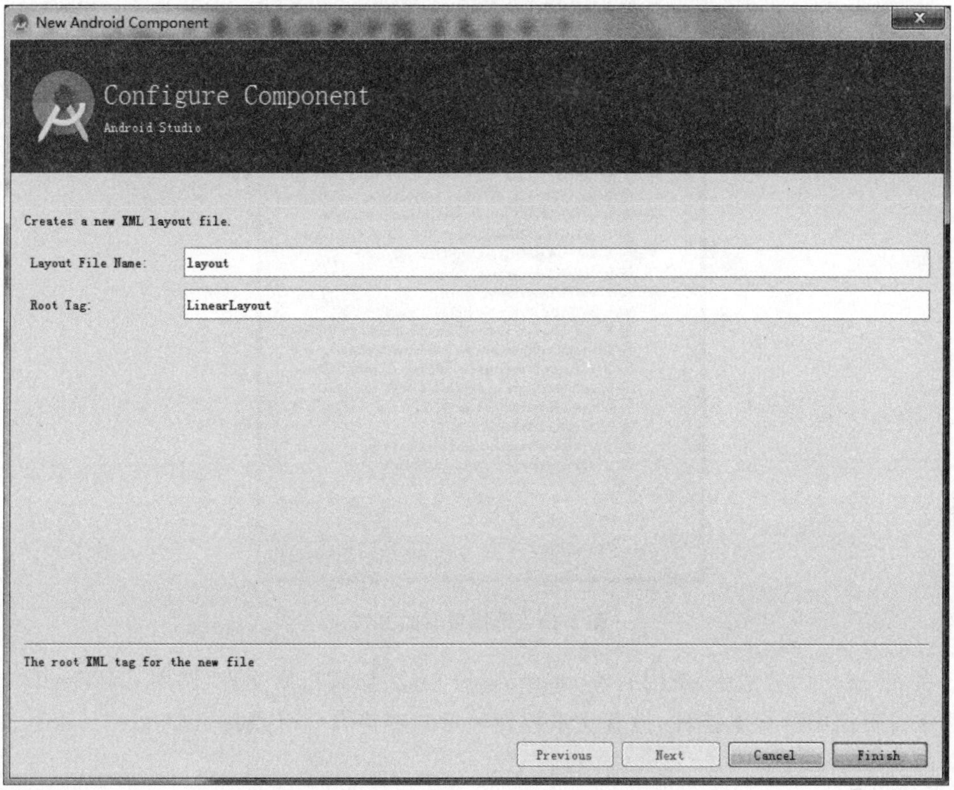

图 2-12　Configure Component 对话框

在新建的布局文件中添加< TextView >控件对象,内容如文件清单 2-5 所示。

文件清单 2-5　activity_my_second.xml

```
<?xml version="1.0" encoding="utf-8"?>
<LinearLayout xmlns:android="http://schemas.android.com/apk/res/android"
    android:layout_width="match_parent"
    android:layout_height="match_parent">
<TextView
        android:layout_width="wrap_content"
        android:layout_height="wrap_content"
        android:text="New Text"
        android:id="@+id/textView" />
</LinearLayout>
```

(3) 重写父类中的一些方法。

选择菜单 Code→Implement Method 命令,弹出所有可以重写的方法,如图 2-13 所示。选择重写 onCreate()方法。

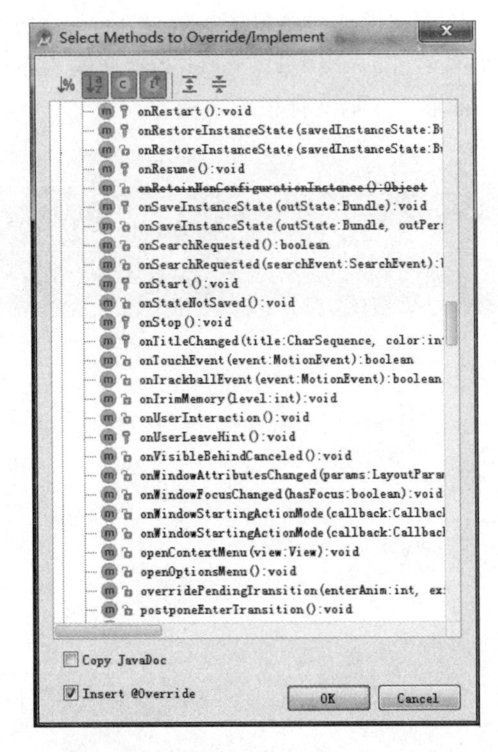

图 2-13 选择重写的方法

在 onCreate()方法中，调用 setContentView()方法加载布局文件，调用 findViewById()方法获得页面中的文本控件，设置文本控件显示的字符串。MySecondActivity. java 文件如文件清单 2-6 所示。

文件清单 2-6 MySecondActivity. java

```
import android.app.Activity;
import android.os.Bundle;
import android.widget.TextView;
import android.app.Activity;
import android.os.Bundle;
import android.widget.TextView;

public class MySecondActivity extends Activity {
    @Override
    protected void onCreate(Bundle savedInstanceState) {
        super.onCreate(savedInstanceState);
        setContentView(R.layout.activity_my_second);
        TextView tv= (TextView) findViewById(R.id.textView);
        tv.setText("Hello World");
    }
}
```

(4) 在 AndroidManifest. xml 文件中对定义的 Activity 进行配置。配置好的 AndroidManifest. xml 如文件清单 2-7 所示。

文件清单 2-7 AndroidManifest. xml

```
<?xml version="1.0" encoding="utf-8"?>
<manifest xmlns:android="http://schemas.android.com/apk/res/android"
    package="com.nsu.zyl.chapter02application">

    <application
        android:allowBackup="true"
        android:icon="@mipmap/ic_launcher"
        android:label="@string/app_name"
        android:supportsRtl="true"
        android:theme="@style/AppTheme">
      <activity android:name=".MainActivity">
        <intent-filter>
          <action android:name="android.intent.action.MAIN" />
          <category android:name="android.intent.category.LAUNCHER" />
        </intent-filter>
        </activity>
        <activity android:name=".MyFirstActivity"></activity>
        <activity android:name=".MySecondActivity"></activity>
    </application>

</manifest>
```

如上所述,完成 MySecondActivity 的创建工作。

2.1.3 Activity 的启动方式

启动 Activity 需要使用 Intent 对象,Android 系统通过 Intent 对象找到需要启动的目标组件。根据被启动 Activity 是否有数据返回,可以将 Activity 的启动方式分为两种,一种是直接启动 Activity,没有返回值;另一种是启动 Activity 后,目标 Activity 有数据返回源 Activity。

1. 直接启动

首先声明 Intent 对象,在 Intent 对象中指定启动源组件和目标组件,然后调用 startActivity(Intent)方法完成 Activity 的启动。

例如,在 Activity A 中启动 Activity B 的代码如下:

```
Intent intent=new Intent(A.this,B.class);
startActivity(intent);
```

上述代码通过 Intent 的构造方法创建了一个 intent 对象。该构造方法接收两个参数,第一个参数 Context 是启动 Activity 的上下文,此处为 A. this;第二个参数 Class 是指定要启动的目标 Activity，此处为 B. class。

2. 启动一个 Activity 并返回结果

如果启动一个 Activity，并且希望返回结果给当前的 Activity，那么可以使用 startActivityForResult()方法启动 Activity，并且当前 Activity 需要重写 onActivityResult()方法处理返回的结果数据。

为了获取被启动 Activity 的返回结果，需要执行以下两个步骤：

(1) 在被启动的 Activity 中调用 setResult(int resultCode, Intent data)方法设置返回的结果数据。

(2) 在当前的 Activity 中重写 onActivityResult(int requestCode, int resultCode, Intent intent)方法处理返回结果，其中 requestCode 代表请求码，用于判断是哪个请求的返回结果；resultCode 代表返回的结果码，判断返回结果的状态。

2.1.4 Activity 之间数据交换

Intent 可以用来启动 Activity，也可以用来在 Activity 之间传递数据。使用 Intent 传递数据只需要调用 Intent 提供的 putExtra(String name, Xxx data)方法，将想要传递的数据以 Key-Value 的形式放到 Intent 中，Xxx 表示存储数据的类型。

例如，从 Activity A 启动 Activity B 时携带 Xxx 类型的数据 data，相应代码如下：

```
Xxx data=" ";          //Xxx 代表数据类型
Intent intent=new Intent(A.this,B.class);
intent.putExtra("key",data);
startActivity(intent);
```

同样，Intent 提供相应的 getXxxExtra(String name)方法，取出 Intent 中 key 为 name 的数据，Xxx 表示存储数据的类型。在被启动的 Activity B 中，把携带的数据取出的代码如下：

```
Intent intent=getIntent();
Xxx data=intent.getXxxExtra();
```

也可以使用 putExtras()方法传递 Bundle 类型的数据，Bundle 是一个数据包，以 Key-Value 的形式存储数据，通过 Key 可以得到 Bundle 存储的 Value 值。

Bundle 提供如下方法放入 Key-Value 形式的数据：

putXxx(String key, Xxx data)：向 Bundle 放入 int、long、String 等类型的数据。

putSerializable(String key, Serializable data)：向 Bundle 放入一个可序列化的对象。

同样，Bundle 提供了根据 Key 获取数据的方法：

getXxx(String name)：取出 Intent 中 key 为 name 的值为 Xxx 类型的数据。

getSerializable(String key, Serializable data)：从 Intent 中获取一个可以序列化的数据。

调用 putExtras()方法传递 Bundle 数据的代码如下：

```
Bundle  bundle=new Bundle();                    //创建 Bundle 对象
bundle.putString("name","Jerry");               //向 bundle 对象放入 Key-Value 形式数据
bundle.putInt("age",24);
Intent intent=new Intent(A.this,B.class); //创建显示意图
intent.putExtras(bundle);                       //将 bundle 对象放入 Intent 中
startActivity(intent);
```

如果要在被启动的 Activity 中取出上述方式传递的数据，可以使用如下代码：

```
Intent intent=getIntent();                      //获取启动该组件的 Intent 对象
Bundle bundle=intent.getExtras();               //获取 Intent 携带的数据包
String stuName=bundle.getString("name"); //读取 bundle 对象中携带的 Key 为"name"的
                                                //数据
int stuAge=bundle.getInt("age");
```

2.1.5 案例

完善 Chapter02Application 项目，在 MainActivity 中分别启动 MyFirstActivity 和 MySecondActivity。

首先对 Chapter02Application 项目中 MainActivity 对应的布局文件 activity_main.xml 进行编辑，编辑后的 activity_main.xml 如文件清单 2-8 所示。

文件清单 2-8 activity_main.xml

```
<?xml version="1.0" encoding="utf-8"?>
<RelativeLayout xmlns:android="http://schemas.android.com/apk/res/android"
    xmlns:tools="http://schemas.android.com/tools"
    android:layout_width="match_parent"
    android:layout_height="match_parent"
    android:paddingBottom="@dimen/activity_vertical_margin"
    android:paddingLeft="@dimen/activity_horizontal_margin"
    android:paddingRight="@dimen/activity_horizontal_margin"
    android:paddingTop="@dimen/activity_vertical_margin"
    tools:context="com.nsu.zyl.chapter02application.MainActivity">
    <TextView
        android:layout_width="wrap_content"
        android:layout_height="wrap_content"
        android:text="用户名："
        android:id="@+id/txtName"
        android:layout_alignParentTop="true"
        android:layout_alignParentLeft="true"
        android:layout_marginTop="42dp" />
    <TextView
        android:layout_width="wrap_content"
```

```
        android:layout_height="wrap_content"
        android:text="密码："
        android:id="@+id/txtPwd"
        android:layout_below="@+id/txtName"
        android:layout_alignParentLeft="true"
        android:layout_marginTop="39dp" />
    <EditText
        android:layout_width="match_parent"
        android:layout_height="wrap_content"
        android:id="@+id/editName"
        android:layout_alignTop="@+id/txtName"
        android:layout_toRightOf="@+id/txtName" />
    <EditText
        android:layout_width="wrap_content"
        android:layout_height="wrap_content"
        android:id="@+id/editPwd"
        android:layout_alignTop="@+id/txtPwd"
        android:layout_alignParentRight="true"
        android:layout_alignLeft="@+id/editName" />
    <Button
        android:layout_width="wrap_content"
        android:layout_height="wrap_content"
        android:text="注册"
        android:id="@+id/btnReg"
        android:layout_below="@+id/editPwd"
        android:layout_toRightOf="@+id/txtPwd"
        android:layout_marginTop="37dp" />
    <Button
        android:layout_width="wrap_content"
        android:layout_height="wrap_content"
        android:text="登录"
        android:id="@+id/btnLogin"
        android:layout_alignTop="@+id/btnReg"
        android:layout_toRightOf="@+id/btnReg"
        android:layout_marginLeft="36dp" />
</RelativeLayout>
```

activity_main.xml 布局文件对应的页面如图 2-14 所示。

当用户在 MainActivity 页面点击"登录"按钮时，启动 MyFirstActivity，当前应用由 MainActivity 页面切换到 MyFirstActivity 对应的页面。

在 Android 中为用户操作提供响应的机制称为事件处理机制。Android 本身提供了强大的事件处理机制，包括基于监听的事件处理机制和基于回调的事件处理机制。基于回调的事件处理机制主要用于 Android 组件内特定的回调方法处理具有通用性的事件；对于某些特定的事件，则主要采用基于监听的事件处理机制。

Android 中基于监听的事件处理机制与 Java AWT、Swing 中的事件处理机制类似，首先在界面中获取组件(事件源)；然后定义实现 XxxListener 接口的事件监听器；最后

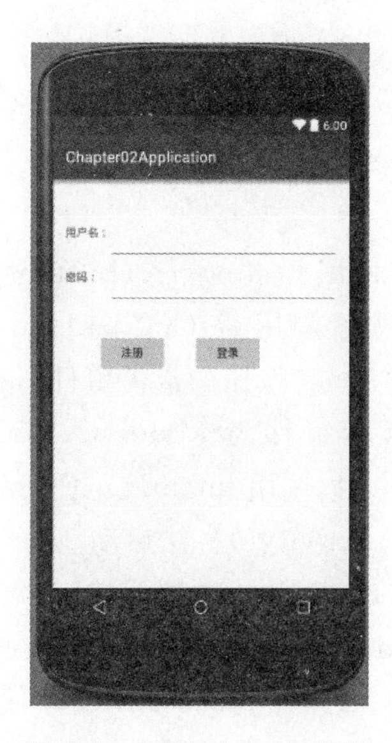

图 2-14 MainActivity 页面

通过调用事件源组件的 setXxxListener()方法将事件监听器对象注册到组件上。

在 MainActivity 中点击“登录”按钮启动 MyFirstActivity 功能，因此需要为“登录”按钮设置 OnClickListener 监听器对象，同时在监听器中重写 onClick(View v)方法，使用 Intent 启动 MyFirstActivity，如文件清单 2-9 所示。

文件清单 2-9 MainActivity. java

```
import android.content.Intent;
import android.support.v7.app.AppCompatActivity;
import android.os.Bundle;
import android.view.View;
import android.widget.Button;

public class MainActivity extends AppCompatActivity {
    private Button btnLogin,btnReg;
    @Override
    protected void onCreate(Bundle savedInstanceState) {
        super.onCreate(savedInstanceState);
        setContentView(R.layout.activity_main);
        btnLogin=(Button)findViewById(R.id.btnLogin);
        btnLogin.setOnClickListener(new View.OnClickListener() {
            @Override
            public void onClick(View v) {
                Intent intent=new Intent(MainActivity.this, MyFirstActivity.class);
                String name=edtName.getText().toString();
                String pwd=edtPwd.getText().toString();
                intent.putExtra("name",name);
                intent.putExtra("pwd",pwd);
                startActivity(intent);
```

```
            }
        });
    }
}
```

在 onCreate()方法中，首先使用 btnLogin=(Button)findViewById(R. id. btnLogin)获得页面中“登录”按钮对象，然后调用 setOnClickListener()为“登录”按钮注册一个 OnClickListener 对象用于监听“登录”按钮的点击事件。在监听器的 onClick()方法中定义 Intent 对象，指定启动的目标组件为 MyFirstActivity，通过 edtName. getText(). toString()获取用户输入的用户名，调用 intent. putExtra()方法将用户输入的账号和密码携带到 Intent，然后调用 startActivity()方法启动 Activity。

在 MainActiviy 中，当用户点击“注册”按钮后跳转至第二个 Activity 即 MySecondActivity，在 MySecondActivity 完成注册后将注册信息返回到 MainActivity，并在文本框和密码框显示返回的值。因此在 MainActivity 中使用 startActivityForResult()方法启动 MySecondActivity。

MainActivity 对应的源代码如文件清单 2-10 所示。

文件清单 2-10　MainActivity. java

```java
import android.content.Intent;
import android.support.v7.app.AppCompatActivity;
import android.os.Bundle;
import android.view.View;
import android.widget.Button;
import android.widget.EditText;

public class MainActivity extends AppCompatActivity {
    private Button btnLogin,btnReg;
    private EditText edtName,edtPwd;
    private final int REQUEST_CODE=101;
    @Override
    protected void onCreate(Bundle savedInstanceState) {
        super.onCreate(savedInstanceState);
        setContentView(R.layout.activity_main);
        btnLogin= (Button)findViewById(R.id.btnLogin);
        btnReg= (Button)findViewById(R.id.btnReg);
        edtName= (EditText)findViewById(R.id.edtName);
        edtPwd= (EditText)findViewById(R.id.edtPwd);
        btnLogin.setOnClickListener(new View.OnClickListener() {
            @Override
            public void onClick(View v) {
                Intent intent=new Intent(MainActivity.this, MyFirstActivity.class);
                String name=edtName.getText().toString();
                String pwd=edtPwd.getText().toString();
                intent.putExtra("name",name);
```

```
                intent.putExtra("pwd",pwd);
                startActivity(intent);            }
        });
        btnReg.setOnClickListener(new View.OnClickListener() {
            @Override
            public void onClick(View v) {
                Intent intent=new Intent(MainActivity.this, MySecondActivity.class);
                startActivityForResult(intent,REQUEST_CODE);
            }
        });

    }
    @Override
    protected void onActivityResult(int requestCode, int resultCode, Intent data) {
        super.onActivityResult(requestCode, resultCode, data);
        if(requestCode==REQUEST_CODE){
            String name=data.getStringExtra("name");
            String pwd=data.getStringExtra("pwd");
            edtName.setText(name);
            edtPwd.setText(pwd);
        }
    }

}
```

在"注册"按钮绑定的点击事件监听器处理方法中，调用 startActivityForResult(intent, REQUEST_CODE)方法启动 MySecondActivity。为了让当前 Activity 处理 MySecondActivity 的返回结果，重写 onActivityResult()方法。当被启动的 MySecondActivity 返回结果时，onActivityResult()方法将被回调处理返回结果。

MySecondActivity 对应的布局文件 activity_my_second.xml 如文件清单 2-11 所示。

文件清单 2-11 activity_my_second.xml

```
<?xml version="1.0" encoding="utf-8"?>
<LinearLayout xmlns:android="http://schemas.android.com/apk/res/android"
    android:layout_width="match_parent"
    android:layout_height="match_parent"
    android:orientation="vertical"
    >
    <LinearLayout
        android:orientation="horizontal"
        android:layout_width="match_parent"
        android:layout_height="73dp">
        <TextView
            android:layout_width="wrap_content"
```

```
        android:layout_height="wrap_content"
        android:text="用户名："
        android:id="@+id/textView" />
    <EditText
        android:layout_width="343dp"
        android:layout_height="wrap_content"
        android:id="@+id/edtName" />
</LinearLayout>

<LinearLayout
    android:orientation="horizontal"
    android:layout_width="match_parent"
    android:layout_height="73dp">
    <TextView
        android:layout_width="wrap_content"
        android:layout_height="wrap_content"
        android:text="密码："
        android:id="@+id/textView2" />
    <EditText
        android:layout_width="match_parent"
        android:layout_height="wrap_content"
        android:id="@+id/edtPwd" />
</LinearLayout>

<LinearLayout
    android:orientation="horizontal"
    android:layout_width="match_parent"
    android:layout_height="wrap_content">
    <TextView
        android:layout_width="wrap_content"
        android:layout_height="wrap_content"
        android:text="重新输入密码"
        android:id="@+id/textView3" />
    <EditText
        android:layout_width="match_parent"
        android:layout_height="wrap_content"
        android:id="@+id/edtRepwd" />
</LinearLayout>
<Button
    android:layout_width="wrap_content"
    android:layout_height="wrap_content"
    android:text="注册"
    android:id="@+id/btnReg" />

</LinearLayout>
```

MySecondActivity 最终内容如文件清单 2-12 所示。

文件清单 2-12 MySecondActivity. java

```
import android.app.Activity;
import android.content.Intent;
import android.os.Bundle;
import android.support.v7.app.AppCompatActivity;
import android.view.View;
import android.widget.Button;
import android.widget.EditText;
import android.widget.TextView;
import android.widget.Toast;

public class MySecondActivity extends AppCompatActivity {
    private Button btnReg;
    private EditText edtName,edtPwd,edtRePwd;
    private static final int RESULT_CODE=101;

    @Override
    protected void onCreate(Bundle savedInstanceState) {
        super.onCreate(savedInstanceState);
        setContentView(R.layout.activity_my_second);
        btnReg= (Button)findViewById(R.id.btnReg);
        edtName= (EditText)findViewById(R.id.edtName);
        edtPwd= (EditText)findViewById(R.id.edtPwd);
        edtRePwd= (EditText)findViewById(R.id.edtRepwd);
        btnReg.setOnClickListener(new View.OnClickListener() {
            @Override
            public void onClick(View v) {
                String name=edtName.getText().toString();
                String pwd=edtPwd.getText().toString();
                String repwd=edtRePwd.getText().toString();
                if(!"".equals(pwd)&&pwd.equals(repwd)){
                    //获得启动该 Activity 的 Intent 对象
                    Intent intent=getIntent();
                    intent.putExtra("name",name);
                    intent.putExtra("pwd", pwd);
                    //设置结果码,并设置结束后返回的 Activity
                    setResult(RESULT_CODE,intent);
                    //结束 RegActivity
                    MySecondActivity.this.finish();
                }else{
                    Toast.makeText(MySecondActivity.this,"密码输入不一致",Toast.
LENGTH_LONG).show();
                }
            }
```

```
        });
    }
}
```

当用户在MySecondActivity对应的页面点击“注册”按钮时，程序将获取用户输入的用户名、密码和重复密码，当密码和重复密码相同时，将携带用户输入的用户名和密码返回启动当前Activity的MainActivity。

2.2 Activity的生命周期

对象从创建到销毁的整个过程称为生命周期，每一个对象都有自己的生命周期。Activity也是有生命周期的。

2.2.1 Activity的状态

Activity的生命周期可以分成四种状态，即运行状态、暂停状态、停止状态和销毁状态。

(1) 运行状态：当Activity显示在屏幕的最前端，能够获得焦点，可以处理用户的交互操作时，该Activity处于运行状态。处于运行状态的Activity具有较高优先权，Android会尽可能地保持它的运行，即使出现内存不足的情况，Android也会杀死处于停止状态的Activity，来确保运行状态Activity的正常运行。

(2) 暂停状态：在某些情况下Activity对用户来说仍然可见，但它不再拥有焦点，不能处理用户对它的操作，这时Activity处于暂停状态。例如，当最上面的Activity没有完全覆盖屏幕或者是半透明时，这时被覆盖的Activity仍然对用户可见，但是已经不能处理用户的操作。当内存不足时，处于暂停状态的Activity可能会被杀死。

(3) 停止状态：当Activity完全不可见时，该Activity就处于停止状态。处于停止状态的Activity仍然保留着当前状态和成员信息。对于处于停止状态的Activity，当内存不足时很容易被杀死。

(4) 销毁状态：当Activity被销毁时，该Activity就处于销毁状态。

Activity在不同状态间切换时，会触发回调方法的执行。在Activity生命周期中会涉及onCreate()、onStart()、onResume()、onPause()、onStop()、onRestart()和onDestroy()等回调方法的执行。

2.2.2 Activity的生命周期及回调方法

图2-15所示为Android官方文档提供的Activity生命周期。

Activity生命周期中涉及的回调方法如表2-1所示。

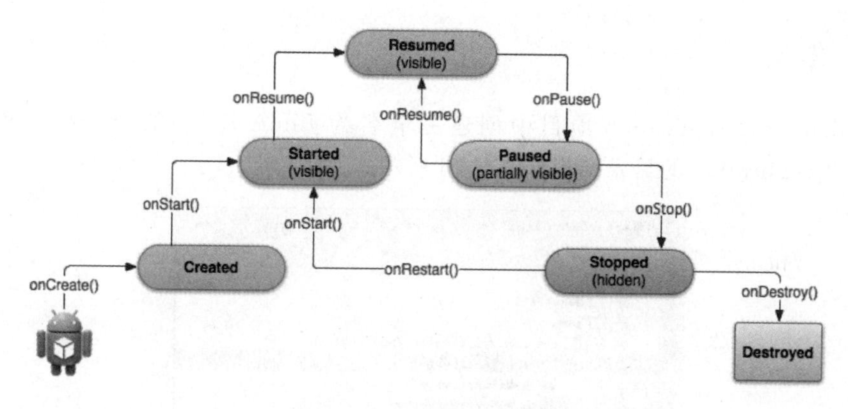

图 2-15 Activity 生命周期

表 2-1 回调方法

方　法	描　述
onCreate(Bundle savedInstanceState)	当 Activity 第一次被创建时调用，可以在该方法体中实现 Activity 的初始化设置
onStart()	当 Activity 正在变为用户所见时被调用
onRestart()	当 Activity 停止后，再次启动前被调用
onResume()	当 Activity 开始与用户进行交互之前被调用。此时 Activity 位于栈顶，用户可见
onPause()	当启动另一个 Activity 或者弹出对话框时调用。此方法主要用于将持久性数据写入存储中，这一切动作需要在短时间内完成，下一个 Activity 必须等到此方法返回才会继续
onStop	当 Activity 不再为用户可见时调用此方法
onDestroy	在 Activity 销毁前调用

Activity 从启动到关闭会依次执行 onCreate()→onStart()→onResume()→onPause()→onStop()→onDestroy()方法。所有的 Activity 都必须实现 onCreate()方法，在该方法中可以对 Activity 进行一些初始化设置。当 Activity 执行到 onPause()方法失去焦点且处于可见状态时，若要重新回到可见且可交互状态，将会回调执行 onResume()方法。当执行到 onStop()方法 Activity 不可见时，再次回到前台可见状态，将会执行 onRestart()方法和 onStart()方法。如果进程被杀死了，若要重新启动 Activity，将会重新执行 onCreate()方法。Activity 被销毁前会执行 onDestroy()方法，释放所有的资源。

Activity 的 7 个回调方法构成了 Activity 完整的生命周期，在完整生命周期中又包含 3 个嵌套的生命周期。

(1) 前台生命周期：始于 onResume()方法调用，止于 onPause()方法调用。处于前台生命周期的 Activity 处于可见且可以交互状态。

(2) 可视生命周期：始于 onStart()方法调用，止于 onStop()方法调用，在此期间用户可以在屏幕看到 Activity。

(3) 完整生命周期：始于第一次调用 onCreate()，止于 onDestroy()方法调用。

2.2.3 案例

在 Chapter02Application 项目中创建一个名为 LifeActivity 的 Activity，创建后，Chapter02Application 的目录结构如图 2-16 所示。

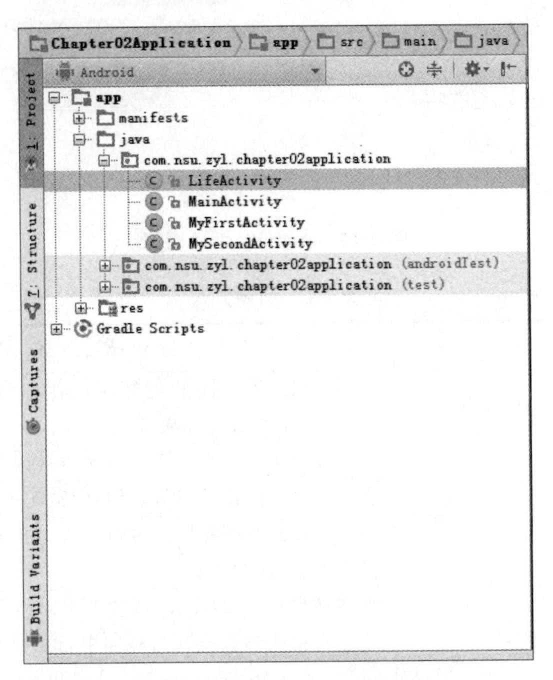

图 2-16 Chapter02Application 项目目录结构

在 LifeActivity 中重写 onCreate( )、onStart( )、onResume( )、onRestart( )、onPause( )、onStop( ) 和 onDestroy() 方法，在这些方法中使用 Log.i() 方法输出一些语句。LifeActivity 如文件清单 2-13 所示。

文件清单 2-13 LifeActivity.java

```
import android.app.Activity;
import android.os.Bundle;
import android.util.Log;
public class LifeActivity extends Activity {
    private static final String TAG="--LifeActivity--";
    @Override
    protected void onCreate(Bundle savedInstanceState) {
        super.onCreate(savedInstanceState);
        setContentView(R.layout.activity_lifie);
        Log.i(TAG,"onCreate is running...");
}
@Override
    protected void onStart() {
        super.onStart();
```

```
        Log.i(TAG,"onStart is running...");
    }@Override
    protected void onResume() {
        super.onResume();
        Log.i(TAG,"onResume is running...");
    }
@Override
    protected void onRestart() {
        super.onRestart();
        Log.i(TAG,"onRestart is running...");
    }
@Override
    protected void onPause() {
        super.onPause();
        Log.i(TAG,"onPause is running...");
    }
@Override
    protected void onStop() {
        super.onStop();
        Log.i(TAG,"onStop is running...");
    }
    @Override
    protected void onDestroy() {
        super.onDestroy();
        Log.i(TAG,"onDestroy is running...");
    }
   }
```

通过修改 AndroidManifest. xml，更改 LifeActivity 作为启动 Activity，修改后的 AndroidManifest. xml 如文件清单 2-14 所示。

文件清单 2-14　AndroidManifest. xml

```
<?xml version="1.0" encoding="utf-8"?>
<manifest xmlns:android="http://schemas.android.com/apk/res/android"
    package="com.nsu.zyl.chapter02application">

    <application
        android:allowBackup="true"
        android:icon="@mipmap/ic_launcher"
        android:label="@string/app_name"
        android:supportsRtl="true"
        android:theme="@style/AppTheme">
        <activity android:name=".MainActivity">
```

```
        </activity>
        <activity android:name=".MyFirstActivity" />
        <activity android:name=".MySecondActivity" />
        <activity android:name=".LifeActivity">
            <intent-filter>
            <action android:name="android.intent.action.MAIN" />

            <category android:name="android.intent.category.LAUNCHER" />
            </intent-filter>
        </activity>
    </application>

</manifest>
```

运行 LifeActivity,在 logcat 管理器中的输出如图 2-17 所示。

图 2-17 运行 LifeActivity logcat 日志输出

点击 Home 按钮回到桌面时,logcat 的日志输出如图 2-18 所示。

```
11-21 01:58:16.924 19445-19445/com.nsu.zyl.chapter02.chapter02project I/--LifeActivity--: onCreate is running...
11-21 01:58:16.924 19445-19445/com.nsu.zyl.chapter02.chapter02project I/--LifeActivity--: onStart is running...
11-21 01:58:16.924 19445-19445/com.nsu.zyl.chapter02.chapter02project I/--LifeActivity--: onResume is running...
11-21 01:58:46.754 19445-19445/com.nsu.zyl.chapter02.chapter02project I/--LifeActivity--: onPause is running...
11-21 01:58:46.784 19445-19445/com.nsu.zyl.chapter02.chapter02project I/--LifeActivity--: onStop is running...
```

图 2-18 点击 Home 按钮 logcat 日志输出

再次回到 LifeActivity 页面,logcat 的日志输出如图 2-19 所示。

```
11-21 01:58:16.924 19445-19445/com.nsu.zyl.chapter02.chapter02project I/--LifeActivity--: onCreate is running...
11-21 01:58:16.924 19445-19445/com.nsu.zyl.chapter02.chapter02project I/--LifeActivity--: onStart is running...
11-21 01:58:16.924 19445-19445/com.nsu.zyl.chapter02.chapter02project I/--LifeActivity--: onResume is running...
11-21 01:58:46.754 19445-19445/com.nsu.zyl.chapter02.chapter02project I/--LifeActivity--: onPause is running...
11-21 01:58:46.784 19445-19445/com.nsu.zyl.chapter02.chapter02project I/--LifeActivity--: onStop is running...
11-21 02:03:14.194 19445-19445/com.nsu.zyl.chapter02.chapter02project I/--LifeActivity--: onRestart is running...
11-21 02:03:14.194 19445-19445/com.nsu.zyl.chapter02.chapter02project I/--LifeActivity--: onStart is running...
11-21 02:03:14.204 19445-19445/com.nsu.zyl.chapter02.chapter02project I/--LifeActivity--: onResume is running...
```

图 2-19 再次运行 LifeActivity logcat 日志输出

点击 Back 按钮，logcat 的日志输出如图 2-20 所示。

图 2-20　点击 Back 按钮 logcat 日志输出

2.3 Activity 启动模式

Android 采用任务栈的方式来管理 Activity 的实例。任务栈默认采用“后进先出”的原则，当启动一个应用时，Android 就会为之创建一个任务栈，启动一个 Activity，该 Activity 就会被压入任务栈中，先启动的 Activity 压在栈底，后启动的 Activity 放在栈顶，当 Activity 结束时，该 Activity 会从该任务栈中弹出(图 2-21)。

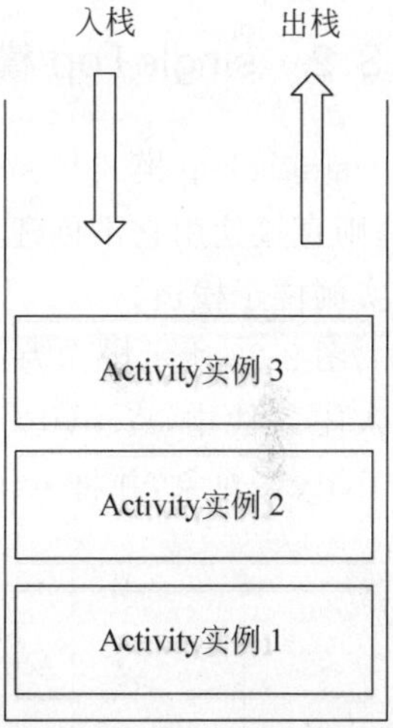

入栈　出栈

Activity实例 3

Activity实例 2

Activity实例 1

图 2-21　任务栈

Android 为 Activity 定义了四种启动模式，即 standard、singleTop、singleTask 和 singleInstance。通过启动模式可以控制 Activity 在任务栈的加载情况。在 AndroidManifest.xml 中，可以通过< activity >标签的 android：launchMode 属性设置 Activity 启动模式。

2.3.1 standard 模式

standard 模式是 Activity 默认的启动模式，在不指定 Activity 启动模式的情况下，所有 Activity 使用的都是 standard 模式。在 standard 模式下，每次启动 Activity 都会创建一个新的 Activity 对象，把它压入任务栈，并处于栈顶的位置。对于使用 standard 模式启动的 Activity，系统不会判断该 Activity 在栈中是否存在，每次启动都会创建一个新的实例。

图 2-22 所示的操作是：启动应用，系统创建一个任务栈，并将创建的 Activity 1 对象压入任务栈中，Activity 1 位于栈顶处于活动状态，用户在手机端看到的是 Activity 1；在

Activity 1 中启动 Activity 2,系统创建 Activity 2 对象,并将其压入任务栈,这时 Activity 2 位于栈顶,处于活动状态,Activity 1 位于栈底处于暂停或者停止状态; Activity 2 中启动 Activity 2,系统会再次创建 Activity 2 对象并压入栈中,新压入栈顶的 Activity 2 处于活动状态,原来 Activity 2 处于暂停或停止状态; 在 Activity 2 中启动 Activity 3,系统新建 Activity 3 对象压入栈中,Activity 3 处于活动状态。这时连续点击 Back 按钮,Activity 3→Activity 2→Activity 2→Activity 1 会按照次序依次退栈。

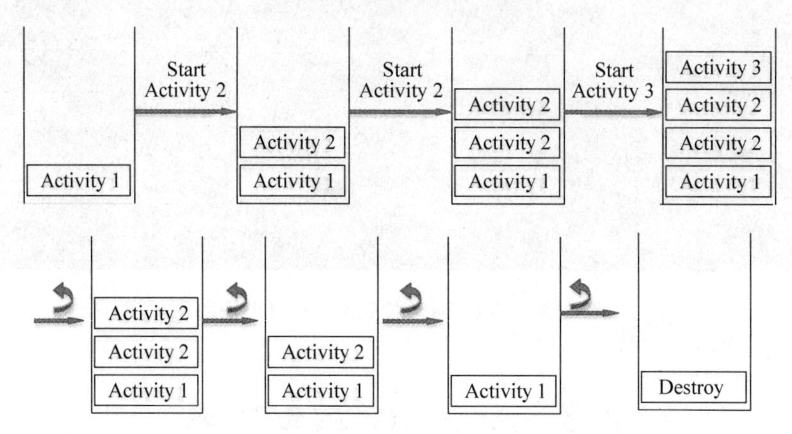

图 2-22 standard 模式任务栈变化

2.3.2 singleTop 模式

singleTop 模式与 standard 模式类似,不同的是,当启动的 Activity 已经位于栈顶时,则直接使用它不创建新的实例。如果启动的 Activity 没有位于栈顶,则创建一个新的实例位于栈顶。

图 2-23 所示操作为: 启动应用,系统创建对应的任务栈,并将启动的 Activity 1 对象压入任务栈中,Activity 1 处于运行状态; 在 Activity 1 中启动 Activity 2,系统新建 Activity 2 对象,并将 Activity 2 压入栈顶,使其处于运行状态; 如果在 Activity 2 位于栈

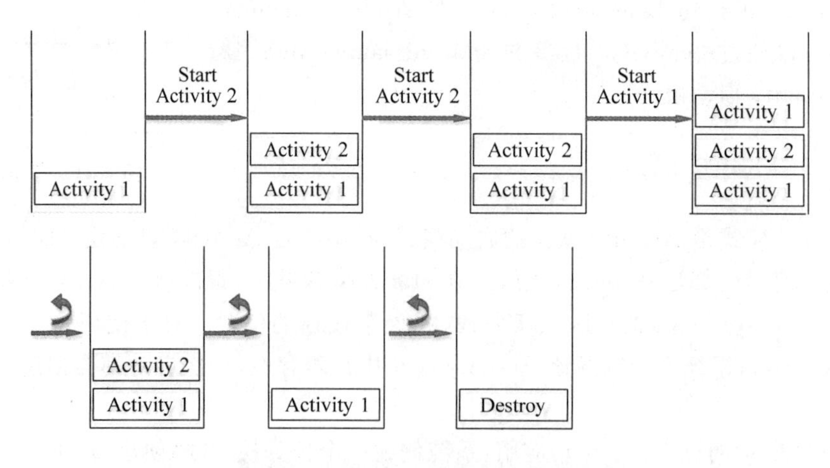

图 2-23 singleTop 模式任务栈变化

顶,处于活动状态时再启动 Activity 2,由于栈顶对象就是 Activity 2,所以不会创建新的 Activity 2 对象;如果启动 Activity 1,由于 Activity 1 位于栈中但不位于栈顶,就会创建 Activity 1 对象压入栈中。这时,用户连续点击 Back 按钮,Activity 1→Activity 2→Activity 1 会按照次序依次退栈。

2.3.3 singleTask 模式

如果希望 Activity 在整个应用中只存在一个实例,可以使用 singleTask 模式,当 Activity 的启动模式指定为 singleTask 时,每次启动该 Activity,系统首先会检查栈中是否存在该 Activity 的实例,如果发现已经存在则直接使用该实例,并将当前 Activity 之上的所有 Activity 出栈,如果没有发现则创建一个新的实例。

图 2-24 所示为任务栈变化,首先启动 Activity 1,Activity 1 压入任务栈,位于栈顶位置,这时 Activity 1 处于运行状态;通过 Activity 1 启动 Activity 2,这时 Activity 2 被压入栈,位于栈顶位置,处于运行状态,Activity 1 由运行状态转换为暂停状态或者停止状态;通过 Activity 2 启动 Activity 2,由于任务栈中存在 Activity 2 实例对象,所以不会创建新的 Activity 2 对象,而使用当前 Activity 2 作为启动对象。通过 Activity 2 启动 Activity 1,由于任务栈包括 Activity 1 对象,所以将任务栈中位于 Activity 1 对象上的 Activity 对象出栈,使 Activity 1 位于栈顶,处于运行状态。

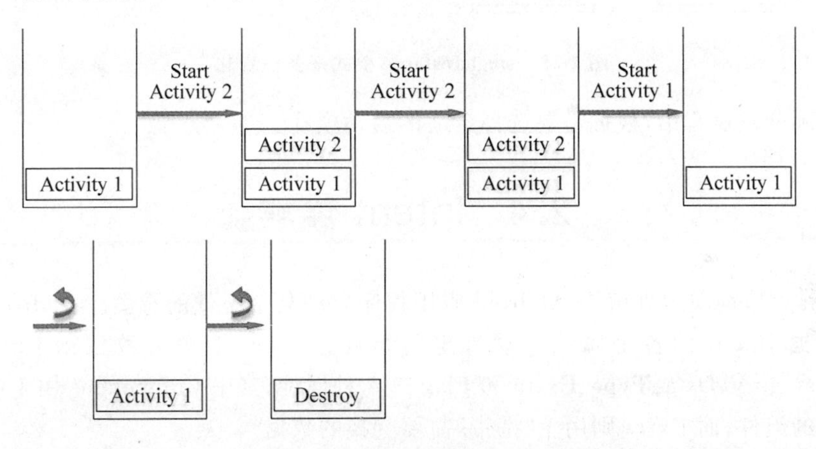

图 2-24 singleTask 模式任务栈变化

2.3.4 singleInstance 模式

在整个程序中,如果需要 Activity 在整个系统中都只有一个实例,就需要用到 singleInstance 模式。不同于上述三种模式,指定为 singleInstance 模式的 Activity 会启动一个新任务栈来管理这个 Activity。

singleInstance 模式加载 Activity 时,无论从哪个任务栈中启动该 Activity,都只会创建一个 Activity 实例,并且使用一个全新的任务栈来装载该 Activity 实例。采用这种模式启动 Activity 有如下两种情况:

(1) 如果要启动的 Activity 不存在,系统会先创建一个新的任务栈,再创建该

Activity 实例,并把该 Activity 加入栈顶。

(2) 如果要启动的 Activity 已经存在,无论位于哪个应用程序或者哪个任务栈,系统都会把该 Activity 所在的任务栈转到前台,从而使该 Activity 显示出来。

图 2-25 所示为任务栈变化,Activity 1 启动 Activity 2 时,将会新建一个 Task 任务栈,并将 Activity 2 压入新建的 Task 任务栈,Activity 2 处于运行状态;在 Activity 2 启动 Activity 1,系统不会新建 Activity 1 对象,将会回到原来的任务 Task 栈,Activity 1 处于运行状态。

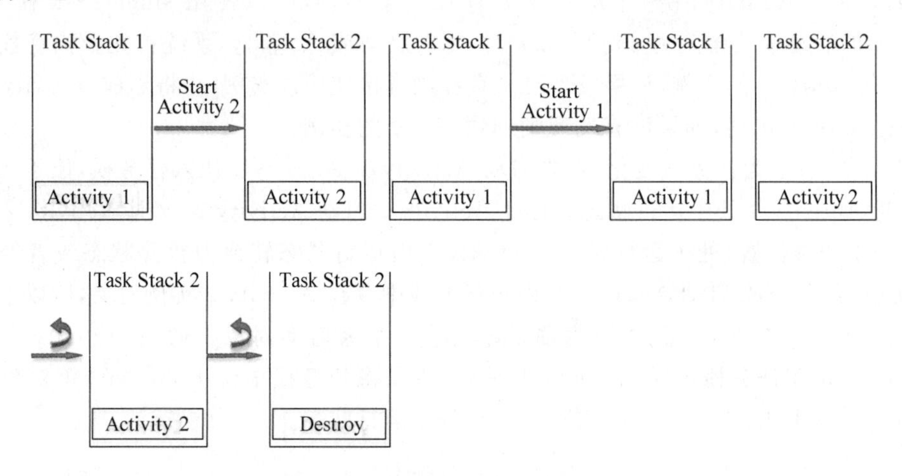

图 2-25 singleInstance 模式任务栈变化

在实际开发过程中,根据需要选择合适的启动模式。

2.4 Intent 详解

Intent 可以简单地理解为 Android 应用程序启动某个组件的意图。Android 应用程序将会根据 Intent 的各个属性启动指定的组件。Intent 对象大致包括 Component、Action、Category、Data、Type、Extra 和 Flag 这 7 种属性,其中 Component 用于明确指定需要启动的组件,而 Extra 则用于"携带"需要交换的数据。

< intent-filter >元素是 AndroidManifest. xml 文件中< activity >等组件元素的子元素,< intent-filter >标签中常用< action >< data >和< category >这些元素,分别对应 Intent 中的 Action、Data 和 Category 属性,用来对 Intent 进行匹配。通过组件的< intent-filter >信息,组件管理服务可以了解各个组件的功能。当组件管理服务接收到调用组件发来的隐式 Intent 的会与所有组件的< intent-filter >匹配,寻找匹配的组件。

2.4.1 Intent 属性与 IntentFilter

Intent 包括 Action(动作)、Data(数据)、Type(类型)、Component(组件)和 Extra(扩展信息)等属性,并且 Intent 提供了设置和获取相应属性的方法。简要说明如表 2-2 所示。

表 2-2 Intent 属性

属　　性	属性名称	数据类型	设置属性的方法	获取属性的方法
Component	组件	ComponentName	setComponent() setClass() setClassName()	getComponent()
Action	动作	String	setAction()	getAction()
Data	数据	URI	setData()	getData()
Category	分类	String	addCategory()	getCategories()
Type	类型	String	setType()	getType()
Extra	扩展信息	Bundle	putExtra()	getXXXExtra() getExtras()
Flag	标记	Integer	setFlags()	getFlags()

1. Component 属性

Component 属性用于指明 Intent 目标组件的类名称，它可以被设置，也可以不被设置。如果不设置 Component 属性，该 Intent 称为隐式 Intent，Android 会根据 Intent 中包含的其他属性信息，如 Action、Data、Type、Category 进行查找，最终找到一个与之匹配的目标组件。如果设置 Component 属性，则该 Intent 成为显式 Intent，会根据组件名查找相应的组件，不再执行上述查找过程。

在开发过程中，显式 Intent 通常用于启动当前应用程序内部组件，隐式 Intent 通常用于启动系统组件或其他应用程序内的组件。

Intent 的 Component 属性需要接受 ComponentName 对象，可以调用构造方法实现 ComponentName 对象的实例化。

ComponentName(String pkg, String cls)：创建指定包 pkg 下的 cls 字符串所对应的组件。

ComponentName(Context context, String cls)：创建 context 对应包下的 cls 字符串所对应的组件。

ComponentName(Context context, Class cls)：创建 context 对应包下的 cls 类所对应的组件。

文件清单 2-15 和 2-16 分别演示如何通过指定 Intent 的 Component 属性来启动另外一个 Activity。第一个 Activity 中的布局文件只包含一个按钮，点击该按钮启动第二个 Activity。

文件清单 2-15　ComponentActivity.java

```
import android.content.ComponentName;
import android.content.Intent;
import android.support.v7.app.AppCompatActivity;
import android.os.Bundle;
import android.view.View;
```

```
import android.widget.Button;

public class ComponentActivity extends AppCompatActivity {
    private Button btnStart;

    @Override
    protected void onCreate(Bundle savedInstanceState) {
        super.onCreate(savedInstanceState);
        setContentView(R.layout.activity_main2);
        btnStart=(Button) findViewById(R.id.btnStart);
        btnStart.setOnClickListener(new View.OnClickListener() {
            @Override
            public void onClick(View v) {
                Intent intent=new Intent();
                ComponentName component = new ComponentName (ComponentActivity.
this,ComponentSecondActivity.class);
                intent.setComponent(component);
                startActivity(intent);
            }
        });

    }
}
```

文件清单 2-16　ComponentSecondActivity.java

```
import android.support.v7.app.AppCompatActivity;
import android.os.Bundle;
import android.widget.TextView;


public class ComponentSecondActivity extends AppCompatActivity {
    private TextView tv;

    @Override
    protected void onCreate(Bundle savedInstanceState) {
        super.onCreate(savedInstanceState);
        setContentView(R.layout.activity_component_second);
        tv=(TextView) findViewById(R.id.txtInfo);
        tv.setText(getIntent().getComponent().getPackageName());
    }
}
```

2. Action 属性

Intent 中 Action 属性值是一个普通的字符串，用于描述 Intent 要完成的动作。Action 要完成的是一个抽象动作，这个动作具体由哪个组件完成取决于组件的<intent-filter>配置，只要某个组件的<intent-filter>配置中包含该 Action，该 Activity 就有可能被启动。并且,Intent类定义了一系列Action属性常量(表2-3),用来标识一套标准动作,如ACTION_CALL、ACTION_EDIT等。

表2-3 Action属性常量

Action常量	行为描述
ACTION_CALL	打电话,即直接呼叫Data中所带电话号码
ACTION_ANSWER	接听电话
ACTION_SEND	由用户指定发送方式进行数据发送操作
ACTION_SENDTO	根据不同的Data类型,通过对应的软件发送数据
ACTION_VIEW	根据不同的Data类型,通过对应软件显示数据
ACTION_EDIT	显示可编辑的数据
ACTION_MAIN	应用程序的入口
ACTION_SYNC	同步服务器与移动设备之间的数据
ACTION_BATTERY_LOW	警告设备电量低
ACTION_HEADSET_PLUG	插入或拔出耳机
ACTION_SCREEN_ON	打开移动设备屏幕
ACTION_TIMEZONE_CHANGED	移动设备时区发生变化

以下通过示例演示Action属性的使用,该示例包含ActionActivity和ActionFilterActivity两个Activity。在ActionActivity中包含一个按钮,当用户点击这个按钮时,程序通过为Intent指定Action属性来启动ActionFilterActivity。ActionActivity的源代码如文件清单2-17所示。

文件清单2-17 ActionActivity.java

```
import android.content.Intent;
import android.support.v7.app.AppCompatActivity;
import android.os.Bundle;
import android.view.View;
import android.widget.Button;
public class ActionActivity extends AppCompatActivity {
    private Button btnAction;
    private static final String NSU_ACTION="cn.eud.nsu.NSU_ACTION";
    @Override
    protected void onCreate(Bundle savedInstanceState) {
        super.onCreate(savedInstanceState);
        setContentView(R.layout.activity_action);
        btnAction=(Button) findViewById(R.id.btnAction);
        btnAction.setOnClickListener(new View.OnClickListener() {
            @Override
            public void onClick(View v) {
                Intent intent=new Intent();
                intent.setAction(NSU_ACTION);
                startActivity(intent);
            }
        });
```

```
        }
    }
```

在上述源代码中，通过调用 intent. setAction()方法为 Intent 对象指定 Action 属性。为了保证 ActionFilterActivity 被启动，要求 ActionFilterActivity 的< intent-filter >配置元素中至少包含一个< action android: name ="cn. eud. nsu. NSU_ACTION"/>的子元素。由于 Intent 在创建时默认启动 Category 属性值为 android. intent. category. Default 的组件，所以在< intent-filter >元素中添加了< category…/>子元素。

在 AndroidManifest. xml 中对 ActionFilterActivity 的配置如下：

```
<activity android:name=".ActionFilterActivity">
    <intent-filter>
        <action android:name="cn.eud.nsu.NSU_ACTION" />
        <category android:name="android.intent.category.DEFAULT" />
    </intent-filter>
</activity>
```

3. Data 属性

Data 属性通常用于向 Action 属性提供操作的数据(表 2-4)。Data 属性由两部分构成，分别是数据的 URI 和数据的 MIME 类型。

表 2-4 Data 属性范例

Data 属性	说　明	范　例
tel://	号码数据格式，后跟电话号码	tel:// 10086
mailto://	邮件数据格式，后跟邮件收件人地址	mailto://developer@163. com
smsto://	短信数据格式，后跟短信接收号码	smsto://123
content://	内容数据格式，后跟需要读取的内容	content://content/people/1
file://	文件数据格式，后跟文件路径	file://sdcard/music. mp3
geo://latitude,longitude	经纬数据格式	geo://180,65

一般 Action 和 Data 匹配使用，不同的 Action 由不同的 Data 数据指定(表 2-5)。

表 2-5 Action 与 Data 配合使用

Action 属性	Data 属性	描　述
ACTION_VIEW	content://contacts/people/1	显示_id 为 1 的联系人信息
ACTION_EDIT	conten://contacts/people/1	编辑_id 为 1 的联系人信息
ACTION_VIEW	tel://10086	显示电话号码为 10086 的联系人信息
ACTION_VIEW	http://www. baidu. com	在浏览器中浏览网页
ACTION_VIEW	file://sdcard/music. mp3	播放 MP3

下面通过示例演示，为 Intent 指定 Action 和 Data 属性来启动浏览器，浏览用户输入网址对应的内容。Activity 对应的布局文件包括一个文本输入框和一个按钮，当用户点

击该按钮就会启动系统浏览器加载在文本框输入的网址对应的内容。

布局文件如文件清单 2-18 所示。

文件清单 2-18 acivity_action_data.xml

```
<?xml version="1.0" encoding="utf-8"?>
<RelativeLayout xmlns:android="http://schemas.android.com/apk/res/android"
    xmlns:tools="http://schemas.android.com/tools"
    android:layout_width="match_parent"
    android:layout_height="match_parent"
    android:paddingBottom="@dimen/activity_vertical_margin"
    android:paddingLeft="@dimen/activity_horizontal_margin"
    android:paddingRight="@dimen/activity_horizontal_margin"
    android:paddingTop="@dimen/activity_vertical_margin"

    tools:context="com.nsu.zyl.chapter02.chapter02project.ActionDataActivity">

    <EditText
        android:layout_width="match_parent"
        android:layout_height="wrap_content"
        android:id="@+id/edtUrl"
        android:layout_alignParentTop="true"
        android:layout_centerHorizontal="true" />

    <Button
        android:layout_width="wrap_content"
        android:layout_height="wrap_content"
        android:text="打开网址"
        android:id="@+id/btnOpen"
        android:layout_marginTop="40dp"
        android:layout_below="@+id/edtUrl"
        android:layout_centerHorizontal="true" />
</RelativeLayout>
```

源代码文件如文件清单 2-19 所示。

文件清单 2-19 ActionDataActivity.java

```
package com.nsu.zyl.chapter02.chapter02project;

import android.content.Intent;
import android.net.Uri;
import android.support.v7.app.AppCompatActivity;
import android.os.Bundle;
import android.view.View;
import android.widget.Button;
import android.widget.EditText;

public class ActionDataActivity extends AppCompatActivity {
    private Button btnOpen;
```

```
    private EditText edtUrl;

    @Override
    protected void onCreate(Bundle savedInstanceState) {
        super.onCreate(savedInstanceState);
        setContentView(R.layout.activity_action_data);
        btnOpen=(Button) findViewById(R.id.btnOpen);
        edtUrl=(EditText)findViewById(R.id.edtUrl);
        btnOpen.setOnClickListener(new View.OnClickListener() {
            @Override
            public void onClick(View v) {
                Intent intent=new Intent();
                String strUrl=edtUrl.getText().toString();
                Uri uri=Uri.parse(strUrl);
                intent.setAction(Intent.ACTION_VIEW);
                intent.setData(uri);
                startActivity(intent);
            }
        });
    }
}
```

由于需要在本实例中访问网络资源，所以需要在 AndroidManifest.xml 中添加允许访问网络资源的权限。具体代码如下：

```
<users-permission android:name="android.permission.INTERNET">
```

4. Category 属性

Category 属性指明一个执行 Action 的分类，一个 Intent 对象最多只能包含一个 Action 属性，但是可以包含多个 Category 属性。调用 Intent.addCategory()方法为 Intent 添加 Category 属性。

Intent 中定义了一系列 Category 属性常量，如表 2-6 所示。

表 2-6 Category 属性

Category 属性	说明
CATEGORY_DEFAULT	默认的执行方式
CATEGORY_HOME	该组件为 Home Activity
CATEGORY_LAUNCHER	优先级最高的 Activity，通常与 ACTION_MAIN 配合使用
CATEGORY_BROWSABLE	可以使用浏览器启动
CATEGORY_GADGET	可以内嵌到另外的 Activity 中

5. Extra 属性

Extra 属性用于添加一些附加的信息，例如发送一封邮件，就可以通过 Extra 属性来添加主题和内容。通过使用 Intent 对象的 putExtra()方法来添加信息。例如，将一个人

的姓名附加到 Intent 对象中,代码如下:

```
Intent intent=new Intent();
intent.putExtra("name","Andy");
```

2.4.2 Activity 中使用 Intent

Intent 最常见的用途是启动应用程序组件,并且在组件之间通信。Intent 一般用于启动 Activity、Service 和发送广播等,承担 Android 应用程序三大核心组件的通信功能。Android 根据 Intent 寻找目标组件的方式可以分为两种,一种是显式意图,另一种是隐式意图。

1. 显式意图

显式意图是指启动 Activity 时需要明确指定激活组件。在 Activity A 中通过显式意图启动 Activity B 有如下几种方式:

(1) 在 Intent 的构造方法中指定被启动的 Activity。

```
Intent intent=new Intent(this,B.class); //创建 Intent 对象,目标组件为 Activity01
startActivity(intent);
```

(2) 根据目标组件的包名、全路径名来指定开启组件。

```
Intent intent=new Intent();
intent.setClassName("包名","全路径名");
startActivity(intent);
```

(3) 通过设置 Intent 的 Component 属性。

```
ComponentName component=new ComponentName(A.this,B.class);
Intent intent=new Intent();
intent.setComponent(component);
startActivity(intent);
```

2. 隐式意图

隐式意图指不明确指定被启动 Activity,只通过设置 Intent 的属性,让 Android 系统根据隐式意图中设置的动作(Action)、类别(Category)、数据(URI 和数据类型)找到最合适的 Activity。

使用隐式意图开启 Activity 的示例代码如下:

```
Intent intent=new Intent();
intent.setAction();
startActivity(intent);
```

在上述代码中，Intent指定了setAction()这个动作，但是并没有指定Category，这是因为清单文件中配置的“android. intent. category. DEFAULT”是一种默认的Category，在调用startActivity()方法时，会自动将这个category添加到Intent中。例如：

```
<activity android:name="">
    <intent-filter>
        <action android:name=""/>
        <category android:name="">
</intent-filter>
</activity>
```

上述代码中，< action >标签指明了当前Activity可以响应的动作，< category >标签包含了一些类别信息，只有当< action >和< category >中的内容同时匹配时，Activity才会被开启。

一个< intent-filter >中可以添加多个< action >子元素，例如：

```
<intent-filter>
    <action android:value="android.intent.VIEW">
<action android:value="android.intent.EDIT">
…
</intent-filter>
```

< intent-filter >列表中的Action属性不能为空，否则所有Intent都会因匹配失败而被阻塞。所以一个< intent-filter >元素下至少需要包含一个< action >子元素，这样系统才能处理Intent消息。

在< intent-filter >中可以添加多个< category >元素，例如：

```
<intent-filter>
    <category android:value="android.intent.category.DEFAULT"/>
    <category android:value="android.intent.category.BROWSABLE"/>
</intent-filter>
```

与Action一样，< intent-filter >列表中的Category属性不能为空。Category属性的默认值android. intent. category. DEFAULT是启动Activity的默认值，在添加其他的Category属性值时，该值必须添加，否则也会匹配失败。

一个< intent-filter >中可以包含多个< data >子元素，用于指定组件可以执行的数据。例如：

```
<intent-filter>
<data
android:mimeType="video/mpeg"
android:scheme="http"
android:host="com.example.android"
```

```
android:path="folder/subfolder/1"
android:port="8888"/>
</intent-filter>
```

显式意图开启组件时必须指定组件的名称，一般只在本应用程序中切换组件时使用。而隐式意图的功能要比显示意图更加强大，不仅可以开启本应用的组件，还可以开启其他应用的组件，例如打开系统的照相机、浏览器等。

本章小结

本章主要讲解Activity与Intent相关知识，包括Activity的创建与配置，Activity的两种启动方式和Activity之间数据的传递，Activity生命周期和Intent的使用。创建Activity需要继承Activity类或它的子类，在创建的类中根据业务需要重写回调方法，然后在AndroidManifest.xml中配置，就可以被其他组件访问。Activity启动方式分为显式启动和隐式启动两种，应用程序不仅可以启动自身Activity还可以启动系统组件，甚至其他应用程序的Activity。

习　题

1. Android四大组件中，Activity是一个用来________的组件。

2. 根据Activity生命周期，可以将Activity分为________、________、________和________四种基本状态。

3. 使用startActivityForResult(Intent intent, int requestCode)方法启动Activity时，需要在源Activity中重写________方法来获取返回值。

4. Intent对象包括以下属性，分别是组件名(ComponentName)、________、________、分类(Category)、扩展信息(Extra)和标志(Flag)。

5. 下面关于Activity栈中Activity与状态说法错误的是(　　)。
   A. Activity被启动且显示给用户，它被压入栈中成为栈顶元素，此时它处于运行状态
   B. 用户点击返回按钮，栈顶的Activity出栈，被终止，它处于终止状态
   C. 栈中非栈顶的Activity，处于停止状态、暂停状态或终止状态
   D. 用户点击Home键，栈顶Activity出栈，被终止，它处于终止状态

6. 下面不能使用Intent激活的组件是(　　)。
   A. Activity　　　　B. Service
   C. BroadcastReceiver　　　　D. ContentProvider

7. 下面不属于启动Activity方法的是(　　)。
   A. startActivity(Intent intent)
   B. startActivities(Intent[] intent[])

C. startActivityForResult(int requestCode,int resultCode, Intent data);

D. startCommand();

8. 简述 Activity 生命周期中的几个状态。

9. 简述显式 Intent 和隐式 Intent 的区别。

10. 编写程序,通过隐式意图打开系统提供的摄像头。

第3章 chapter 3

Android UI 开发

主要内容：Android UI 布局，常用控件，对话框，菜单，导航，Adapter 与 AdapterView

建议课时：12 课时

知识目标：(1) 掌握 Android 常用布局的使用；

(2) 掌握 Android 常用控件的用法及常用的交互策略；

(3) 掌握 Adapter 和 AdapterView 的使用。

能力目标：(1) 初步具备用户界面交互设计的能力；

(2) 具备 Android UI 开发的能力。

在学习本章之前，需要在 Android Studio 中创建 Chapter03UI 项目。本章相关实例均创建在 Chapter03UI 项目中。MainActivity 是 Chapter03UI 项目的主 Activity，通过对 MainActivity 页面中按钮的点击操作可以将各个实例串联起来。Chapter03UI 启动后展示的首页如图 3-1 所示。

图 3-1　Chapter03UI 首页

MainActivity 的布局文件为 activity_main. xml,如文件清单 3-1 所示。

文件清单 3-1 activity_main. xml

```
<?xml version="1.0" encoding="utf-8"?>
<RelativeLayout xmlns:android="http://schemas.android.com/apk/res/android"
    xmlns:tools="http://schemas.android.com/tools"
    android:id="@+id/activity_main"
    android:layout_width="match_parent"
    android:layout_height="match_parent"
    android:paddingBottom="@dimen/activity_vertical_margin"
    android:paddingLeft="@dimen/activity_horizontal_margin"
    android:paddingRight="@dimen/activity_horizontal_margin"
    android:paddingTop="@dimen/activity_vertical_margin"
    tools:context="com.nsu.zyl.Chapter03ui.MainActivity">

    <TextView
        android:id="@+id/textView1"
        android:layout_width="wrap_content"
        android:layout_height="wrap_content"
        android:layout_alignParentTop="true"
        android:layout_centerHorizontal="true"
        android:layout_marginTop="10dp"
        android:textSize="25sp"
        android:textAllCaps="false"
        android:text="Android UI 开发" />

    <Button
        android:id="@+id/btn_basic_layout"
        android:layout_width="match_parent"
        android:layout_height="wrap_content"
        android:layout_centerHorizontal="true"
        android:layout_below="@id/textView1"
        android:textAllCaps="false"
        android:text="Android UI 布局" />

    <Button
        android:id="@+id/btn_basic_view"
        android:layout_width="match_parent"
        android:layout_height="wrap_content"
        android:layout_below="@id/btn_basic_layout"
        android:layout_centerHorizontal="true"
        android:text="常用控件" />

    <Button
        android:id="@+id/btn_basic_dialog"
        android:layout_width="match_parent"
        android:layout_height="wrap_content"
        android:layout_below="@+id/btn_basic_view"
```

```
        android:layout_centerHorizontal="true"
        android:text="对话框用法讲解" />

    <Button
        android:id="@+id/btn_basic_toast"
        android:layout_width="match_parent"
        android:layout_height="wrap_content"
        android:layout_below="@+id/btn_basic_dialog"
        android:layout_centerHorizontal="true"
        android:textAllCaps="false"
        android:text="Toast 的使用" />

    <Button
        android:id="@+id/btn_basic_menu"
        android:layout_width="match_parent"
        android:layout_height="wrap_content"
        android:layout_below="@+id/btn_basic_toast"
        android:layout_centerHorizontal="true"
        android:text="菜单的使用" />

    <Button
        android:id="@+id/btn_basic_guide"
        android:layout_width="match_parent"
        android:layout_height="wrap_content"
        android:layout_below="@+id/btn_basic_menu"
        android:layout_centerHorizontal="true"
        android:text="导航栏的使用" />

    <Button
        android:id="@+id/btn_basic_adapter"
        android:layout_width="match_parent"
        android:layout_height="wrap_content"
        android:layout_below="@+id/btn_basic_guide"
        android:layout_centerHorizontal="true"
        android:textAllCaps="false"
        android:text="Adapter 与 Adapter View" />
</RelativeLayout>
```

MainActivity 页面中不同按钮对应不同的演示示例，为 MainActivity 中的按钮设置点击事件监听器，根据页面中被点击按钮的不同，跳转到不同的页面。MainActivity 初始内容如文件清单 3-2 所示。

文件清单 3-2　MainActivity.java

```
package com.nsu.zyl.Chapter03ui;
import android.content.Intent;
import android.support.v7.app.AppCompatActivity;
import android.os.Bundle;
```

```
import android.view.View;
import android.widget.Button;
public class MainActivity extends AppCompatActivity {
    private Button btn_basicLayout;
    private Button btn_basicView;
    private Button btn_basicDialog;
    private Button btn_basicToast;
    private Button btn_menu;
    private Button btn_guide;
    private Button btn_adapter;

    @Override
    protected void onCreate(Bundle savedInstanceState) {
        super.onCreate(savedInstanceState);
        setContentView(R.layout.activity_main);
        initView();
    }

    private void initView() {
        btn_basicLayout = (Button)findViewById(R.id.btn_basic_layout);
        btn_basicView = (Button)findViewById(R.id.btn_basic_view);
        btn_basicDialog = (Button) findViewById(R.id.btn_basic_dialog);
        btn_basicToast = (Button) findViewById(R.id.btn_basic_toast);
        btn_menu = (Button) findViewById(R.id.btn_basic_menu);
        btn_guide = (Button) findViewById(R.id.btn_basic_guide);
        btn_adapter = (Button) findViewById(R.id.btn_basic_adapter);
        btn_basicLayout.setOnClickListener(new MyClickListener());
        btn_basicView.setOnClickListener(new MyClickListener());
        btn_basicDialog.setOnClickListener(new MyClickListener());
        btn_basicToast.setOnClickListener(new MyClickListener());
        btn_menu.setOnClickListener(new MyClickListener());
        btn_guide.setOnClickListener(new MyClickListener());
        btn_adapter.setOnClickListener(new MyClickListener());
    }

    private class MyClickListener implements View.OnClickListener {
        @Override
        public void onClick(View v) {
            switch (v.getId()) {
                case R.id.btn_basic_layout:

                    break;
                case R.id.btn_basic_view:

                    break;
                case R.id.btn_basic_dialog:

                    break;
```

```
                case R.id.btn_basic_toast:

                    break;
                case R.id.btn_basic_menu:

                    break;
                case R.id.btn_basic_guide:

                    break;
                case R.id.btn_basic_adapter:

                    break;
            }
        }
    }
}
```

这样 Chapter03UI 项目的准备工作就完成了。下面进行 Android UI 编程的学习，完成演示案例，并关联到 MainActivity。

3.1 Android UI 布局

Android UI 布局(Layout)是用户界面结构的描述，定义了界面中所有元素的结构和相互关系。本节将详细讲解 Android UI 布局的相关知识。

3.1.1 Android 布局概述

在 Android 应用程序中，用户界面由 View 和 ViewGroup 对象构建，Android 中有很多种 View 和 ViewGroup，它们都继承自 View 类，View 与 ViewGroup 之间的关系如图 3-2 所示。View 对象是 Android 平台上表示用户界面的基本单元。ViewGroup 是 View 的子类，既可作为 View 使用，也可向其中添加 View。

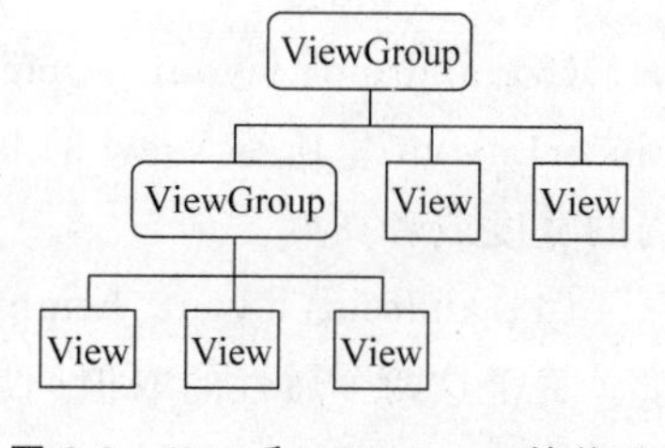

图 3-2 View 和 ViewGroup 的关系

一般布局方式是指一组 View 元素如何布局，准确地说是一个 ViewGroup 中包含的一些 View 怎样布局。以下介绍的关于 View 布局方式的类，都是直接或间接继承自 ViewGroup 类，如图 3-3 所示。

Android 中所有的布局方式都可以归类为 ViewGroup 的 6 个直接子类，其他的一些布局都扩展自这 6 个类。

Android 程序的界面布局有两种声明方法：

(1) 使用 XML 文件描述界面布局；

(2) 在程序运行时动态添加或修改界面布局。

用户既可以独立使用任何一种声明界面布局的方式，也可以同时使用两种方式。

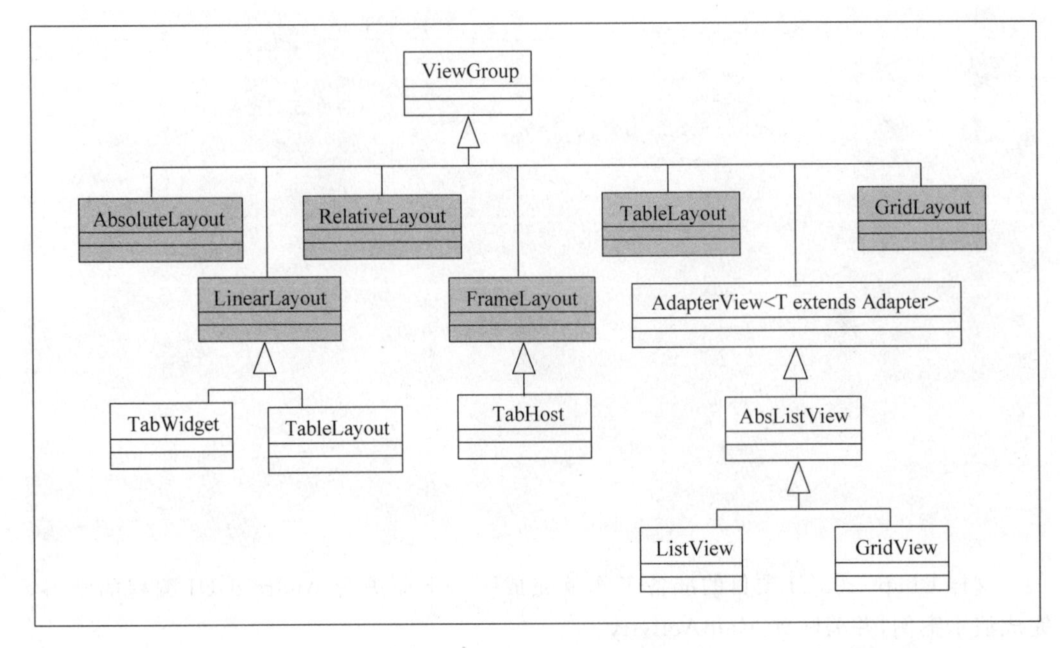

图 3-3 ViewGroup 继承体系

3.1.2 线性布局

线性布局(LinearLayout)是一种常用的界面布局方式,在线性布局中,所有的子元素都按照垂直或水平的顺序在界面上排列,每一个子元素都位于前一个子元素的后面,当超过边界时,超过的那部分将不显示。

LinearLayout 相关属性说明如下:

(1) android:orientation:指定布局中 View 控件的排列方式,取值为 vertical 时,表示垂直排列,取值为 horizontal 时,表示水平排列。

(2) android:layout_gravity:指定 View 控件在容器中的对齐方式。

(3) android:layout_weight:指定 View 控件在容器中所占的权重,例如将 LinearLayout 里所有 View 的 layout_weight 都设为 1,那么这些 View 将平分该线性布局的宽度或者高度。

(4) android:layout_height\android:layout_width:指定 LinearLayout 的高度/宽度。关于这两个属性的取值问题,常取 fill_parent、match_parent、wrap_content 和具体像素值等。

取值说明如下:

fill_parent 表示控件的宽度或高度与父容器相同。

match_parent 与 fill_parent 完全相同,从 Android 2.2 版本以后推荐使用此属性。

wrap_content 表示控件的大小刚好包裹它的内容即可。

由于 Android 设备分辨率差别太大,布局时一般不推荐使用具体像素值。

在 Chapter03UI 项目中新建布局文件 activity_linear_layout.xml,该布局文件使用

线性布局,其源代码如文件清单 3-3 所示。

文件清单 3-3　activity_linear_layout.xml

```
<?xml version="1.0" encoding="utf-8"?>
<LinearLayout xmlns:android="http://schemas.android.com/apk/res/android"
    android:layout_width="match_parent"
    android:layout_height="match_parent"
    android:orientation="vertical">
    <TextView
        android:text="用户名"
        android:layout_width="wrap_content"
        android:layout_height="wrap_content"
        android:id="@+id/textView" />
    <EditText
        android:layout_width="wrap_content"
        android:layout_height="wrap_content"
        android:inputType="textPersonName"
        android:hint="请输入用户名"
        android:ems="8"
        android:id="@+id/editText2"/>
    <Button
        android:text="确认"
        android:layout_width="wrap_content"
        android:layout_height="wrap_content"
        android:id="@+id/button" />
    <Button
        android:text="取消"
        android:layout_width="wrap_content"
        android:layout_height="wrap_content"
        android:id="@+id/button2"/>
</LinearLayout>
```

这段布局文件的页面显示效果如图 3-4 所示。因为布局方式 android:orientation 取值为 vertical,所以布局里的各 View 垂直分布,每行仅包含一个 View。相反,如果将 android:orientation 的值改为 horizontal,保持其他代码不变,页面的显示效果将变成水平分布,如图 3-5 所示。需要注意的是 LinearLayout 不会自动换行,如果布局中的元素太多,则不会显示多余的 View。

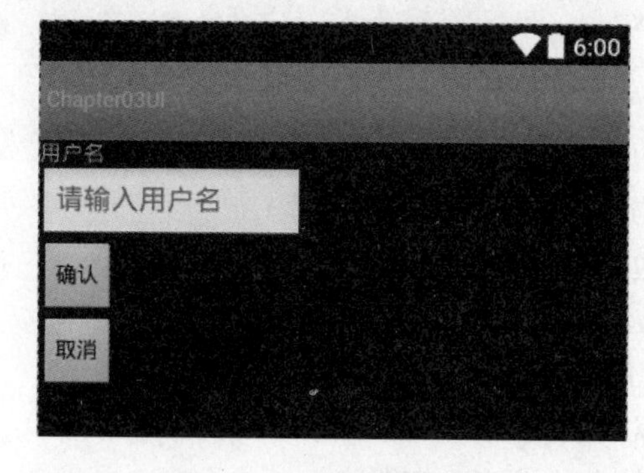

图 3-4　线性布局(纵向)

图 3-5　线性布局(横向)

线性布局(LinearLayout)作为一种广泛使用的布局方式,关于它更多的属性用法,需要大家在实践中去反复体会。

3.1.3 相对布局

相对布局(RelativeLayout)利用控件之间的相对位置关系来进行布局,相对位置关系主要是控件与父容器、控件与其他控件之间的相对关系。相对布局的属性很多,可以归纳为以下3类:

1. 属性值为true或false

android:layout_centerHorizontal:水平居中;
android:layout_centerVertical:垂直居中;
android:layout_centerInParent:相对于父元素完全居中;
android:layout_alignParentBottom:紧贴父元素的下边缘;
android:layout_alignParentLeft:紧贴父元素的左边缘;
android:layout_alignParentRight:紧贴父元素的右边缘;
android:layout_alignParentTop:紧贴父元素的上边缘;
android:layout_alignWithParentIfMissing:如果对应的兄弟元素找不到,就以父元素做参照物。

2. 属性值必须为id的引用名"@+id/id-name"

android:layout_below:在某元素的下方;
android:layout_above:在某元素的上方;
android:layout_toLeftOf:在某元素的左边;
android:layout_toRightOf:在某元素的右边;
android:layout_alignTop:本元素的上边缘和某元素的上边缘对齐;
android:layout_alignLeft:本元素的左边缘和某元素的左边缘对齐;
android:layout_alignBottom:本元素的下边缘和某元素的下边缘对齐;
android:layout_alignRight:本元素的右边缘和某元素的右边缘对齐。

3. 属性值为具体的像素值,如30dp,40px

android:layout_marginBottom:离某元素底边缘偏移一定距离;
android:layout_marginLeft:离某元素左边缘偏移一定距离;
android:layout_marginRight:离某元素右边缘偏移一定距离;
android:layout_marginTop:离某元素上边缘偏移一定距离;
android:layout_paddingBottom:往内部元素的底边填充一定距离;
android:layout_paddingLeft :往内部元素的左边填充一定距离;
android:layout_paddingRight:往内部元素的右边填充一定距离;
android:layout_paddingTop:往内部元素的上边填充一定距离。

在 Chapter03UI 项目中新建布局文件 activity_relative_layout. xml,该布局文件使用相对布局,源代码如文件清单 3-4 所示。

文件清单 3-4 activity_relative_layout. xml

```
<?xml version="1.0" encoding="utf-8"?>
<RelativeLayout xmlns:android="http://schemas.android.com/apk/res/android"
    android:layout_width="match_parent"
    android:layout_height="match_parent">
    <TextView
        android:text="用户名"
        android:layout_width="wrap_content"
        android:layout_height="wrap_content"
        android:layout_alignParentTop="true"
        android:layout_alignParentStart="true"
        android:id="@+id/textView3" />
<EditText
        android:layout_width="match_parent"
        android:layout_height="wrap_content"
        android:inputType="textPersonName"
        android:layout_below="@+id/textView3"
        android:layout_alignParentStart="true"
        android:hint="请输入用户名"
        android:layout_marginTop="13dp"
        android:id="@+id/editText" />
    <Button
        android:text="取消"
        android:layout_width="wrap_content"
        android:layout_height="wrap_content"
        android:layout_below="@+id/editText"
        android:layout_alignParentEnd="true"
        android:layout_marginTop="23dp"
        android:id="@+id/button3" />

    <Button
        android:text="确定"
        android:layout_width="wrap_content"
        android:layout_height="wrap_content"
        android:layout_alignTop="@+id/button3"
        android:layout_toStartOf="@+id/button3"
        android:layout_marginEnd="33dp"
        android:id="@+id/button4" />
</RelativeLayout>
```

这段布局文件非常简单,定义了 4 个控件,以相对布局的方式,确定每个控件的位置,页面显示效果如图 3-6 所示。

图 3-6 相对布局

3.1.4 帧布局

帧布局(FrameLayout),也叫框架布局,是一个非常简单的界面布局。这个布局就是直接在屏幕上开辟一块空白的区域,当向帧布局添加控件时默认放到左上角,帧布局没有任何定位方式,不过可以为 View 组件添加 layout_gravity 属性,从而指定组件的对齐方式。

在帧布局中,如果帧布局有多个控件,那么后放置的控件将遮挡先放置的控件,布局的大小则由子控件中最大的那个控件决定。如果所有控件都一样大,同一时刻我们就只能看到最上面的那个控件。可以使用 Android SDK 中提供的层级观察器(Hierarchy Viewer)进一步分析界面布局,层级观察器能够对用户界面进行分析和调试,并以图形化的方式展示树形结构的界面布局。

FrameLayout 常用属性说明如下：

(1) android:foreground：设置该帧布局容器的前景图像(永远处于帧布局最上面,直接面对用户的图像,即不会被覆盖的图片)。

(2) android:foregroundGravity：设置前景图像显示的位置。

(3) android:layout_gravity ：指定子元素在 FrameLayout 中的对齐方式,常取 center、bottom、left、right、top 等值。如果不指定 View 组件的这个属性,则默认 View 组件都将位于 FrameLayout 空间的左上角。

在 Chapter03UI 项目中新建布局文件 activity_frame_layout.xml,该布局文件使用帧布局,源代码如文件清单 3-5 所示。

文件清单 3-5 activity_frame_layout.xml

```
<?xml version="1.0" encoding="utf-8"?>
<FrameLayout xmlns:android="http://schemas.android.com/apk/res/android"
    android:layout_width="match_parent"
    android:layout_height="match_parent">

    <TextView
        android:id="@+id/tv_frame1"
        android:layout_width="300dp"
```

```
        android:layout_height="300dp"
        android:layout_gravity="center"
        android:background="#FF6143"
        android:text="第一个 TextView" />

    <TextView
        android:id="@+id/tv_frame2"
        android:layout_width="250dp"
        android:layout_height="250dp"
        android:layout_gravity="center"
        android:background="#7BFE00"
        android:text="第二个 TextView" />

    <TextView
        android:id="@+id/tv_frame3"
        android:layout_width="200dp"
        android:layout_height="200dp"
        android:layout_gravity="center"
        android:background="#FFFF00"
        android:text="第三个 TextView" />

    <TextView
        android:id="@+id/tv_frame4"
        android:layout_width="150dp"
        android:layout_height="150dp"
        android:layout_gravity="center"
        android:background="#0000FF"
        android:text="第四个 TextView" />
</FrameLayout>
```

这段布局文件定义了 4 个 TextView 控件，以 FrameLayout 的布局方式来排列这 4 个控件。这 4 个 TextView 大小不一，都位于父容器的正中心，我们以不同的颜色标示了每个 TextView。其页面显示效果如图 3-7 所示，可以看到后绘制的 TextView 覆盖在前一个 TextView 的上面，实现了遮挡的效果。

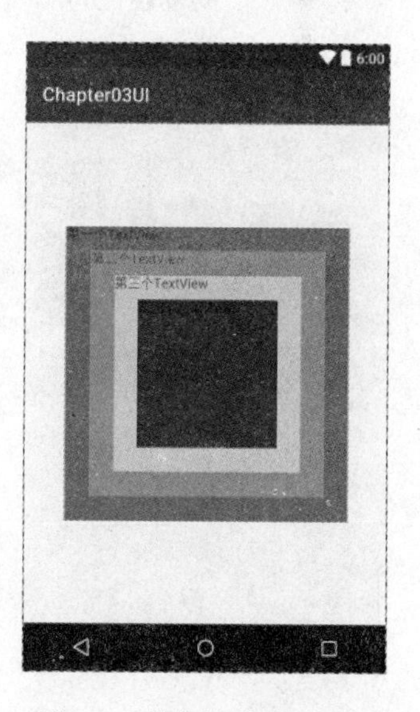

图 3-7　帧布局(框架布局)

3.1.5　绝对布局

绝对布局(AbsoluteLayout)能通过指定界面元素的坐标位置，来确定用户界面的整体布局，又可以称为坐标布局。

AbsoluteLayout 常用属性说明如下：

(1) android:layout_x：指定当前子类控件在 Android 坐标系 x 轴的位置。

(2) android:layout_y：指定当前子类控件在 Android 坐标系 y 轴的位置。

在 Android 手机中，坐标系以手机屏幕左上角的顶点为坐标原点，从该点向右为 x 轴的正方向，从该点向下则为 y 轴的正方向。

需要说明的是，由于 Android 设备的分辨率种类繁多，小到 320×240，大到 1920×1080 等，差异较大，而 AbsoluteLayout 通过 x 轴和 y 轴确定界面元素位置后，Android 系统不能够根据不同屏幕对界面元素的位置进行调整，从而降低了界面布局对不同类型和尺寸屏幕的适应能力。因此，现在绝对布局是一种不推荐使用的界面布局。但是从学习知识的角度，我们还是有必要来了解这种布局方式。

在 Chapter03UI 项目中新建布局文件 activity_absolute_layout.xml，该布局文件使用绝对布局，源代码如文件清单 3-6 所示。

文件清单 3-6　activity_absolute_layout.xml

```
<?xml version="1.0" encoding="utf-8"?>
<AbsoluteLayout xmlns:android="http://schemas.android.com/apk/res/android"
    android:layout_width="match_parent"
    android:layout_height="match_parent">

    <TextView
        android:text="用户名"
        android:layout_width="wrap_content"
        android:layout_height="wrap_content"
        android:layout_x="90dp"
        android:layout_y="129dp"
        android:id="@+id/textView4" />

    <EditText
        android:layout_width="wrap_content"
        android:layout_height="wrap_content"
        android:inputType="textPersonName"
        android:hint="请输入用户名"
        android:ems="10"
        android:layout_x="90dp"
        android:layout_y="172dp"
        android:id="@+id/editText3" />

    <Button
        android:text="确定"
        android:layout_width="wrap_content"
        android:layout_height="wrap_content"
        android:layout_x="90dp"
        android:layout_y="255dp"
        android:id="@+id/button5" />

    <Button
        android:text="取消"
```

```
        android:layout_width="wrap_content"
        android:layout_height="wrap_content"
        android:layout_x="206dp"
        android:layout_y="257dp"
        android:id="@+id/button6" />
</AbsoluteLayout>
```

这段布局文件以绝对布局的方式,包含了4个控件,通过指定每个控件在屏幕上的(x,y)坐标,来确定控件的具体位置,页面显示效果如图3-8所示。

图3-8 绝对布局

3.1.6 表格布局

表格布局(TableLayout)将屏幕划分表格,通过指定行和列可以将界面元素添加到表格中,表格的边界对用户是不可见的。表格布局的每一行是一个TableRow的对象,当然也可以是一个View的对象。如果直接往TableLayout中添加组件,那么这个组件将占满一行。如果一行上有多个组件,就要添加一个TableRow容器,把组件都放到里面。

表格布局还支持嵌套,可以将另一个表格布局放置在前一个表格布局的表格中,也可以在表格布局中添加其他界面布局,例如线性布局、相对布局等。

TableLayout常用属性说明如下:

(1) android:collapseColumns = "1,2":设置需要被隐藏的列序号(序号从0开始)。列之间必须用逗号隔开,例如,1,2,5。

(2) android:shrinkColumns="1,2":设置允许被收缩的列的序号(序号从0开始)。

(3) android:stretchColumns = "1,2":设置允许被拉伸的列的序号(序号从0开始),以填满剩下的多余空白空间,列之间必须用逗号隔开。

对于<TableRow></TableRow>内的控件而言,下标从0开始,即:

android:layout_column = "1"表示该控件显示在第2列;

android:layout_span = "2"表示该控件占据2列。

在Chapter03UI项目中新建布局文件activity_table_layout.xml,该布局文件使用表格布局,源代码如文件清单3-7所示。

文件清单3-7 activity_table_layout.xml

```
<?xml version="1.0" encoding="utf-8"?>
<TableLayout xmlns:android="http://schemas.android.com/apk/res/android"
```

```
    android:layout_width="match_parent"
    android:layout_height="match_parent"
    android: stretchColumns="0" >

    <TableRow>
        <TextView
            android:text="用户名"
            android:layout_width="wrap_content"
            android:layout_height="wrap_content"
            android:gravity="right"
            android:layout_weight="1"/>

        <EditText
            android:layout_width="wrap_content"
            android:layout_height="wrap_content"
            android:inputType="textPersonName"
            android:layout_weight="1"
            android:hint="请输入用户名"/>
    </TableRow>

    <TableRow>
        <Button
            android:text="确定"
            android:layout_width="wrap_content"
            android:layout_height="wrap_content"
            android:layout_weight="1"/>

        <Button
            android:text="确定"
            android:layout_width="wrap_content"
            android:layout_height="wrap_content"
            android:layout_weight="1"/>
    </TableRow>
</TableLayout>
```

该表格布局的设计如图 3-9 所示，共包含两个 TableRow，第一个 TableRow 里存放一个 TextView 控件和一个 EditText 控件，第二个 TableRow 里存放两个 Button 控件。具体实现时，在布局文件中设置了 android：stretchColumns＝“0”，表示允许第 1 列被拉伸，以填满剩下的多余空白空间，最终页面效果如图 3-10 所示。

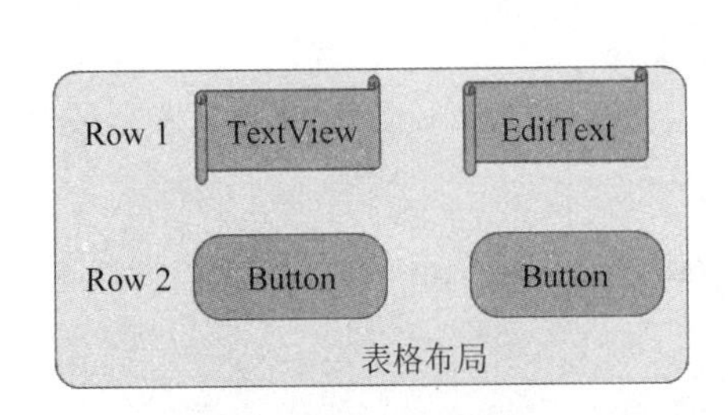

图 3-9　表格布局设计

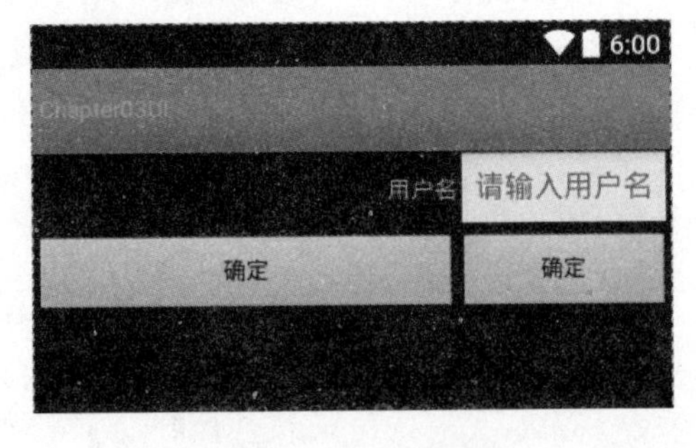

图 3-10　表格布局效果

3.1.7 网格布局

网格布局(GridLayout)类似于表格布局,是 Android 4.0 及以上版本新增加的布局,主要以网格的形式来布局窗口控件。使用虚细线将布局划分为行、列和单元格,也支持一个控件横跨多行或多列。GridLayout 分为水平和垂直两种方式,默认是水平布局,一个控件挨着一个控件从左到右依次排列。

GridLayout 常用属性说明如下:

(1) android:orientation、android:layout_gravity 用于设置布局排列对齐。

android:orientation="vertical|horizontal"设置组件的排列方式,当值为 vertical 时代表采用的是竖直的排列方式,当值为 horizontal 时代表采用的是水平的排列方式。

android:layout_gravity="top|left|right|bottom|center|center_vertical|center_horizontal|fill|fill_vertical|fill_horizontal|start|end",该属性用于设置控件相对于容器的对齐方式,这些值是多选的,用"|"分割,例如:android:layout_gravity=" bottom|left"。

(2) android:rowCount、android:columnCount 用于设置布局为几行几列。

android:rowCount="2",设置网格布局有 2 行。

android:columnCount="2",设置网格布局有 2 列。

(3) android:layout_row、android:layout_column 用于设置某个组件位于第几行第几列(注:都是从 0 开始算计)。

android:layout_row = "1",设置组件位于第 2 行。

android:layout_column = "2",设置该组件位于第 3 列。

(4) android:layout_rowSpan、android:layout_columnSpan 用于设置某个组件跨越几行几列。

android:layout_rowSpan = "3",纵向跨 3 行。

android:layout_columnSpan = "3",横向跨 3 列。

设置某控件跨越多行或多列,只须将该子控件的 android:layout_rowSpan 或者 layout_columnSpan 属性设置为数值,再设置其 layout_gravity 属性为 fill 即可,前一个设置表明该控件跨越的行数或列数,后一个设置表明该控件填满所跨越的整行或整列。

在 Chapter03UI 项目中新建布局文件 activity_grid_layout.xml,该布局文件使用网格布局,源代码如文件清单 3-8 所示。

文件清单 3-8 activity_grid_layout.xml

```
<?xml version="1.0" encoding="utf-8"?>
<GridLayout xmlns:android="http://schemas.android.com/apk/res/android"
    android:layout_width="wrap_content"
    android:layout_height="wrap_content"
    android:columnCount="4"
android:rowCount="6">

<TextView
       android:layout_columnSpan="2"
```

```
            android:layout_gravity="fill"
            android:textSize="25sp"
            android:text="0"/>
        <Button android:text="C"/>
        <Button
            android:layout_column="3"
            android:text="/"/>
        <Button android:text="1"/>
        <Button android:text="2"/>
        <Button android:text="3"/>
        <Button android:text=" * "/>
        <Button android:text="4"/>
        <Button android:text="5"/>
        <Button android:text="6"/>
        <Button android:text="-"/>
        <Button android:text="7"/>
        <Button android:text="8"/>
        <Button android:text="9"/>
        <Button
            android:layout_gravity="fill"
            android:layout_rowSpan="3"
            android:text="+" />
        <Button android:text="0"/>
        <Button
            android:layout_gravity="fill"
            android:layout_columnSpan="2"
            android:text="00"/>
        <Button
            android:layout_gravity="fill"
            android:layout_columnSpan="3"
            android:text="="/>
    </GridLayout>
```

这段布局文件以 GridLayout 的布局方式，实现了一个简易计算器的页面，页面效果如图 3-11 所示。该布局中 android:columnCount="4"，android:rowCount="6"，指定这是一个 6 行 4 列的网格布局。TextView 控件属性 android:layout_columnSpan="2"表明文本框跨越 2 列，Button 属性 android:layout_gravity="fill"，android:layout_rowSpan="3"，指定"+"这个按钮跨越 3 行。

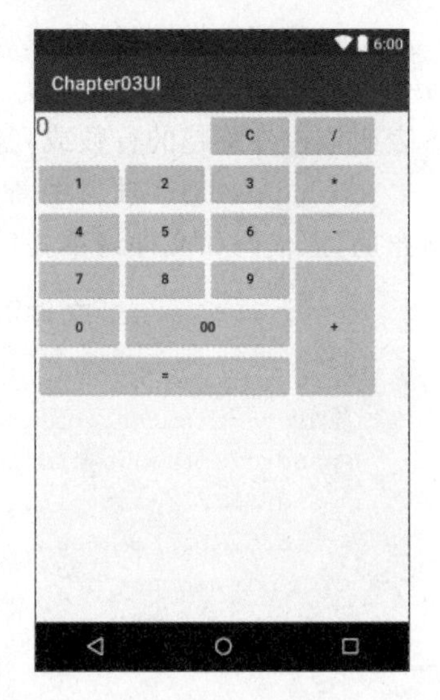

图 3-11　网格布局

3.1.8　约束性布局

ConstraintLayout(约束性布局)是 Google 在 2016 年的 I/O 大会上推出的一个新的布局，根据布局中其他元素或视图，确定 View 在屏幕中的位置，受到其他视图、父容器和基准线三类约束。利用其他布局编写界面时，复杂的布局伴随着多层嵌套，从而降低了程序的性能。ConstraintLayout 使用约束的方式来指定各个控件的位置和关系，可以有效解决布局嵌套过多的问题，并且 ConstraintLayout 相比其他布局更适合使用可视化的方式编写界面。

ConstraintLayout 是一个新的 Support 库，支持 Android 2.3（API Level 9）以及以后的版本。使用 ConstraintLayout 需要确保在 Android Studio 2.2 及以后版本，并且在 Android Studio 中使用 ConstraintLayout 之前需要先下载最新的 ConstraintLayout 库，步骤如下：

(1) 在 Android Studio 中选择菜单 Tools→Android→SDK Manager 命令，打开对话框，见图 3-12。

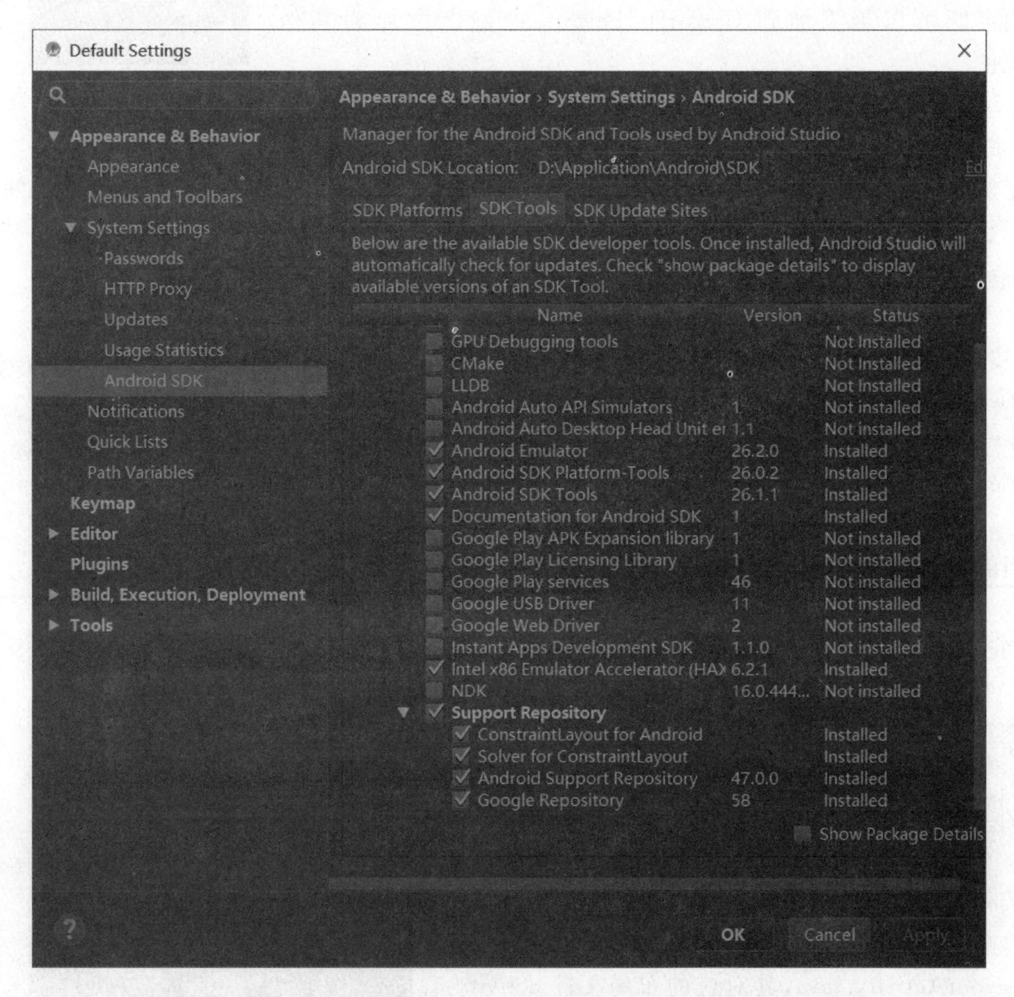

图 3-12　安装 ConstraintLayout for Android 和 Solver for ConstraintLayout

(2) 点击 SDK Tools 标签页，滚动到最下面找到 Support Repository 部分，然后勾选 ConstraintLayout for Android 和 Solver for ConstraintLayout，点击 OK 按钮安装需要的最新版本。安装完以后，在 app/build.gradle 文件中添加 ConstraintLayout 的依赖：

```
dependencies {
    compile 'com.android.support.constraint:constraint-layout:1.0.1'
}
```

在 Android Studio 2.3 及以上版本中，默认支持 ConstraintLayout 布局。ConstraintLayout 的使用比较简单，直接拖曳需要的控件到页面中，然后添加约束。控件的约束都分为垂直和水平两类，一共可以在 4 个方向为控件添加约束，如图 3-13 所示。

图 3-13 中 Button 上下左右 4 个圆圈就是用来添加约束的。约束既可以是添加到 ConstraintLayout，也可以添加到其他控件。

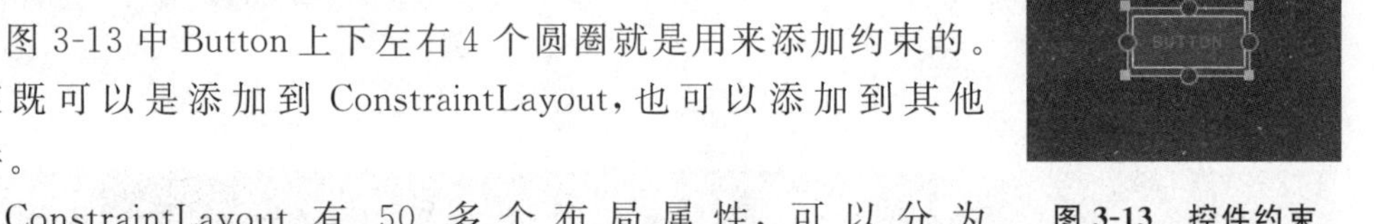

图 3-13 控件约束

ConstraintLayout 有 50 多个布局属性，可以分为 ConstraintLayout 本身使用的属性、Guideline 使用的属性、相对定位属性、Margin 属性、居中和偏移属性、子 View 的尺寸控制属性、UI 编辑器使用的属性。

常用属性有：

layout_constraintTop_toTopOf：表示当前 View 顶部与另一个 View 顶部对齐。

layout_constraintTop_toBottomOf：表示当前 View 顶部与另一个 View 的底部对齐。

layout_constraintBottom_toTopOf：表示当前 View 底部与另一个 View 的顶部对齐。

layout_constraintBottom_toBottomOf：表示当前 View 底部与另一个 View 底部对齐。

layout_constraintLeft_toLeftOf：表示当前 View 左边与另一个 View 左边对齐。

layout_constraintLeft_toRightOf：表示当前 View 左边与另一个 View 的右边对齐。

layout_constraintRight_toLeftOf：表示当前 View 右边与另一个 View 的左边对齐。

layout_constraintRight_toRightOf：表示当前 View 右边与另一个 View 的右边对齐。

在 Chapter03UI 项目中新建 ConstraintLayoutActivity.java，其对应的布局文件 activity_constraint_layout.xml 采用默认的 ConstraintLayout 布局方式，拖曳控件到该布局文件，并且编辑控件的约束关系，如图 3-14 所示。

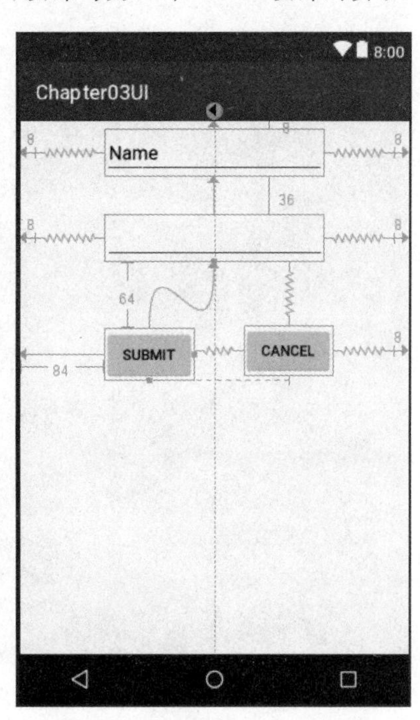

图 3-14 ConstraintLayoutActivity 页面预览

activity_constraint_layout.xml 对应的源文件如文件清单 3-9 所示。

文件清单 3-9 activity_constraint_layout.xml

```
<?xml version="1.0" encoding="utf-8"?>
<android.support.constraint.ConstraintLayout xmlns:android="http://schemas.
android.com/apk/res/android"
    xmlns:app="http://schemas.android.com/apk/res-auto"
    xmlns:tools="http://schemas.android.com/tools"
    android:layout_width="match_parent"
    android:layout_height="match_parent"
    tools:context="com.nsu.zyl.Chapter03ui.layout.ConstraintLayoutActivity">
    <Button
        android:id="@+id/btnSubmit"
        android:layout_width="wrap_content"
        android:layout_height="50dp"
        android:layout_marginStart="84dp"
        android:layout_marginTop="64dp"
        android:text="Submit"
        app:layout_constraintStart_toStartOf="parent"
        app:layout_constraintTop_toBottomOf="@+id/editText5"
        tools:ignore="MissingConstraints" />

    <EditText
        android:id="@+id/editText4"
        android:layout_width="wrap_content"
        android:layout_height="wrap_content"
        android:layout_marginEnd="8dp"
        android:layout_marginStart="8dp"
        android:layout_marginTop="8dp"
        android:ems="10"
        android:inputType="textPersonName"
        android:text="Name"
        app:layout_constraintEnd_toEndOf="parent"
        app:layout_constraintStart_toStartOf="parent"
        app:layout_constraintTop_toTopOf="parent" />

    <EditText
        android:id="@+id/editText5"
        android:layout_width="wrap_content"
        android:layout_height="wrap_content"
        android:layout_marginEnd="8dp"
        android:layout_marginStart="8dp"
        android:layout_marginTop="36dp"
        android:ems="10"
        android:inputType="textPassword"
        app:layout_constraintEnd_toEndOf="parent"
        app:layout_constraintHorizontal_bias="0.503"
        app:layout_constraintStart_toStartOf="parent"
        app:layout_constraintTop_toBottomOf="@+id/editText4" />
```

```
    <Button
        android:id="@+id/btnCancel"
        android:layout_width="wrap_content"
        android:layout_height="wrap_content"
        android:layout_marginBottom="8dp"
        android:layout_marginEnd="8dp"
        android:layout_marginStart="8dp"
        android:layout_marginTop="64dp"
        android:text="Cancel"
        app:layout_constraintBottom_toBottomOf="@+id/btnSubmit"
        app:layout_constraintEnd_toEndOf="parent"
        app:layout_constraintHorizontal_bias="0.37"
        app:layout_constraintStart_toEndOf="@+id/btnSubmit"
        app:layout_constraintTop_toBottomOf="@+id/editText5" />

    <android.support.constraint.Guideline
        android:id="@+id/guideline3"
        android:layout_width="wrap_content"
        android:layout_height="wrap_content"
        android:orientation="vertical"
        app:layout_constraintGuide_begin="192dp" />

</android.support.constraint.ConstraintLayout>
```

为了便于理解这些布局方式的特点，在Chapter03UI项目中新建BasicLayoutActivity和ShowLayoutActivity，用户在BasicLayoutActivity中选择具体的布局，ShowLayoutActivity将根据选择的布局进行展示。编辑文件清单3-2 MainActivity.java，在MyClickListener的onClick()处理方法中，添加用户点击"Android UI布局"按钮时，应用跳转至BasicLayoutActivity的逻辑。

```
    switch (v.getId()) {
            case R.id.btn_basic_layout:
                Intent intent1 =new Intent();
                intent1.setClass(MainActivity.this, BasicLayoutActivity.class);
                startActivity(intent1);
                break;
......
}
```

BasicLayoutActivity页面显示如图3-15所示，点击页面中的按钮程序跳转至ShowLayoutActivity，ShowLayoutActivity根据点击按钮的不同以不同的布局方式显示内容。由于图3-15对应页面布局比较简单，在此就不给出布局文件详细内容。

BasicLayoutActivity.java对应的源代码如文件清单3-10所示。

图 3-15 BasicLayoutActivity 页面

文件清单 3-10 BasicLayoutActivity. java

```
package com.nsu.zyl.Chapter03ui;

import android.content.Intent;
import android.support.v7.app.AppCompatActivity;
import android.os.Bundle;
import android.view.View;
import android.widget.Button;

import com.nsu.zyl.Chapter03ui.layout.ShowLayoutActivity;

public class BasicLayoutActivity extends AppCompatActivity {
    private Button btnLinear, btnRelative, btnFrame, btnGrid, btnTable, btnAbsolute,
btnConstraint;

    @Override
    protected void onCreate(Bundle savedInstanceState) {
        super.onCreate(savedInstanceState);
        setContentView(R.layout.activity_basic_layout);
        btnLinear=(Button)findViewById(R.id.btnLinear);
        btnRelative=(Button)findViewById(R.id.btnRelative);
        btnFrame=(Button)findViewById(R.id.btnFrame);
        btnGrid=(Button)findViewById(R.id.btnGrid);
```

```
        btnTable=(Button)findViewById(R.id.btnTable);
        btnAbsolute=(Button)findViewById(R.id.btnAbsolute);
        btnConstraint=(Button)findViewById(R.id.btnConstraint);
        MyClickListener myClickListener=new MyClickListener();
        btnLinear.setOnClickListener(myClickListener);
        btnRelative.setOnClickListener(myClickListener);
        btnFrame.setOnClickListener(myClickListener);
        btnGrid.setOnClickListener(myClickListener);
        btnTable.setOnClickListener(myClickListener);
        btnAbsolute.setOnClickListener(myClickListener);
        btnConstraint.setOnClickListener(myClickListener);
    }
    class MyClickListener implements View.OnClickListener {

        @Override
        public void onClick(View v) {
            Intent intent=new Intent(BasicLayoutActivity.this, ShowLayoutActivity.
class);
            int layout=R.layout.activity_linear_layout;
            switch (v.getId()){
                case R.id.btnLinear:
                    layout=R.layout.activity_linear_layout;
                    break;
                case R.id.btnRelative:
                    layout=R.layout.activity_relative_layout;
                    break;
                case R.id.btnFrame:
                    layout=R.layout.activity_frame_layout;
                    break;
                case R.id.btnGrid:
                    layout=R.layout.activity_grid_layout;
                    break;
                case R.id.btnTable:
                    layout=R.layout.activity_table_layout;
                    break;
                case R.id.btnAbsolute:
                    layout=R.layout.activity_absolute_layout;
                    break;
                case R.id.btnConstraint:
                    layout=R.layout.activity_constraint_layout;
                    break;
            }
            intent.putExtra("layout",layout);
            startActivity(intent);
        }
    }
}
```

ShowLayoutActivity 将调用 getIntent(). getIntExtra()获得传递过来的参数,根据参数以不同的布局文件显示,ShowLayoutActivity. java 详细内容如文件清单 3-11 所示。

文件清单 3-11 ShowLayoutActivity. java

```
package com.nsu.zyl.Chapter03ui.layout;
import android.support.v7.app.AppCompatActivity;
import android.os.Bundle;
import com.nsu.zyl.Chapter03ui.R;

public class ShowLayoutActivity extends AppCompatActivity {

    @Override
    protected void onCreate(Bundle savedInstanceState) {
        super.onCreate(savedInstanceState);
        int layout=getIntent().getIntExtra("layout", R.layout.activity_show_
layout);
        setContentView(layout);
    }
}
```

3.2 常用控件的使用

Android 提供了大量的 UI 控件,合理地使用这些控件就可以轻松地编写出不错的界面,这些是 Android 学习的基础。本节主要涉及的控件包括文本类控件、按钮类控件、图片控件、进度条控件等。

3.2.1 TextView 与 EditText

TextView 是界面设计中最为常见的控件,也是很多其他控件的父类,例如 Button、EditText 等。本节介绍 TextView 控件及其子类的常用属性和用法。

1. TextView

TextView 是一种用于显示文本信息的控件,它不能被编辑,其重要属性如表 3-1 所示。

表 3-1 TextView 属性说明

属性	功能说明
android:layout_width	控件宽度,可取值 fill_parent、match_parent、wrap_content,或者具体像素值(单位 dp)
android:layout_height	控件高度,取值同 android:layout_width
android:lines	设置文本的行数,设置 2 行就显示 2 行,即使第 2 行没有数据

续表

属　　性	功能说明
android:singleLine	设置单行显示,取值 true 或 false。如果和 layout_width 一起使用,当文本不能全部显示时,后面用"…"表示
android:maxLength	限制显示的文本长度,超出部分不显示
android:gravity	设置文本显示位置,如设置成 center,文本将居中显示
android:text	设置显示的文本信息,推荐使用@string/xx 的方式
android:textSize	设置文字大小,推荐度量单位 sp,如 15sp
android:textColor	设置文本颜色
android:drawableBottom	在 text 的下方输出一个 drawable,如图片。如果指定一个颜色,会把 text 的背景设为该颜色,和 background 同时使用时会覆盖后者
android:drawableLeft	在 text 的左边输出一个 drawable,如图片
android:drawableRight	在 text 的右边输出一个 drawable
android:drawableTop	在 text 的正上方输出一个 drawable
android:drawablePadding	设置 text 与 drawable 的间隔,与 drawableLeft、drawableRight、drawableTop、drawableBottom 一起使用,可设置为负数,单独使用没有效果
android:autoLink	设置是否当文本为 URL 链接/email/电话号码/map 时,文本显示为可点击的链接。可选值(none /web /email /phone /map /all)
android:ellipsize	设置当文字过长时,该控件该如何显示。有如下值设置:start——省略号显示在开头;end——省略号显示在结尾;middle——省略号显示在中间;marquee——以跑马灯的方式显示(动画横向移动)
android:marqueeRepeatLimit	在 ellipsize 指定 marquee 的情况下,设置重复滚动的次数,当设置为 marquee_forever 时表示无限次

通过代码体验一下这些属性的用法,代码如下:

```
<?xml version="1.0" encoding="utf-8"?>
<LinearLayout xmlns:android="http://schemas.android.com/apk/res/android"
    android:layout_width="match_parent"
    android:layout_height="match_parent"
    android:orientation="vertical">

    <TextView
        android:text="1、我是一个普通的 TextView"
        android:layout_width="wrap_content"
        android:layout_height="wrap_content"
        android:textSize="20sp"
        android:id="@+id/tv_main_1" />

    <TextView
        android:text="2、我是一个超链接 TextView,电话: 18032580286,\n 网页: http://
www.baidu.com"
        android:layout_width="wrap_content"
        android:layout_height="wrap_content"
```

```
        android:textSize="20sp"
        android:id="@+id/tv_main_2"
        android:autoLink="all" />

    <TextView
        android:text="3、我是一个跑马灯效果的 TextView"
        android:layout_width="wrap_content"
        android:layout_height="wrap_content"
        android:textSize="20sp"
        android:id="@+id/tv_main_3"
        android:marqueeRepeatLimit="marquee_forever"
        android:ellipsize="marquee"
        android:focusable="true"
        android:focusableInTouchMode="true"
        android:singleLine="true"/>
</LinearLayout>
```

这段布局代码中使用了 3 个 TextView 控件，第一个是普通的 TextView，设置了文本框的宽度、高度及字体大小；第二个 TextView 中设置了超链接，android:autoLink 属性取值为 all，表示将识别 web、email、phone、map 等超链接；第三个 TextView 中设置了一个跑马灯的效果，页面效果如图 3-16 所示。

图 3-16 TextView 运行效果

2. EditText

EditText 是一个具有编辑功能的 TextView，是用来输入和编辑字符串的控件。除了具备 TextView 的属性外，EditText 还具有表 3-2 所示的常用属性。

表 3-2 EditText 常用属性

属　性	功 能 说 明
android:editable	设置是否可编辑。默认为 true，当值为 false 时仍然可以获取光标，但是无法输入
android:hint	Text 为空时显示的文字提示信息，可通过 textColorHint 设置提示信息的颜色
android:inputType	设置文本的类型，用于帮助输入法显示合适的键盘类型
android:imeOptions	指定输入法窗口中 Enter 键的功能，可选值为 normal、actionNext、actionDone、actionSearch 等
android:digits	指定要支持的字符
android:cursorVisible	设置光标是否显示

以下通过代码体验这些属性的用法：

```
<?xml version="1.0" encoding="utf-8"?>
<LinearLayout xmlns:android="http://schemas.android.com/apk/res/android"
    android:layout_width="match_parent"
    android:layout_height="match_parent"
    android:orientation="vertical">

    <TextView
        android:layout_width="wrap_content"
        android:layout_height="wrap_content"
        android:text="请输入个人信息："
        android:textSize="25sp" />

    <LinearLayout
        android:orientation="horizontal"
        android:layout_width="match_parent"
        android:layout_height="wrap_content">

        <TextView
            android:layout_width="0dp"
            android:layout_height="wrap_content"
            android:text="姓名："
            android:id="@+id/textView"
            android:textSize="22sp"
            android:layout_weight="1" />

        <EditText
            android:layout_width="0dp"
            android:layout_height="wrap_content"
            android:inputType="textPersonName"
            android:ems="10"
            android:id="@+id/edT1"
            android:layout_weight="4"
            android:singleLine="true"
            android:hint="请输入您的姓名" />
    </LinearLayout>

    <LinearLayout
        android:orientation="horizontal"
        android:layout_width="match_parent"
        android:layout_height="wrap_content">

        <TextView
            android:layout_width="0dp"
            android:layout_height="wrap_content"
            android:text="年龄："
            android:id="@+id/textView2"
```

```
            android:textSize="22sp"
            android:layout_weight="1" />

        <EditText
            android:layout_width="0dp"
            android:layout_height="wrap_content"
            android:inputType="number"
            android:ems="10"
            android:id="@+id/edT2"
            android:layout_weight="4"
            android:singleLine="true"
            android:hint="请输入您的年龄" />
    </LinearLayout>

    <LinearLayout
        android:orientation="horizontal"
        android:layout_width="match_parent"
        android:layout_height="wrap_content">

        <TextView
            android:layout_width="0dp"
            android:layout_height="wrap_content"
            android:text="电话："
            android:id="@+id/textView3"
            android:textSize="22sp"
            android:layout_weight="1" />

        <EditText
            android:layout_width="0dp"
            android:layout_height="wrap_content"
            android:inputType="phone"
            android:ems="10"
            android:id="@+id/edT3"
            android:layout_weight="4"
            android:singleLine="true"
            android:hint="请输入您的电话号" />
    </LinearLayout>

    <Button
        android:layout_width="wrap_content"
        android:layout_height="wrap_content"
        android:text="确定"
        android:id="@+id/btn"
        android:layout_gravity="center_horizontal"
        android:onClick="show" />

    <TextView
        android:layout_width="wrap_content"
        android:layout_height="wrap_content"
```

```
        android:id="@+id/txv"
        android:layout_gravity="center_horizontal"
        android:textSize="25sp" />
</LinearLayout>
```

页面运行效果如图 3-17 和图 3-18 所示。这里应重点理解 hint、inputType、singleLine、ems 等属性,其他属性可自行尝试。

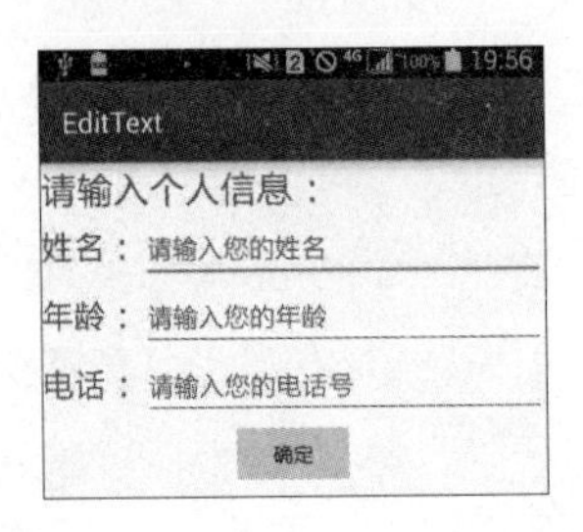

图 3-17　EditText 输入前效果

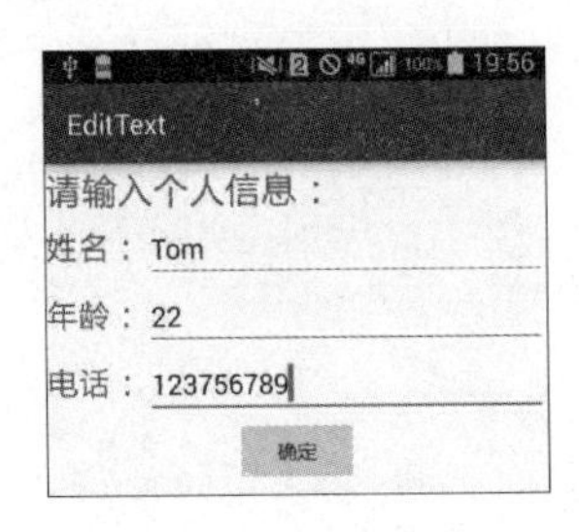

图 3-18　EditText 输入效果

3.2.2　Button

Button 继承自 TextView,它的功能就是提供一个按钮,这个按钮可以供用户点击。当用户对按钮进行操作时,触发相应事件,如点击、触摸等。Button 的相关属性说明如表 3-3 所示。

表 3-3　Button 属性

属　　性	功能说明
android:clickable	取值 true 或者 false,设置是否允许点击
android:background	通过资源文件设置背景色
android:drawableTop	在 Button 组件上放置图片,图片在上,文字在下。类似的属性还有 drawableLeft、drawableRight、drawableBottom
android:text	设置文字
android:textColor	设置文字颜色
android:onClick	设置按钮的监听器,点击按钮时将调用对应的方法,方法应为 public void XXXX(View v)

1. 事件处理

以下重点讲解 Button 点击事件和触摸事件的处理,其他事件的处理方式与此类似,只是触发的时机不同而已。

对 Button 的点击事件处理,需要对 Button 添加 View. OnClickListener 监听器,并且需要实现监听器中的 onClick(View v)方法,其中 v 为触发当前事件的控件,例如:

```
button.setOnClickListener(new View.OnClickListener() {
    @Override
    public void onClick(View v) {
        // TODO Auto-generated method stub
    }
});
```

同理，对 Button 的触摸事件的处理，需要对 Button 添加 View. OnTouchListener 监听器，并且实现 onTouch(View v, MotionEvent event)方法，其中 v 为当前触发事件的控件，event 包括了触摸时的具体内容，例如移动、按下等，例如：

```
playBtn.setOnTouchListener(new View.OnTouchListener() {
    @Override
    public boolean onTouch(View v, MotionEvent event) {
        // TODO Auto-generated method stub
        return false;
    }
});
```

2. 图文混排

在实际项目中，经常需要将按钮展示设置为图文混排的效果，这样可以通过简短的文字说明将图标的功能展示给用户。对于 Button 控件，图文混排需要用到 android: drawableXxx 属性(Xxx 为图片所在按钮的方向)，这个属性配合 android: text，就可以实现图文混排的效果，如图 3-19 所示。

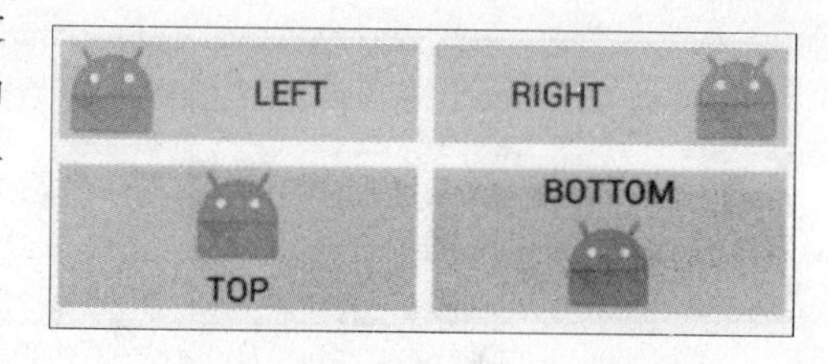

图 3-19 Button 图文混排运行效果

图 3-19 所示布局的代码如下：

```
<LinearLayout
        android:layout_width="match_parent"
        android:layout_height="wrap_content"
        android:orientation="horizontal">

        <Button
            android:layout_width="0dp"
            android:layout_height="wrap_content"
            android:text="Left"
            android:id="@+id/btnLeft"
            android:paddingLeft="5dp"
            android:paddingRight="5dp"
            android:paddingTop="5dp"
            android:paddingBottom="5dp"
```

```
            android:drawableLeft="@mipmap/ic_launcher"
            android:layout_weight="1" />

        <Button
            android:layout_width="0dp"
            android:layout_height="wrap_content"
            android:text="Right"
            android:id="@+id/btnRight"
            android:paddingLeft="5dp"
            android:paddingRight="5dp"
            android:paddingTop="5dp"
            android:paddingBottom="5dp"
            android:drawableRight="@mipmap/ic_launcher"
            android:layout_weight="1" />

    </LinearLayout>

    <LinearLayout
        android:orientation="horizontal"
        android:layout_width="match_parent"
        android:layout_height="wrap_content">

        <Button
            android:layout_width="0dp"
            android:layout_height="wrap_content"
            android:text="Top"
            android:id="@+id/btnTop"
            android:paddingLeft="5dp"
            android:paddingRight="5dp"
            android:paddingTop="5dp"
            android:paddingBottom="5dp"
            android:drawableTop="@mipmap/ic_launcher"
            android:layout_weight="1" />

        <Button
            android:layout_width="0dp"
            android:layout_height="wrap_content"
            android:text="Bottom"
            android:id="@+id/btnBottom"
            android:paddingLeft="5dp"
            android:paddingRight="5dp"
            android:paddingTop="5dp"
            android:paddingBottom="5dp"
            android:drawableBottom="@mipmap/ic_launcher"
            android:layout_weight="1"
            android:layout_gravity="center_vertical" />
    </LinearLayout>
```

3.2.3 ImageView 和 ImageButton

在 Android 应用开发中,经常需要显示图片。与图片相关的 View 控件主要有 ImageView 和 ImageButton。二者虽然都可显示图片,但用法也存在差异,本节深入理解这两个控件。

1. ImageView

ImageView 可以加载各种来源的图片(如资源或图片库),用于在页面中显示图片(图片的浏览)。ImageView 在页面中显示时,需要计算图像的尺寸,并提供缩放和着色(渲染)等各种显示选项。ImageView 的相关属性如表 3-4 所示。

表 3-4 ImageView 属性说明

属 性	功能说明
android: adjustViewBounds	是否保持宽高比。需要与 maxWidth、maxHeight 一起使用,单独使用没有效果
android:maxHeight	设置 View 的最大高度,单独使用无效,需要与 setAdjustViewBounds 一起使用
android:maxWidth	设置 View 的最大宽度,同上
android:cropToPadding	是否截取指定区域用空白代替
android:scaleType	设置图片的填充方式,可有以下取值: matrix: 用矩阵来绘图; fitXY: 拉伸图片(不按比例)以填充 View 的宽高; fitStart: 按比例拉伸或压缩图片,处理后图片的高度或宽度为 View 的高度或宽度,且显示在 View 的左边; fitCenter: 按比例拉伸或压缩图片,处理后图片的高度或宽度为 View 的高度或宽度,且显示在 View 的中间; fitEnd: 按比例拉伸或压缩图片,处理后图片的高度为 View 的高度或宽度,且显示在 View 的右边; center: 按原图大小显示图片,但图片宽高大于 View 的宽高时,截取图片中间部分显示; centerCrop: 按比例放大原图直至等于某边 View 的宽高显示; centerInside: 当原图宽高小于等于 View 的宽高时,按原图大小居中显示,效果同 center; 反之,将原图按比例缩放至 View 的宽高居中显示,效果同 fitCenter
android:src	设置 View 的 drawable(如图片,也可以是颜色,但是需要指定 View 的大小)

通过下面这段代码来体验一下 ImageView 的使用要点:

```
<?xml version="1.0" encoding="utf-8"?>
<RelativeLayout xmlns:android="http://schemas.android.com/apk/res/android"
    xmlns:app="http://schemas.android.com/apk/res-auto"
```

```
    android:layout_width="match_parent"
    android:layout_height="match_parent">

    <ImageView
        android:layout_width="match_parent"
        android:layout_height="match_parent"
        app:srcCompat="@mipmap/lena"
        android:layout_centerVertical="true"
        android:layout_centerHorizontal="true"
        android:id="@+id/imageView2"
        android:scaleType="matrix" />
</RelativeLayout>
```

在布局文件中，设置ImageView显示在屏幕正中，通过修改android:scaleType的值，图片的分辨率及位置会发生相应的变化，页面显示效果如图3-20所示。

图3-20 ImageView不同填充方式运行效果

需要说明的是，ImageView中XML属性background会根据ImageView组件给定的长宽拉伸，而src存放的就是原图的大小，不会拉伸。src是图片内容(前景)，background是背景，可以同时使用，scaleType只对src起作用，background可设置透明。

如果想设置图片固定大小，又想保持图片宽高比，需要如下设置：

(1) 设置setAdjustViewBounds为true；

(2) 设置maxWidth、maxHeight；

(3) 设置layout_width和layout_height为wrap_content。

2. ImageButton

ImageButton(图片按钮)继承自ImageView，可以在ImageButton中显示一个图片展示给用户看，需要注意其Text属性是无效的，其他功能与Button一样。ImageButton属性基本与ImageView类似，只不过它默认是可以获得焦点的。

可以通过点击实现切换按钮图片的效果，来看下面这段布局代码：

```
<ImageButton
      android:layout_width="wrap_content"
      android:layout_height="wrap_content"
      android:background="@drawable/img_btn_status"
      android:id="@+id/imageButton" />
```

ImageButton 的 background 属性指定的并不是一张图片，而是位于 drawable 文件夹下的一个名为 img_btn_status.xml 的选择器文件。代码如下：

```
<?xml version="1.0" encoding="utf-8"?>
<selector xmlns:android="http://schemas.android.com/apk/res/android">
    <item android:state_pressed="false" android:drawable="@mipmap/icon_home" />
    <item android:state_focused="true" android:drawable="@mipmap/icon_home" />
    <item android:state_pressed="true" android:drawable="@mipmap/icon_home_
light" />
</selector>
```

执行这段代码，正如 selector 中指定的一样，当 ImageButton 在点击前，获得焦点时显示效果如图 3-21 所示，显示的是 icon_home.png 图片；而当点击时，显示的则是 icon_home_light.png 图片，效果如图 3-22 所示。

图 3-21 ImageButton 点击前

图 3-22 ImageButton 点击时

（注：图标变成了红色）

ImageView 与 ImageButton 的用法区别：ImageView 会根据设置的具体宽高尺寸变化，而 ImageButton 只会显示图片的原始像素大小。给 ImageButton 设置 scaleType 属性可以完成 ImageView 的效果，但是那样会使图片失真。

3.2.4 ToggleButton、RadioButton 和 CheckBox

ToggleButton、RadioButton 和 CheckBox 都是继承自 android.widget.CompoundButton 的组件。CompoundButton 有两个状态，分别是 checked 和 not checked。在 Android 应用开发中，可以根据实际需要，灵活选择相应的控件来实现需要的功能。

1. ToggleButton

ToggleButton(开关按钮)继承自 CompoundButton，是一个具有选中和未选中两种状态的按钮，并且需要为不同的状态设置不同的显示文本，常用于表示开-关场景中。

ToggleButton 的相关属性如表 3-5 所示。

ToggleButton 的常用方法：

表 3-5　ToggleButton 属性说明

属　　性	功 能 说 明
android：disabledAlpha	设置按钮在禁用时的透明度。数字对应具体效果如下图所示：
android：textOff	未选中时按钮的文本
android：textOn	选中时按钮的文本
android：checked	设置该按钮是否选中

public CharSequence getTextOff ()：返回按钮未选中时的文本。

public CharSequence getTextOn ()：返回按钮选中时的文本。

public void setChecked (boolean checked)：改变按钮的选中状态，参数 checked 值为 true 让按钮选中，为 false 让按钮不选中。

可以为 ToggleButton 设置 OnCheckedChangeListener 监听器来监听开关按钮的状态变化。

在 Chapter03UI 项目中创建 ToggleButtonActivity 用于体验 ToggleButton 的用法，ToggleButtonActivity 对应的布局文件 toggle_button_layout. xml 如文件清单 3-12 所示，相对布局中有两个控件，一个 ImageView 显示图片，一个 ToggleButton 控制图片的变化。

文件清单 3-12　toggle_button_layout. xml

```
<?xml version="1.0" encoding="utf-8"?>
<RelativeLayout xmlns:android="http://schemas.android.com/apk/res/android"
    android:layout_width="match_parent"
    android:layout_height="match_parent">

    <ImageView
        android:layout_width="wrap_content"
        android:layout_height="wrap_content"
        android:src="@mipmap/close"
        android:layout_marginTop="71dp"
        android:id="@+id/img_light_on"
        android:layout_alignParentTop="true"
        android:layout_centerHorizontal="true" />
```

```
    <ToggleButton
        android:text="ToggleButton"
        android:layout_width="wrap_content"
        android:layout_height="wrap_content"
        android:layout_below="@+id/img_light_on"
        android:layout_centerHorizontal="true"
        android:layout_marginTop="63dp"
        android:id="@+id/toggleButton" />
</RelativeLayout>
```

对应的 Activity 核心代码如文件清单 3-13 所示。

文件清单 3-13　ToggleButtonActivity. java

```
public class ToggleButtonActivity extends Activity {

    private ImageView imageView;
    private ToggleButton toggleButton;

    @Override
    protected void onCreate(Bundle savedInstanceState) {
        super.onCreate(savedInstanceState);
        setContentView(R.layout.toggle_button_layout);

        imageView = (ImageView) findViewById(R.id.img_light_on);
        toggleButton = (ToggleButton) findViewById(R.id.toggleButton);
        toggleButton.setTextOn("关灯");
        toggleButton.setTextOff("开灯");
        toggleButton.setChecked(false);

        toggleButton.setOnCheckedChangeListener(new
CompoundButton.OnCheckedChangeListener() {
            @Override
            public void onCheckedChanged(CompoundButton buttonView,
boolean isChecked) {
                if (isChecked){
                    imageView.setImageResource(R.mipmap.open);
                }else {
                    imageView.setImageResource(R.mipmap.close);
                }
            }
        });
    }
}
```

该案例实现一个模拟的开/关灯效果，如图 3-23 所示。当 ToggleButton 未选中时，显示一个未发光的灯泡图片；而当选中时，则显示一个发光的灯泡图片。

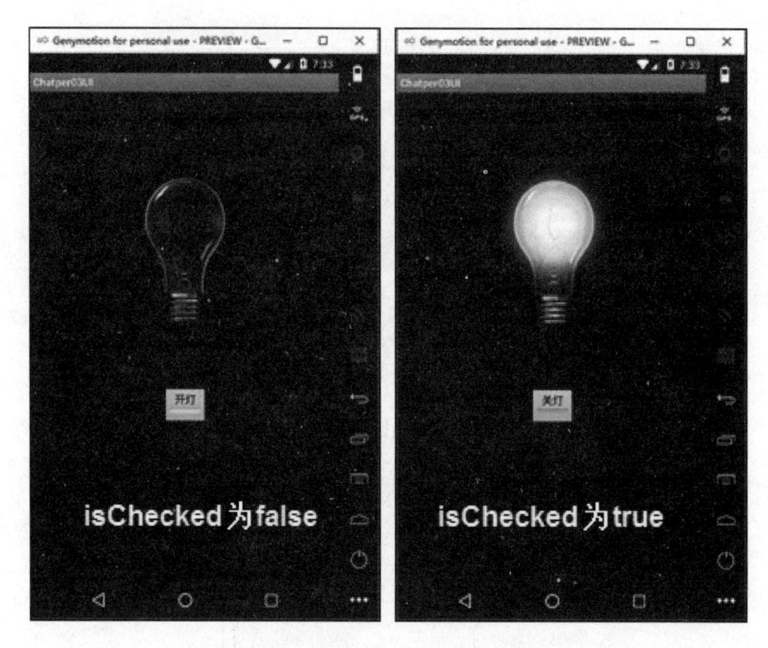

图 3-23 ToggleButton 案例运行效果

2. RadioButton

RadioButton(单选按钮)继承自 CompoundButton,是一种具有选中和未选中两种状态的按钮。在单选按钮没有被选中时,用户能够按下或点击它来选中单选按钮,用户一旦选中就不能够取消选中。

RadioButton 通常配合 RadioGroup 使用,表示一个单选按钮组,包含几个单选按钮,选中其中一个的同时,将取消其他选中的单选按钮。

可以为 RadioGroup 设置 OnCheckedChangeListener 事件监听器,监听单选按钮的变化;也可以针对具体的 RadioButton 来设置 OnCheckedChangeListener 事件监听器。

通过如下代码体验 RadioButton 的用法:

```
<LinearLayout
        android:orientation="vertical"
        android:layout_width="match_parent"
        android:layout_height="match_parent">

        <TextView
            android:text="你的性别是?"
            android:layout_width="match_parent"
            android:layout_height="wrap_content"
            android:textSize="25sp"
            android:id="@+id/tv_radio_title" />

        <RadioGroup
```

```
        android:id="@+id/radio_sex"
        android:layout_width="match_parent"
        android:layout_height="wrap_content"
        android:orientation="horizontal">

        <RadioButton
            android:text="男"
            android:layout_width="wrap_content"
            android:layout_height="wrap_content"
            android:id="@+id/radioButton_m"
            android:layout_weight="1" />

        <RadioButton
            android:text="女"
            android:layout_width="wrap_content"
            android:layout_height="wrap_content"
            android:id="@+id/radioButton_f"
            android:layout_weight="1" />

        <RadioButton
            android:text="其他"
            android:layout_width="wrap_content"
            android:layout_height="wrap_content"
            android:id="@+id/radioButton_o"
            android:layout_weight="1" />
    </RadioGroup>

    <TextView
        android:text="TextView"
        android:layout_width="match_parent"
        android:layout_height="match_parent"
        android:gravity="center"
        android:textSize="25sp"
        android:id="@+id/tv_radio_result" />
</LinearLayout>
```

对应的 Activity 文件的核心代码如下：

```
//初始化控件
sexChoice = (RadioGroup) findViewById(R.id.radio_sex);
sexResult = (TextView) findViewById(R.id.tv_radio_result);
//设置监听器
sexChoice.setOnCheckedChangeListener(new
                          RadioGroup.OnCheckedChangeListener() {
    @Override
    public void onCheckedChanged(RadioGroup group, int checkedId) {
            RadioButton choice = (RadioButton) findViewById(
sexChoice.getCheckedRadioButtonId());
            sexResult.setText("你选择的性别是"+choice.getText().toString());
```

```
        }
});
```

最终运行效果如图3-24所示。点击RadioButton,对应的控件将变成选中状态,可以通过getText()方法获取RadioButton的文字内容。

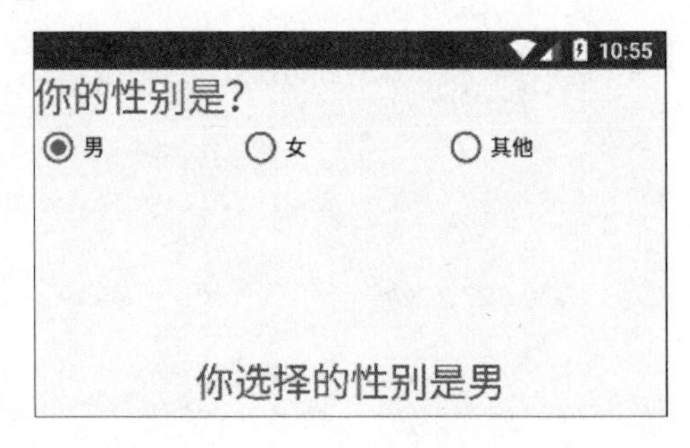

图3-24 RadioButton案例运行效果

3. CheckBox

CheckBox(复选框)与RadioButton相同,也继承自CompoundButton,是一种有两种状态的特殊按钮,可以选中或者不选中。可以先在布局文件中定义多选按钮,然后对每一个多选按钮设置事件监听setOnCheckedChangeListener,通过isChecked属性来判断选项是否被选中,做出相应的事件响应。

通过如下代码体验CheckBox的用法,监听单个CheckBox的选中与未选中,同时从外面获取所有CheckBox的选中状态。布局文件核心代码如下:

```
<LinearLayout
    android:orientation="vertical"
    android:layout_width="match_parent"
    android:layout_height="match_parent"
    android:layout_weight="1">

    <TextView
        android:text="下列课程中,你喜欢的有哪些?"
        android:layout_width="match_parent"
        android:layout_height="wrap_content"
        android:textSize="25sp"
        android:id="@+id/tv_cbx_title" />

    <CheckBox
        android:text="Android应用开发"
          android:layout_width="match_parent"
          android:layout_height="wrap_content"
          android:id="@+id/cbx_a" />

      <CheckBox
          android:text="软件测试基础"
          android:layout_width="match_parent"
          android:layout_height="wrap_content"
          android:id="@+id/cbx_b" />

      <CheckBox
          android:text="面向对象分析与设计"
```

```
        android:layout_width="match_parent"
        android:layout_height="wrap_content"
        android:id="@+id/cbx_c" />

    <CheckBox
        android:text="Java 程序设计"
        android:layout_width="match_parent"
        android:layout_height="wrap_content"
        android:id="@+id/cbx_d" />

    <Button
        android:text="提交"
        android:layout_width="wrap_content"
        android:layout_height="wrap_content"
        android:id="@+id/btn_submit" />

    <TextView
        android:text="TextView"
        android:layout_width="match_parent"
        android:layout_height="match_parent"
        android:gravity="center"
        android:id="@+id/tv_cbx_result" />
</LinearLayout>
```

要想在外面获取所有 CheckBox 的选中状态，需要在 Activity 中先声明一个全局变量 cbxList，这里我们用一个列表：

```
private List<CheckBox> cbxList = new ArrayList<CheckBox> ();
//初始化控件
cbxA = (CheckBox) findViewById(R.id.cbx_a);
cbxB = (CheckBox) findViewById(R.id.cbx_b);
cbxC = (CheckBox) findViewById(R.id.cbx_c);
cbxD = (CheckBox) findViewById(R.id.cbx_d);
btnSubmit = (Button) findViewById(R.id.btn_submit);
answerResult = (TextView) findViewById(R.id.tv_cbx_result);
//将 4 个 CheckBox 加入列表
cbxList.add(cbxA);
cbxList.add(cbxB);
cbxList.add(cbxC);
cbxList.add(cbxD);
//监听单个 CheckBox 的选中状态
for (final CheckBox cbx:cbxList) {
    cbx.setOnCheckedChangeListener(new CompoundButton.OnCheckedChangeListener() {
      @Override
      public void onCheckedChanged(CompoundButton buttonView, boolean isChecked)
{
          if (isChecked){
```

```
                Toast.makeText(BasicViewActivity.this,cbx.getText().toString()+
                    "已选中",
Toast.LENGTH_SHORT).show();
              }else {
                Toast.makeText(BasicViewActivity.this,cbx.getText().toString()+
                    "已取消",
Toast.LENGTH_SHORT).show();
              }
          }
      });
}
//监听所有 CheckBox 的选中状态
btnSubmit.setOnClickListener(new View.OnClickListener() {
        @Override
        public void onClick(View v) {
            StringBuffer sb =new StringBuffer();
            //遍历集合中的 CheckBox,判断是否选择,获取选中的文本
            for (CheckBox cbx : cbxList) {
                if (cbx.isChecked()){
                    sb.append(cbx.getText().toString() +" ");
                }
            }
              if (sb!=null && "".equals(sb.toString())){
                  Toast.makeText(BasicViewActivity.this, "请至少选择一个",
Toast.LENGTH_SHORT).show();
              }else{
                  answerResult.setText("你的选择是 "+sb.toString());
              }
        }
});
```

最终运行效果如图 3-25 所示。选中或取消选中 CheckBox,以 Toast 消息提示。点击“提交”按钮,获取所有已选中的 CheckBox 的文本内容。

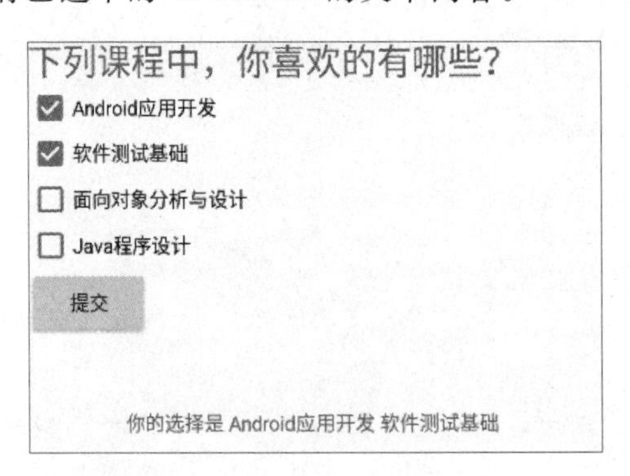

图 3-25　CheckBox 案例运行效果

3.2.5 ProgressBar、SeekBar 和 RatingBar

ProgressBar 是 Android 的进度条，当作耗时操作时，可以使用 ProgressBar 给用户提供进度提示。ProgressBar 派生出两个子类：SeekBar 和 RatingBar。下面详细介绍 ProgressBar、SeekBar 和 RatingBar。

1. ProgressBar

默认为圆形进度条，如图 3-26 所示。圆形进度条根据规格大小可分为大、中、小三种规格，可通过 style 来设置进度条规格，默认中等尺寸规格。

```
style="?android:attr/progressBarStyleLarge"设置为大尺寸。
style="?android:attr/progressBarStyleSmall"设置为小尺寸。
```

也可以通过设置 style 属性将圆形进度条更改为水平进度条，如图 3-27 所示。

```
style="@android:style/Widget.ProgressBar.Horizontal"
style="?android:attr/progressBarStyleHorizontal"
```

以上两种方式是等价的。

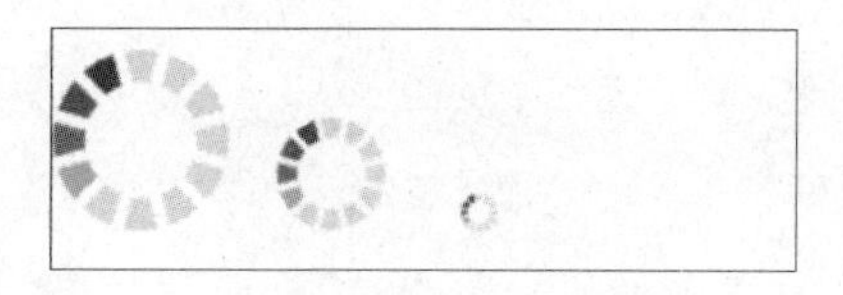

图 3-26 圆形进度条

图 3-27 水平进度条

在 android. R. attr 中定义了 ProgressBar 各种样式的参数，有 progressBarStyle-Horizontal、progressBarStyleSmall、progressBarStyle、progressBarStyleLarge；也可以在表示圆形进度条的样式参数后面加上 Inverse，即 progressBarStyleSmallInverse、progressBarStyleInverse、progressBarStyleLargeInverse，就可以得到反转的圆形进度条。

通过下面这个例子深入理解 ProgressBar 的用法。布局文件如下：

```
<LinearLayout
    android:orientation="vertical"
    android:layout_width="match_parent"
    android:layout_height="wrap_content">

    <TextView
        android:layout_width="wrap_content"
        android:layout_height="wrap_content"
        android:text="ProgressBar:"
        android:id="@+id/textView"
        android:textSize="25sp" />

    <!--设置大圆形进度条-->
```

```
    <ProgressBar
        style="?android:attr/progressBarStyleLarge"
        android:layout_width="wrap_content"
        android:layout_height="wrap_content"
        android:id="@+id/progressBar"
        android:max="100"
        android:visibility="gone"/>

    <!--设置常规圆形进度条-->
    <ProgressBar
        android:layout_width="wrap_content"
        android:layout_height="wrap_content"
        android:id="@+id/progressBar2"
        android:max="100"
        android:visibility="gone"/>

    <!--设置小圆形进度条-->
    <ProgressBar
        style="?android:attr/progressBarStyleSmall"
        android:layout_width="wrap_content"
        android:layout_height="wrap_content"
        android:id="@+id/progressBar3"
        android:max="100"
        android:visibility="gone"/>

    <!--设置水平条状-->
    <ProgressBar
        style="?android:attr/progressBarStyleHorizontal"
        android:layout_width="match_parent"
        android:layout_height="wrap_content"
        android:id="@+id/progressBar4"
        android:max="100"
        android:visibility="gone"/>

    <Button
        android:layout_width="wrap_content"
        android:layout_height="wrap_content"
        android:text="启动"
        android:id="@+id/btn" />

    <TextView
        android:layout_width="wrap_content"
        android:layout_height="wrap_content"
        android:textAppearance="?android:attr/textAppearanceLarge"
        android:id="@+id/txv"
        android:layout_gravity="center_horizontal"
        android:textSize="25sp" />
</LinearLayout>
```

给按钮添加点击事件监听器，当用户点击“增加进度”按钮时，动态修改 ProgressBar 的进度值。页面运行效果如图 3-28 所示。

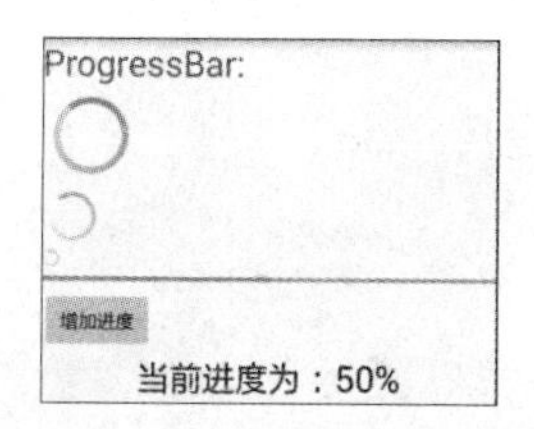

图 3-28 ProgressBar 运行效果

按钮的点击事件监听器代码如下：

```
btn.setOnClickListener(new View.OnClickListener() {
      @Override
      public void onClick(View v) {
          btn.setText("增加进度");
          if (pro ==0) {
              progressBar.setVisibility(View.VISIBLE);       //设置进度条为可见
                progressBar2.setVisibility(View.VISIBLE);
                progressBar3.setVisibility(View.VISIBLE);
                progressBar4.setVisibility(View.VISIBLE);
            } else if (pro <100) {
                progressBar.setProgress(pro);                //设置进度条进度
                progressBar2.setProgress(pro);
                progressBar3.setProgress(pro);
                progressBar4.setProgress(pro);
            } else {                                   //当进度条满 100% 后，进度条消失
                progressBar.setVisibility(View.GONE);
                progressBar2.setVisibility(View.GONE);
                progressBar3.setVisibility(View.GONE);
                progressBar4.setVisibility(View.GONE);
                btn.setVisibility(View.GONE);
            }
            txv.setText("当前进度为: "+pro+"% ");
            pro +=10;
      }
});
```

除了以上基本用法之外，Android 系统也提供了便利的方法来改变进度条的外观，可以自定义 drawable 文件，使用以下属性指定即可：

```
android:progressDrawable="@drawable/my_bar"
android:indeterminateDrawable="@drawable/progress_image"
```

2. SeekBar

SeekBar（拖动条）是 ProgressBar 的扩展，在其基础上增加了一个可拖动的 thumb。

用户可以触摸 thumb 并向左或向右拖动，也可以使用方向键设置当前的进度等级。要在布局文件中拖放 SeekBar 控件，对 SeekBar 作如下设置：

```
<SeekBar
        android:id="@+id/seekbar"
        android:layout_width="fill_parent"
        android:layout_height="wrap_content"
        android:max="255"
        android:progress="55" />
```

其中，max 表示拖动条的最大进度，progress 表示拖动条的当前进度。

要监听 SeekBar 的滑动消息，需要实现 SeekBar. OnSeekBarChangeListener 接口，该接口包含 3 个方法，即 onStartTrackingTouch（ ）、onStopTrackingTouch（ ）和 onProgressChanged()，分别代表按住 SeekBar 时触发、松开 SeekBar 时触发和 SeekBar 改变时触发。代码如下：

```
seekBar.setOnSeekBarChangeListener(new SeekBar.OnSeekBarChangeListener() {
        @Override           //进度条发生改变时触发
        public void onProgressChanged (SeekBar seekBar, int progress, boolean
fromUser) {
            txv2.setText("当前进度为："+progress+"% ");
        }

        @Override           //按住 SeekBar 时触发
        public void onStartTrackingTouch(SeekBar seekBar) {
        }

        @Override           //松开 SeekBar 时触发
        public void onStopTrackingTouch(SeekBar seekBar) {
        }
});
```

页面运行效果如图 3-29 所示，拖动 SeekBar，动态显示当前进度。

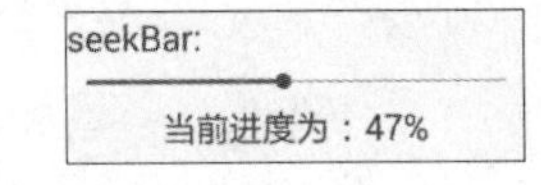

图 3-29　SeekBar 运行效果

3. RatingBar

RatingBar(星级评分条)以五角星来展示进度值，常用于一些游戏及应用的等级评分中。要在布局文件中拖放 RatingBar 控件，对 RatingBar 作如下设置：

```
<RatingBar
        android:id="@+id/ratingbar"
        android:layout_width="wrap_content"
        android:layout_height="wrap_content"
        android:numStars="5"
        android:rating="3"
        android:stepSize="0.2" />
```

RatingBar 的相关属性说明如下：

(1) android：numStars 表示总级别、总分数、星星个数；

(2) android：rating 表示当前级别、分数、星星个数；

(3) android：stepSize 表示每次变化的步长。

针对 RatingBar 控件，可以通过设置 OnRatingBarChangeListener 监听评分变化。

```
ratingBar.setOnRatingBarChangeListener(new RatingBar.OnRatingBarChangeListener() {
      @Override
      public void onRatingChanged (RatingBar ratingBar, float rating, boolean
fromUser) {
                txv3.setText("当前获得" +rating+ "星！");
      }
});
```

程序运行效果如图 3-30 所示，点击星星，进行评分。

图 3-30　RatingBar 运行效果

3.3　对话框的使用

对话框既能引起用户的注意，也可以接收用户的输入。在提示重要信息或提供用户选项方面，对话框是一个不错的选择。在 Android 应用开发中，一般涉及以下几种对话框：

(1) AlertDialog：功能最为丰富，实际应用最为广泛的对话框。

(2) ProgressDialog：进度条对话框，是对简单进度条的封装。

(3) DataPickerDialog：日期选择对话框。

(4) TimePickerDialog：时间选择对话框。

(5) 自定义对话框：对话框布局自定义，并设置监听事件。

下面在 Chapter03UI 项目中创建 BasicDialogActivity，该 Activity 对应的页面如图 3-31 所示，在该页面中每个按钮对应一种类别的对话框，当用户点击按钮时将弹出该按钮对应类别的对话框。

BasicDialogActivity 对应的布局文件比较简单，这里就不给出具体内容。

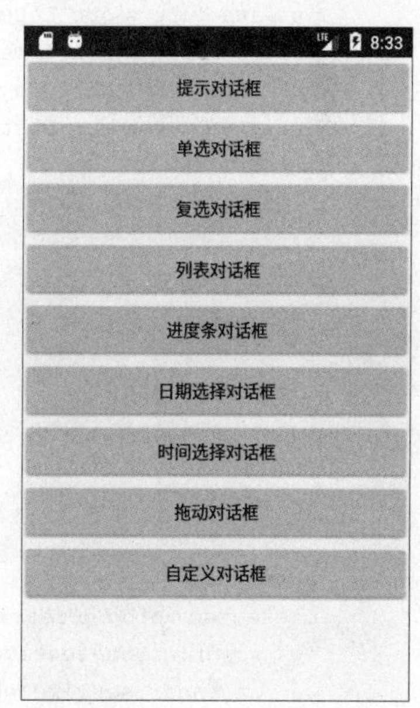

图 3-31　BasicDialogActivity 页面

BasicDialogActivity 详细代码如文件清单 3-14 所示。在监听器处理方法中，会根据用户点击按钮的不同弹出不同类别的对话框。

文件清单 3-14 BasicDialogActivity. java

```
package com.nsu.zyl.Chapter03ui;
import android.app.Activity;
import android.app.DatePickerDialog;
import android.app.ProgressDialog;
import android.app.TimePickerDialog;
import android.content.DialogInterface;
import android.os.Bundle;
import android.support.v7.app.AlertDialog;
import android.view.View;
import android.widget.Button;
import android.widget.DatePicker;
import android.widget.EditText;
import android.widget.SeekBar;
import android.widget.TextView;
import android.widget.TimePicker;
import android.widget.Toast;
import java.util.Calendar;
public class BasicDialogActivity extends Activity{

    private Button btn_AlertDialog;
    private Button btn_RadioDialog;
    private Button btn_CheckBoxDialog;
    private Button btn_ListDialog;
    private Button btn_ProgressDialog;
    private Button btn_DatePickerDialog;
    private Button btn_TimePickerDialog;
    private Button btn_DragDialog;
    private Button btn_CustomDialog;

    private boolean[] user_choice;
    private int index;
    String results ="你选择的是: ";
    int year;
    int month;
    int dayOfMonth;
    int hourOfDay;
    int minute;
    Calendar calendar =Calendar.getInstance();

    @Override
    protected void onCreate(Bundle savedInstanceState) {
        super.onCreate(savedInstanceState);
        calendar.setTimeInMillis(System.currentTimeMillis());
        setContentView(R.layout.basic_dialog);
```

```
        initView();
    }

    private void initView() {
        btn_AlertDialog = (Button)findViewById(R.id.btn_AlertDialog);
        btn_AlertDialog.setOnClickListener(new MyButtonClickListener());
        btn_RadioDialog = (Button)findViewById(R.id.btn_RadioDialog);
        btn_RadioDialog.setOnClickListener(new MyButtonClickListener());
        btn_CheckBoxDialog = (Button)findViewById(R.id.btn_CheckBoxDialog);
        btn_CheckBoxDialog.setOnClickListener(new MyButtonClickListener());
        btn_ListDialog = (Button)findViewById(R.id.btn_ListDialog);
        btn_ListDialog.setOnClickListener(new MyButtonClickListener());
        btn_ProgressDialog = (Button)findViewById(R.id.btn_ProgressDialog);
        btn_ProgressDialog.setOnClickListener(new MyButtonClickListener());
        btn_DatePickerDialog = (Button)findViewById(R.id.btn_DatePickerDialog);
        btn_DatePickerDialog.setOnClickListener(new MyButtonClickListener());
        btn_TimePickerDialog = (Button)findViewById(R.id.btn_TimePickerDialog);
        btn_TimePickerDialog.setOnClickListener(new MyButtonClickListener());
        btn_DragDialog = (Button)findViewById(R.id.btn_DragDialog);
        btn_DragDialog.setOnClickListener(new MyButtonClickListener());
        btn_CustomDialog = (Button)findViewById(R.id.btn_CustomDialog);
        btn_CustomDialog.setOnClickListener(new MyButtonClickListener());
    }

    private class MyButtonClickListener implements View.OnClickListener,Runnable {
        @Override
        public void onClick(View v) {
            switch (v.getId()) {
                case R.id.btn_AlertDialog://弹出提示对话框

                    break;
                case R.id.btn_RadioDialog://弹出单选对话框

                    break;
                case R.id.btn_CheckBoxDialog://弹出复选对话框

                    break;
                case R.id.btn_ListDialog://弹出列表对话框

                    break;
                case R.id.btn_ProgressDialog://弹出进度条对话框

                    break;
                case R.id.btn_DatePickerDialog://弹出日期选择对话框

                    break;
                case R.id.btn_TimePickerDialog://弹出时间选择对话框
```

```
                    break;
                case R.id.btn_DragDialog://弹出拖动对话框

                    break;
                case R.id.btn_CustomDialog://弹出自定义对话框

                    break;
            }
        }

    }
}
```

3.3.1 AlertDialog

AlertDialog 生成的对话框分为 4 个区域：图标区、标题区、内容区和按钮区，如图 3-32 所示。

图 3-32　AlertDialog 对话框结构

通常，可以按以下步骤创建一个 AlertDialog 对话框：

(1) 使用 AlertDialog. Builder 创建对象；

(2) 调用 AlertDialog. Builder 的 setTitle()或 setCustomTitle()方法设置标题；

(3) 调用 AlertDialog. Builder 的 setIcon()方法设置图标；

(4) 调用 AlertDialog. Builder 的相关设置方法设置对话框内容；

(5) 调用 AlertDialog. Builder 的 setPositiveButton()、setNegativeButton()或 setNeutralButton()方法添加多个按钮；

(6) 调用 AlertDialog. Builder 的 create()方法创建 AlertDialog 对象，再调用 AlertDialog 对象的 show()方法将该对话框显示出来。

针对第(4)步，设置对话框的内容，可根据不同场景细分为以下 5 种：

(1) 设置提示消息：

```
setMessage(CharSequence message)
```

(2) 设置单选列表：

```
setSingleChoiceItems(int itemsId, int checkedItem, OnClickListener listener)
```

(3) 设置多选列表：

```
setMultiChoiceItems(CharSequence[] a, boolean[] b,OnClickListener listener)
```

(4) 设置普通列表：

```
setItems(CharSequence[] a,OnClickListener listener)
```

(5) 设置自定义视图：

```
setView(View v)
```

1. 消息提示对话框

当 AlertDialog 的内容区为一条提示消息时，称为消息提示对话框，如图 3-33 所示。对应使用的方法为 setMessage(CharSequence message)。

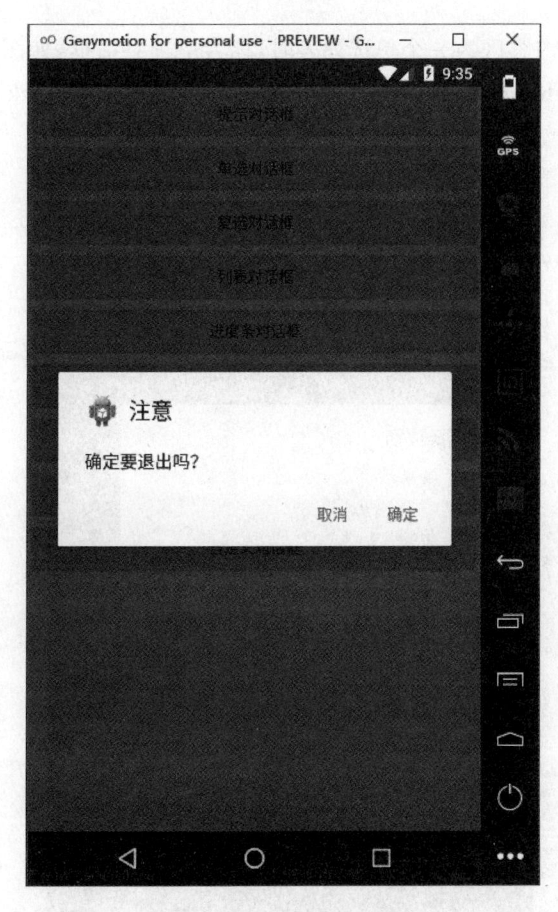

图 3-33　消息提示对话框

因此在文件清单 3-14 中添加创建消息提示对话框的代码，如下所示：

```
new AlertDialog.Builder(BasicDialogActivity.this)
        .setIcon(R.drawable.launcher_icon)
        .setTitle("注意")
        .setMessage("确定要退出吗？")
        .setPositiveButton("确定", new DialogInterface.OnClickListener() {
                @Override
                public void onClick(DialogInterface dialog, int which) {
                     finish();
                }
          })
          .setNegativeButton("取消", null)
          .create()
          .show();
```

2. 单选对话框

当 AlertDialog 的内容区为一个单选列表时，称为单选对话框，如图 3-34 所示。对应使用的方法为 setSingleChoiceItems(int itemsId, int checkedItem, OnClickListener listener)。

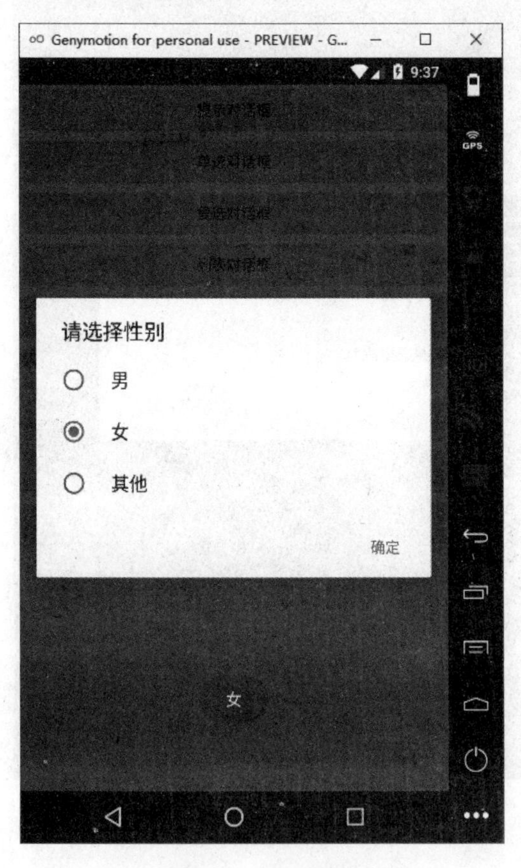

图 3-34 单选对话框

因此在文件清单 3-14 中添加创建单选对话框的代码，如下所示：

```
AlertDialog.Builder builder =new AlertDialog.Builder(BasicDialogActivity.this);
builder.setTitle("请选择性别");
builder.setCancelable(false);
builder.setSingleChoiceItems (R.array.sex_array, -1, new DialogInterface.
OnClickListener() {
      @Override
      public void onClick(DialogInterface dialog, int which) {
      index =which;
      String content =getResources().getStringArray(R.array.sex_array)[which];
      Toast.makeText(getApplicationContext(), content, Toast.LENGTH_SHORT).show();
}
});
builder.setPositiveButton("确定", new DialogInterface.OnClickListener() {
        @Override
        public void onClick(DialogInterface dialog, int which) {
        String content =getResources().getStringArray(R.array.sex_array)[index];
        Toast.makeText(getApplicationContext(), "你的选择是" +content,
Toast.LENGTH_SHORT).show();
        }
});
builder.create().show();
```

3. 复选对话框

当 AlertDialog 的内容区为一个多选列表时，称为复选（多选）对话框，如图 3-35 所示。创建多选对话框对应使用的方法为 setMultiChoiceItems(CharSequence[] a, boolean[] b,OnClickListener listener)。

因此，在文件清单 3-14 中添加创建复选对话框的代码，如下所示：

```
//布尔数组,记录每一个城市的选中状态
user_choice =new boolean[getResources().getStringArray(R.array.city_array).length];
new AlertDialog.Builder(BasicDialogActivity.this)
    .setTitle("复选对话框")
    .setMultiChoiceItems(R.array.city_array, null, new DialogInterface.
OnMultiChoiceClickListener() {
          @Override
          public void onClick(DialogInterface dialog, int which, boolean isChecked) {
                  user_choice[which] =isChecked;
          }
      })
    .setPositiveButton("确定", new DialogInterface.OnClickListener() {
            @Override
            public void onClick(DialogInterface dialog, int which) {
```

```
                for (int i =0; i <user_choice.length; i++) {
                    if (user_choice[i]) {
                        results =results +getResources().
getStringArray(R.array.city_array)[i];
                    }
                }
                Toast.makeText(getApplicationContext(), results,
Toast.LENGTH_SHORT).show();
                results ="你选择的是：";
            }
        })
        setNegativeButton("取消", new DialogInterface.OnClickListener() {
            @Override
            public void onClick(DialogInterface dialog, int which) {
            }
        })
        .create().show();
```

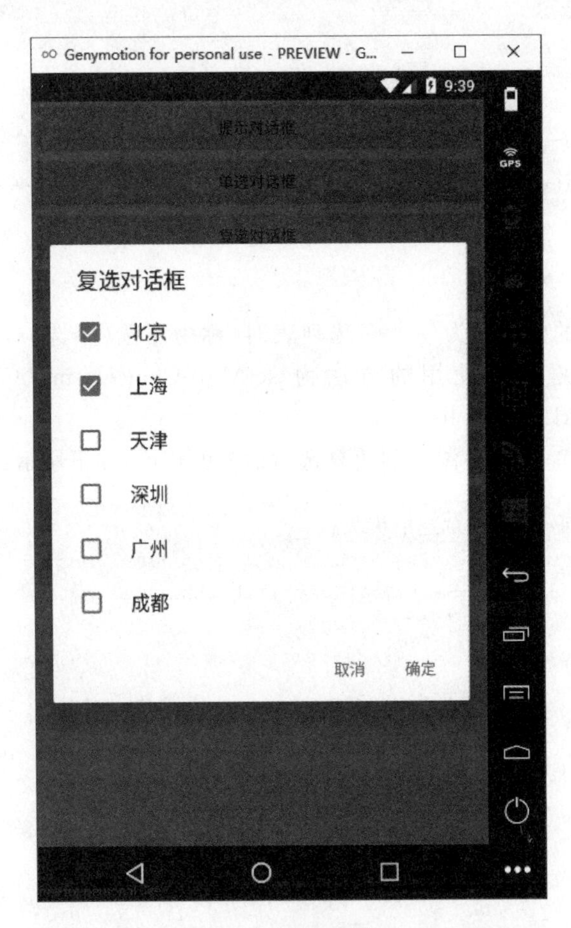

图 3-35　复选对话框

4. 列表对话框

当 AlertDialog 的内容区为一个普通列表时，称为列表对话框，如图 3-36 所示。对应使用的方法为 setItems（CharSequence[] a，OnClickListener listener），其中参数 CharSequence[] a 为列表项内容，OnClickListener listener 监听器用于处理列表项被点击的事件。

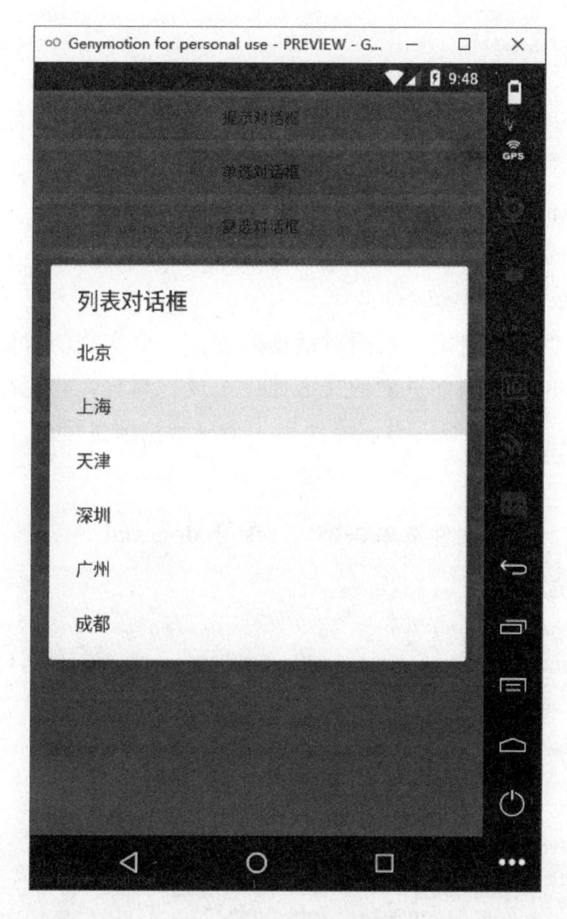

图 3-36 列表对话框

因此，在文件清单 3-14 中添加创建列表对话框的代码，如下所示：

```
new AlertDialog.Builder(BasicDialogActivity.this)
      .setTitle("列表对话框")
      .setItems(R.array.city_array, new DialogInterface.OnClickListener() {
            @Override
            public void onClick(DialogInterface dialog, int which) {
            String content = "你点击的是：" +
getResources().getStringArray(R.array.city_array)[which];
```

```
                Toast.makeText(getApplicationContext(), content, Toast.LENGTH_
SHORT)
.show();
            }
        })
        .create().show();
```

5. 自定义视图对话框

当 AlertDialog 的内容区为用户自己定义的一个 View 视图时，称为自定义视图对话框。创建自定义视图对话框的步骤如下：

(1) 自定义一个布局文件 xxx.xml；

(2) 获取该布局的实例 xxxLayout；

(3) 设置 setView(xxxLayout)。

下面通过一个实例演示自定义视图对话框的开发。要实现的对话框如图 3-35 所示，对话框由进度条控件和显示当前进度的文本控件组成。首先，自定义一个布局文件 seek_dialog.xml，布局中包含一个 SeekBar 控件和一个显示当前进度的 TextView 控件，代码如文件清单 3-15 所示。

文件清单 3-15 seek_dialog.xml

```
<?xml version="1.0" encoding="utf-8"?>
<RelativeLayout xmlns:android="http://schemas.android.com/apk/res/android"
    android:layout_width="match_parent"
    android:layout_height="match_parent">

    <SeekBar
        android:id="@+id/seekBar1"
        android:layout_width="match_parent"
        android:layout_height="wrap_content"
        android:layout_alignParentTop="true"
        android:layout_alignParentLeft="true"
        />

    <TextView
        android:id="@+id/tv_seekbar"
        android:layout_width="wrap_content"
        android:layout_height="wrap_content"
        android:layout_centerHorizontal="true"
        android:layout_below="@+id/seekBar1"
        android:text="" />
</RelativeLayout>
```

然后，通过 LayoutInflater 的 inflate()方法获取上述布局文件的对象，通过布局文件对象获取布局文件中的 SeekBar 控件对象和 TextView 控件对象，并为它们赋值，在文件清单 3-14 中添加如下代码：

```
View seekView =getLayoutInflater().inflate(R.layout.seek_dialog, null);
SeekBar sbar = (SeekBar) seekView.findViewById(R.id.seekBar1);
sbar.setMax(100);
final TextView tv_seekbar = (TextView) seekView.findViewById(R.id.tv_seekbar);
tv_seekbar.setText("当前的进度为：" +sbar.getProgress());
sbar.setOnSeekBarChangeListener(new SeekBar.OnSeekBarChangeListener(){
        @Override
        public void onStopTrackingTouch(SeekBar seekBar){
        }
        @Override
        public void onStartTrackingTouch(SeekBar seekBar){
        }
        @Override
            public void onProgressChanged (SeekBar seekBar, int progress, boolean
fromUser){
                tv_seekbar.setText("当前的进度为：" +seekBar.getProgress());
        }
});
```

最后，在 AlertDialog 中，调用 setView()方法设置自定义的视图。运行该实例，显示效果如图 3-37 所示。

```
new AlertDialog.Builder(BasicDialogActivity.this)
        .setTitle("拖动对话框")
        .setView(seekView)
        .setPositiveButton("确定", new DialogInterface.OnClickListener() {
                @Override
                public void onClick(DialogInterface dialog, int which) {

                }
        })
        .create().show();
```

提示：在 Android 中获得 LayoutInflater 实例有以下 3 种方式：

(1) LayoutInflater inflater = getLayoutInflater();//调用 Activity 的 getLayoutInflater();

(2) LayoutInflater inflater = LayoutInflater. from(context);

(3) LayoutInflater inflater = LayoutInflater. context. getSystemService (Context. LAYOUT_INFLATER_SERVICE)。

这三种方式本质是相同的，在实际使用中可根据需要灵活选择。

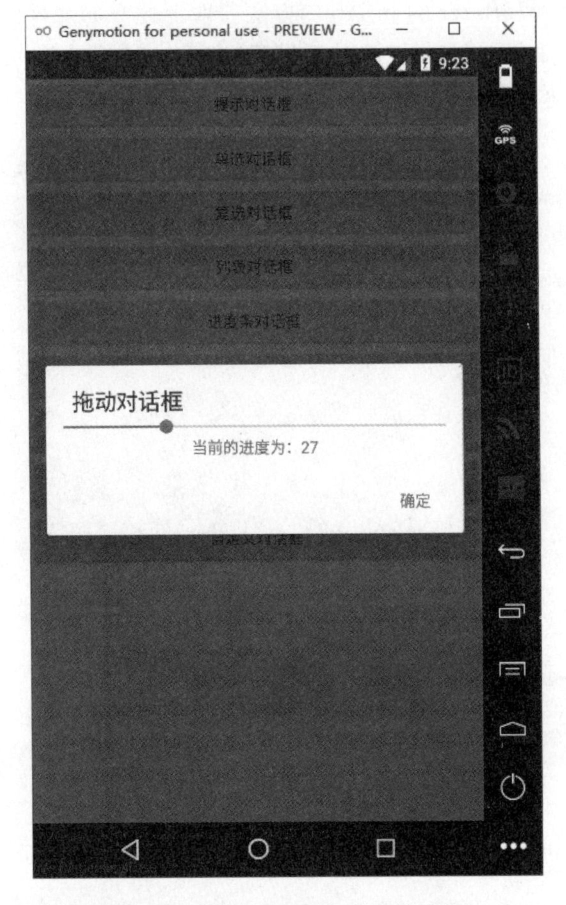

图 3-37　自定义视图对话框(拖动条)

3.3.2　ProgressDialog

ProgressDialog 继承自 AlertDialog，将进度条简单包裹起来，如图 3-38 所示。在进度条对话框中可以设置进度条的样式，ProgressDialog 的样式有两种，一种是圆形状态，另一种是水平进度条状态。

因此，在文件清单 3-14 中添加创建进度条对话框的代码，如下所示：

```
ProgressDialog pDialog =new ProgressDialog(BasicDialogActivity.this);
pDialog.setTitle("文件下载中");
pDialog.setIcon(R.drawable.launcher_icon);
pDialog.setMax(100);
pDialog.setMessage("文件已下载");
    //设置进度条风格，STYLE_SPINNER为圆形、旋转进度条，STYLE_HORIZONTAL为长形进度条
pDialog.setProgressStyle(ProgressDialog.STYLE_HORIZONTAL); //是否可以按Back键取消
pDialog.setCancelable(true);
pDialog.show();
new Thread(this).start();
```

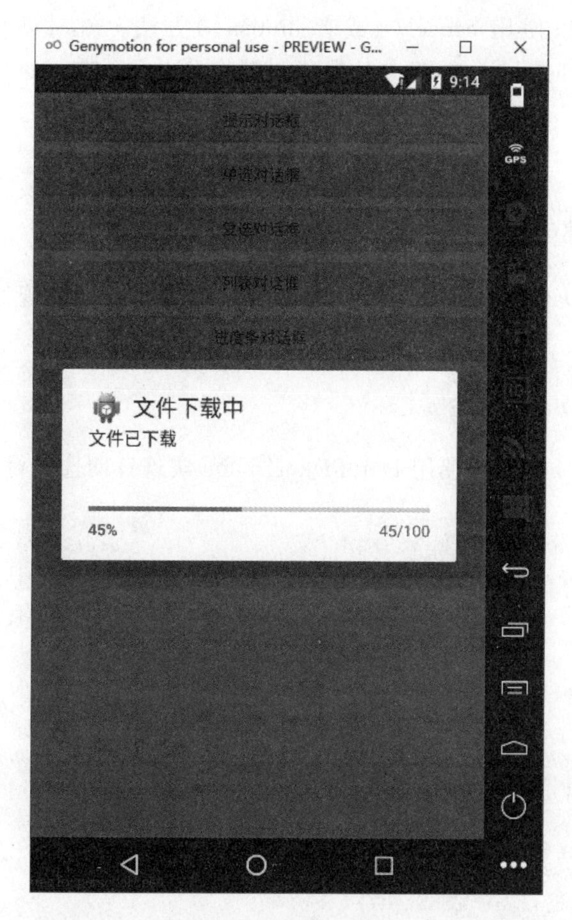

图 3-38 进度条对话框

本例模拟耗时任务的进行，需要另起线程去进行，下载完成后，对话框自动消失。代码如下：

```
@Override
public void run() {
      int progress =0;
      while (progress <100) {
            try {
                Thread.sleep(100);
                progress++;
                pDialog.incrementProgressBy(5);
            } catch (InterruptedException e) {
                e.printStackTrace();
            }
      }
}
```

取消对话框可以使用 cancel()或者 dimiss()方法。就对话框而言，cancel()和 dismiss()方法本质都是一样的，都是从屏幕中删除 Dialog。唯一的区别是：如果注册了 DialogInterface.OnCancelListener，调用 cancel()方法会回调此方法，而 dismiss()方法则不会回调。

3.3.3 DatePickerDialog 和 TimePickerDialog

Android 提供 DatePickerDialog 和 TimePickerDialog 用于实现日期选择对话框和时间选择对话框。

1. DatePickerDialog

在文件清单 3-14 中添加使用 DatePickerDialog 实现日期选择对话框的代码如下：

```
year =calendar.get(Calendar.YEAR);                          //获取年
month =calendar.get(Calendar.MONTH);                        //获取月,从 0 开始
dayOfMonth =calendar.get(Calendar.DAY_OF_MONTH);            //获取日
//监听日期设置
DatePickerDialog.OnDateSetListener listener1 =new
DatePickerDialog.OnDateSetListener() {
          @Override
          public void onDateSet(DatePicker view, int m_year, int m_month,
int m_dayOfMonth) {
               year =m_year;
               month =m_month +1;
               dayOfMonth =m_dayOfMonth;
               Toast.makeText(BasicDialogActivity.this, "你设置的时间是：" +year +
"年" +month +"月" +dayOfMonth +"日",
Toast.LENGTH_SHORT).show();
          }
};

DatePickerDialog dpDialog =new DatePickerDialog(BasicDialogActivity.this,
                    listener1, year, month, dayOfMonth);
dpDialog.setIcon(R.drawable.launcher_icon);
dpDialog.setMessage("请选择日期");
dpDialog.show();
```

注意：在使用 Calendar 获取月份时，取值为 0～11，表示 1～12 月。运行效果如图 3-39 所示，用户可以在对话框中进行日期选择。

2. TimePickerDialog

在文件清单 3-14 中添加使用 TimePickerDialog 创建时间选择对话框的代码如下：

图 3-39 日期选择对话框

```
TimePickerDialog.OnTimeSetListener listener2 =new
TimePickerDialog.OnTimeSetListener() {
      @Override
      public void onTimeSet(TimePicker view, int m_hourOfDay, int m_minute) {
              hourOfDay =m_hourOfDay;
              minute =m_minute;
              Toast.makeText(BasicDialogActivity.this, "你设置的时间是：" +
hourOfDay +" : " +minute, Toast.LENGTH_SHORT).show();
        }
};

TimePickerDialog tpDialog = new TimePickerDialog (BasicDialogActivity. this,
listener2,
         calendar.get(Calendar.HOUR_OF_DAY), calendar.get(Calendar.MINUTE),
true);
tpDialog.setIcon(R.drawable.launcher_icon);
tpDialog.setMessage("请设置时间");
tpDialog.show();
```

运行显示图 3-40 所示的时间选择对话框，当用户在对话框中进行时间选择后，以 Toast 消息提示的方式显示用户设置的时间。

图 3-40 时间选择对话框

3.3.4 自定义 Dialog

前面介绍的对话框都是使用系统已经封装好的接口，基本上能满足绝大多数的开发需求。如果在定义对话框时希望有更大的自由度，可以通过继承 Dialog 来实现自定义的 Dialog。

接下来实现图 3-41 所示的自定义对话框(图 3-41)。

(1) 自定义对话框布局 user_dialog_layout. xml，如文件清单 3-16 所示。

文件清单 3-16 user_dialog_layout. xml

```
<?xml version="1.0" encoding="utf-8"?>
<LinearLayout xmlns:android="http://schemas.android.com/apk/res/android"
    android:layout_width="match_parent"
    android:layout_height="match_parent"
```

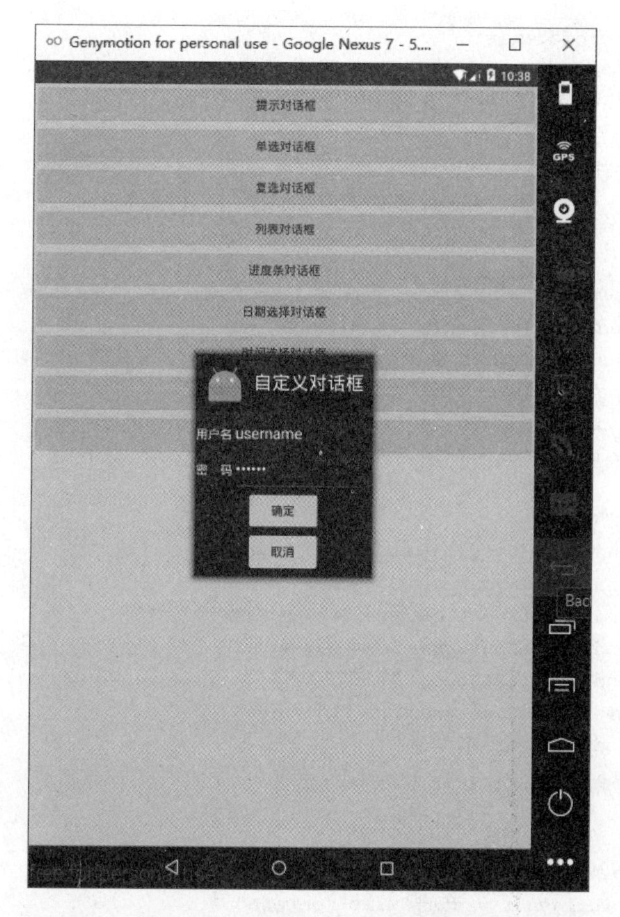

图 3-41　自定义 Dialog 对话框

```
    android:orientation="vertical">

    <LinearLayout
        android:layout_width="match_parent"
        android:layout_height="wrap_content" >

        <TextView
            android:id="@+id/tv_username_dialog"
            android:layout_width="wrap_content"
            android:layout_height="wrap_content"
            android:textColor="#FFFFFF"
            android:text="用户名" />
        <EditText
            android:id="@+id/et_username_dialog"
            android:layout_width="wrap_content"
            android:layout_height="wrap_content"
            android:textColor="#FFFFFF"
            android:layout_weight="1"
```

```
            android:ems="20" >
        </EditText>
    </LinearLayout>

    <LinearLayout
        android:layout_width="match_parent"
        android:layout_height="wrap_content" >

        <TextView
            android:id="@+id/textView2"
            android:layout_width="wrap_content"
            android:layout_height="wrap_content"
            android:textColor="#FFFFFF"
            android:text="密    码" />

        <EditText
            android:id="@+id/et_pwd_dialog"
            android:layout_width="wrap_content"
            android:layout_height="wrap_content"
            android:inputType="textPassword"
            android:textColor="#FFFFFF"
            android:layout_weight="1"
            android:ems="20" />
    </LinearLayout>

    <Button
        android:id="@+id/btn_ok_dialog"
        android:layout_width="wrap_content"
        android:layout_height="wrap_content"
        android:layout_gravity="center"
        android:text="确定" />

    <Button
        android:id="@+id/btn_cancel_dialog"
        android:layout_width="wrap_content"
        android:layout_height="wrap_content"
        android:layout_gravity="center"
        android:text="取消" />
</LinearLayout>
```

(2) 在 res/values/styles.xml 中自定义对话框的样式。

```
<style name="MyDialog" parent="android:Theme.Dialog">
    <!--设置背景颜色以及背景透明程度 -->
    <item name="android:windowBackground">@android:color/transparent</item>
    <!--设置是否浮现-->
    <item name="android:windowIsFloating">true</item>
    <!--设置窗体是否半透明 -->
```

```
    <item name="android:windowIsTranslucent">true</item>
    <!--设置背景模糊的透明度-->
    <item name="android:backgroundDimAmount">0.1</item>
    <!--设置背景是否模糊 -->
    <item name="android:backgroundDimEnabled">true</item>
</style>
```

(3) 继承 Dialog 类,实现构造方法和 onCreate()方法,如文件清单 3-17 所示。

文件清单 3-17　MyDialog.java

```
public class MyDialog extends AlertDialog {
    private Context context;

    protected MyDialog(Context context) {
        super(context , R.style.MyDialog);
        this.context =context;
    }

    @Override
    protected void onCreate(Bundle savedInstanceState) {
        //通过 LayoutInflater 来获取布局文件对象
        LayoutInflater inflater =LayoutInflater.from(context);
        View userDialog =inflater.inflate(R.layout.user_dialog_layout, null);
        setView(userDialog);
        super.onCreate(savedInstanceState);
    }
}
```

(4) 在文件清单 3-14 中添加创建对话框,并进行事件监听的代码如下:

```
//创建自定义的 Dialog 对象
final MyDialog userDialog =new MyDialog(BasicDialogActivity.this);
//设置对话框的图标
userDialog.setIcon(R.mipmap.ic_launcher);
//设置标题
userDialog.setTitle("自定义对话框");
//显示对话框
userDialog.show();
//通过 MyDialog 对象找到相关控件
Button btn_ok = (Button) userDialog.findViewById(R.id.btn_ok_dialog);
Button btn_cancel = (Button) userDialog.findViewById(R.id.btn_cancel_dialog);
final EditText et_userName = (EditText) userDialog
.findViewById(R.id.et_username_dialog);
final EditText et_pwd = (EditText) userDialog.findViewById(R.id.et_pwd_dialog);
//设置按钮的监听事件
btn_ok.setOnClickListener(new View.OnClickListener() {
     @Override
     public void onClick(View v) {
           String userName =et_userName.getText().toString();
           String userPwd =et_pwd.getText().toString();
```

```
            //弹出一个短消息
            Toast.makeText(BasicDialogActivity.this, "用户名"+userName+"密码"
+userPwd, Toast.LENGTH_SHORT).show();
        }
});
btn_cancel.setOnClickListener(new View.OnClickListener() {
    @Override
    public void onClick(View v) {
        //回收对话框
        userDialog.dismiss();
    }
});
```

3.4 Toast 的使用

Toast 是 Android 中用来显示信息的一种机制，在应用程序上浮动显示一些帮助或提示信息给用户。Toast 具有不会获得焦点、不影响用户的输入和显示时间有限等特点。

下面在 Chapter03UI 项目中创建 BasicToastActivity，该 Activity 对应的页面如图 3-42 所示，在该页面中每个按钮对应一种类别的 Toast，当点击按钮时将弹出该按钮对应类别的 Toast。BasicToastActivity 对应的布局文件比较简单，这里就不给出详细内容了。

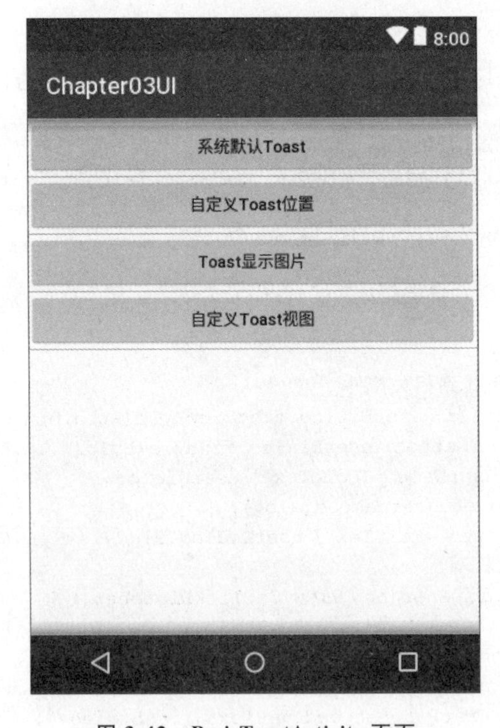

图 3-42 BasicToastActivity 页面

BasicToastActivity 初始化的内容如文件清单 3-18 所示。

文件清单 3-18 BasicToastActivity. java

```
package com.nsu.zyl.Chapter03ui;
import android.app.Activity;
import android.os.Bundle;
import android.view.Gravity;
import android.view.LayoutInflater;
import android.view.View;
import android.widget.Button;
import android.widget.ImageView;
import android.widget.LinearLayout;
import android.widget.TextView;
import android.widget.Toast;

public class BasicToastActivity extends Activity implements View.OnClickListener{
private Button btnNormalToast;
private Button btnGravityToast;
private Button btnPicToast;
private Button btnDiyToast;

@Override
protected void onCreate(Bundle savedInstanceState) {
    super.onCreate(savedInstanceState);
    setContentView(R.layout.basic_toast);
    btnNormalToast = (Button) findViewById(R.id.btn_normal_toast);
    btnGravityToast = (Button) findViewById(R.id.btn_gravity_toast);
    btnPicToast = (Button) findViewById(R.id.btn_pic_toast);
    btnDiyToast = (Button) findViewById(R.id.btn_diy_toast);
    btnNormalToast.setOnClickListener(this);
    btnGravityToast.setOnClickListener(this);
    btnPicToast.setOnClickListener(this);
    btnDiyToast.setOnClickListener(this);
}

@Override
public void onClick(View v) {
    switch (v.getId()){
        case R.id.btn_normal_toast:     //普通 Toast

        break;
        case R.id.btn_gravity_toast:    //修改位置 Toast

        break;
        case R.id.btn_pic_toast:        //带图片 Toast

            break;
        case R.id.btn_diy_toast:        //自定义 Toast

            break;
        }
    }
}
```

3.4.1 系统默认 Toast 的用法

Toast 最常见的创建方式是使用静态方法 Toast. makeText(Context context, CharSequence text,int duration),该方法的 3 个参数说明如下:

Context context: 当前的上下文环境,可用 getApplicationContext()或者 XXXActivity . this。

CharSequence text: 要显示的字符串,也可是 R. string 中字符串的 ID。

int duration: 显示的时间长短,Toast 默认的时间取值有两个,即 Toast. LENGTH_LONG(长)和 Toast. LENGTH_SHORT(短),也可以使用具体的数值表示多少毫秒。

需要注意: 创建 Toast 后,一定要调用 show()方法显示 Toast 的信息。

在程序中使用 Toast 的示例代码:

```
//获取 Toast 对象
Toast toast=Toast.makeText(getApplicationContext(), "这是一个默认的 Toast 消息",
Toast.LENGTH_SHORT);
//创建好 Toast 对象后,需要调用 show()方法来显示 toast 信息。
toast.show();
```

运行效果如图 3-43 所示,点击"系统默认 Toast"按钮,弹出 Toast 消息。

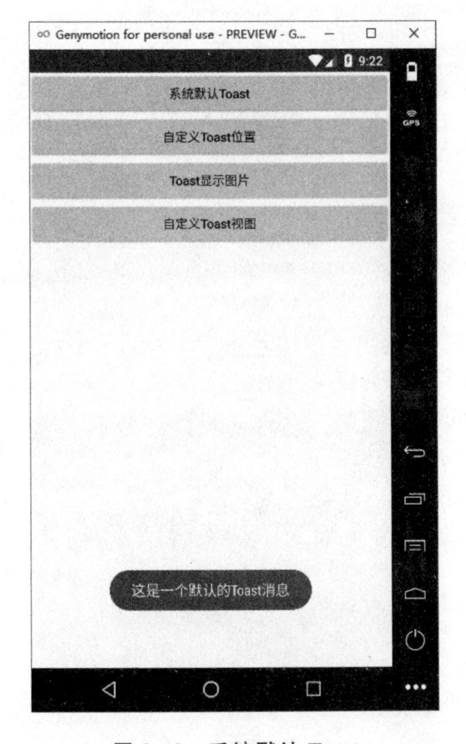

图 3-43 系统默认 Toast

3.4.2 自定义 Toast

Android 应用开发中系统默认的 Toast 就可以满足大多数情况下的基本需求,当然作为一个开放的平台，Android 系统也给用户提供了个性化设计的方法，例如可以自定义 Toast 的显示位置，给 Toast 消息加上图片，以及完全自定义 Toast 的布局样式等。

1. 自定义 Toast 显示位置

```
Toast toast= Toast.makeText(getApplicationContext(), "这是一个自定义位置的 Toast 消息",
                    Toast.LENGTH_SHORT);
```

获取 Toast 对象后，调用 setGravity(int gravity,int xOffset,int yOffset)方法设置 Toast 在屏幕上的位置。

int gravity：设置 Toast 消息在屏幕中显示的位置。

int xOffset：相对于第一个参数位置，设置 toast 位置的横向 X 轴的偏移量，正数表示向右偏移，负数表示向左偏移。

int yOffset：相对于第一个参数位置，设置 toast 位置的纵向 Y 轴的偏移量，正数表示向下偏移，负数表示向上偏移。

如果设置的偏移量超过了当前屏幕的范围，Toast 将在屏幕内靠近超出的边界显示。例如：

toast. setGravity(Gravity. TOP | Gravity. CENTER，－50，100)；设置是居中靠顶，向左偏移 50，向下偏移 100，运行效果如图 3-44 所示。

toast. setGravity(Gravity. CENTER，0，0)；设置屏幕居中显示，X 轴和 Y 轴偏移量都是 0。

设置 Toast 的位置后，调用 show()方法来显示 Toast 信息。

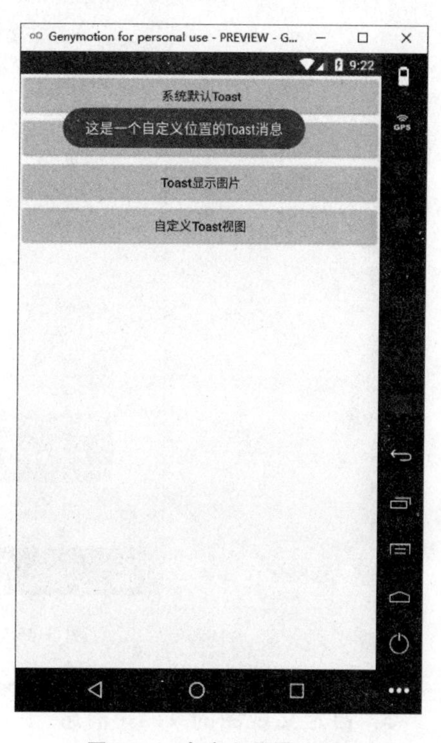

图 3-44 自定义位置 Toast

2. 带图片显示的 Toast 消息

```
Toast toast =Toast.makeText(getApplicationContext(),"这是一个显示图片的 Toast 消息",
Toast.LENGTH_SHORT);
                                        //创建 ImageView 对象
ImageView imageView=new ImageView(getApplicationContext());
                                        //设置 ImageView 显示的图片
imageView.setImageResource(R.mipmap.ic_launcher);
                                        //获得 toast 所在的布局
LinearLayout toastView = (LinearLayout) toast.getView();
                                        //设置此布局为横向线性布局
toastView.setOrientation(LinearLayout.HORIZONTAL);
                                        //将 ImageView 加入到此布局中的第一个位置
toastView.addView(imageView, 0);
toast.show();
```

运行效果如图 3-45 所示，点击按钮，弹出带图片的 Toast 消息，也可以结合前面所讲的内容，设置 Toast 的位置。

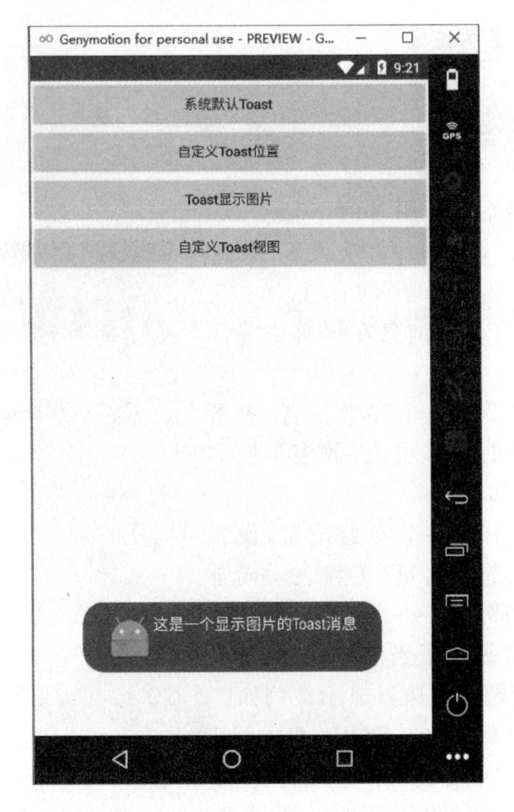

图 3-45 带图片的 Toast 消息

3. 自定义视图的 Toast 消息

除了显示图片外，Toast 也可以显示任意的布局视图。自定义 Toast 消息的视图时，需要设计一个视图布局，然后使用 LayoutInflater 类来创建视图对象，调用 setView()方法将此视图对象设置为 Toast 的视图。核心代码如下：

```
LayoutInflater inflater =getLayoutInflater();
                                //通过制定 XML 文件及布局 ID 来填充一个视图对象
View view =inflater.inflate(R.layout.toast_diy_layout,null);
ImageView imgToast = (ImageView) view.findViewById(R.id.img_diy_toast);
imgToast.setImageResource(R.mipmap.ic_launcher);
TextView tvToast = (TextView) view.findViewById(R.id.tv_diy_toast);
tvToast.setText("这是一个自定义视图的 Toast 消息");
Toast toast3 =new Toast(this);
toast3.setGravity(Gravity.CENTER, 0, 0);
toast3.setDuration(Toast.LENGTH_SHORT);
toast3.setView(view);
toast3.show();
```

自定义的布局文件为 toast_diy_layout.xml,如文件清单 3-19 所示。

文件清单 3-19 toast_diy_layout.xml

```
<?xml version="1.0" encoding="utf-8"?>
<LinearLayout xmlns:android="http://schemas.android.com/apk/res/android"
    xmlns:app="http://schemas.android.com/apk/res-auto"
    android:layout_width="match_parent"
    android:layout_height="match_parent"
    android:orientation="vertical">

    <ImageView
        android:layout_width="match_parent"
        android:layout_height="wrap_content"
        app:srcCompat="@mipmap/ic_launcher"
        android:id="@+id/img_diy_toast" />

    <TextView
        android:text="TextView"
        android:gravity="center"
        android:layout_width="match_parent"
        android:layout_height="wrap_content"
        android:id="@+id/tv_diy_toast" />
</LinearLayout>
```

最终运行效果如图 3-46 所示。

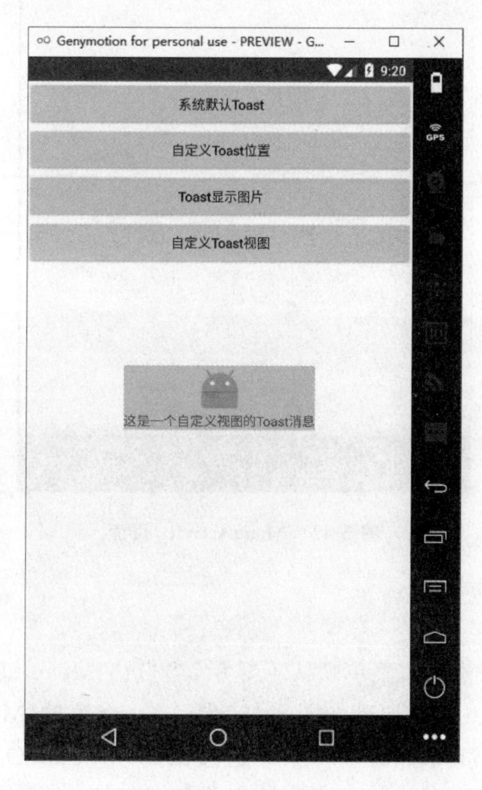

图 3-46 自定义 Toast

3.5 菜单的用法

菜单是用户界面中最常见的元素之一，是许多应用程序不可或缺的一部分。在Android中，菜单可分为三种：选项菜单(OptionsMenu)、上下文菜单(ContextMenu)和弹出式菜单(PopupMenu)。

下面在Chapter03UI项目中创建MenuActivity、MenuContentActivity、ActionBarActivity和ToolBarActivity。MenuActivity对应的页面如图3-47所示。当点击页面上的按钮时，分别跳转至MenuContentActivity、ActionBarActivity和ToolBarActivity。MenuActivity本身和对应的布局文件比较简单，这里就不给出详细内容了。我们将在MenuContentActivity、ActionBarActivity和ToolBarActivity中演示菜单的用法。

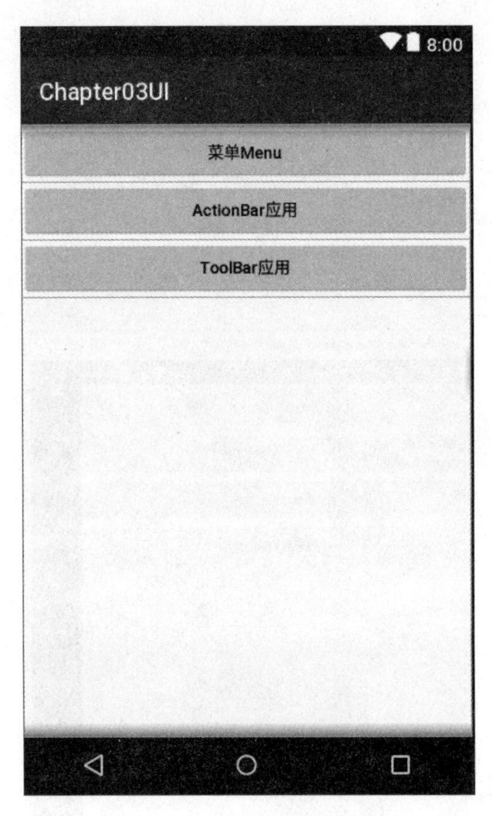

图3-47 MenuActivity页面

3.5.1 选项菜单

OptionsMenu(选项菜单)，菜单默认不显示，当点击Menu键时，系统才显示应用关联的菜单。OptionsMenu工作原理如图3-48所示。一个菜单(Menu)可包含多个菜单项(MenuItem)与子菜单(SubMenu)，Activity通过回调onCreateOptionsMenu()方法创建菜单，通过onOptionsItemSelected()方法处理菜单项事件。

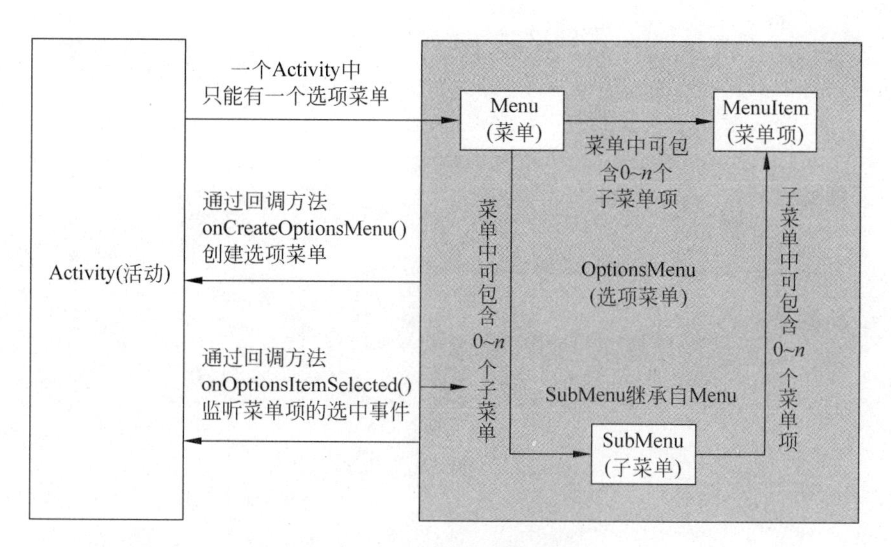

图 3-48 OptionsMenu 工作原理

在 Android 应用开发中，添加菜单或子菜单的步骤如下：

(1) 重写 Activity 的 onCreateOptionsMenu(Menu menu)方法，在该方法中调用 Menu 对象的方法来添加菜单项或子菜单。

方法 add(int groupId, int itemId, int order, CharSequence title)用于添加菜单项，方法 addSubMenu(int groupId, int itemId, int order, CharSequence title)用于添加子菜单。

Menu 可以包含多个 SubMenu，SubMenu 可以包含多个 MenuItem，但是 SubMenu 不能包含 SubMenu(子菜单不能嵌套)。同时，子菜单可以添加菜单头标题、图标，但菜单项不能显示图标。

在 MenuContentActivity 中重写 onCreateOptionsMenu()，详细内容如下：

```
@Override
public boolean onCreateOptionsMenu(Menu menu) {
    try {
        Class<?>menuClass =
                Class.forName("com.android.internal.view.menu.MenuBuilder");
        Method menuMethod =menuClass
                .getDeclaredMethod("setOptionalIconsVisible", boolean.class);
        menuMethod.setAccessible(true);
        menuMethod.invoke(menu, true);
    }catch (Exception e) {
        e.printStackTrace();
    }
    MenuItem show_item =menu.add(0, Menu.FIRST, 0, "显示")
.setIcon(R.mipmap.ic_launcher);
    //子菜单
```

```
        SubMenu shareMenu =menu.addSubMenu(0, Menu.FIRST+1, 0, "分享")
.setIcon(R.mipmap.ic_launcher);
        shareMenu.setHeaderIcon(R.mipmap.ic_launcher);
        shareMenu.setHeaderTitle("分享到...");
        shareMenu.add(0, 100, 0, "微信");
        shareMenu.add(0, 101, 0, "QQ");
        shareMenu.add(0, 102, 0, "新浪微博");

        MenuItem detail_item =menu.add(0, Menu.FIRST+2, 0, "详细");
        MenuItem save_item =menu.add(0, Menu.FIRST+3, 0, "保存")
.setIcon(R.mipmap.ic_launcher);

        return true;
}
```

(2) 如果希望应用程序响应菜单项的点击事件,可重写 Activity 的 onOptionsItemSelected(MenuItem item)方法,调用 item. getItemId()获得被点击菜单项的 ID,做出不同的响应。

在 MenuContentActivity 中重写 onOptionsItemSelected(),详细内容如下:

```
@Override
public boolean onOptionsItemSelected(MenuItem item) {
    switch (item.getItemId()) {
        case Menu.FIRST:
            Toast.makeText(MenuContentActivity.this, "你点击了第一个菜单",
Toast.LENGTH_SHORT).show();
            break;

          case 100:
              Toast.makeText(MenuContentActivity.this, "你打算分享到微信哇",
Toast.LENGTH_SHORT).show();
                break;
        //其他菜单项的监听请自行添加 case 语句
    }
    return true;
}
```

在 Android 早期版本中,菜单超过 6 个 MenuItem 时,第 6 个显示为 more,之后的子菜单式样不再显示图标。在 Android 4.0 版本后,菜单默认不显示图标,导致 setIcon 方法给菜单添加图标无效。原因在于 4.0 系统中,涉及菜单的源码类 MenuBuilder 做了改变,mOptionalIconsVisible 成员初始值默认为 false。为了解决这个问题,采用 Java 反射机制,在代码运行创建菜单时通过反射调用 setOptionalIconsVisible 方法设置

mOptionalIconsVisible 为 true,然后再给菜单添加 icon,这样就可以在菜单中显示添加的图标了,具体可参考前一页方框中的代码。

另外,如果想要动态改变选项菜单的内容,须重写 onPrepareOptionsMenu(Menu),其中的内容可参考 onCreateOptionsMenu(Menu menu)。不同的是,onCreateOptionsMenu 只会在 Menu 显示之前调用一次,之后就不会再去调用;而 onPrepareOptionsMenu 是每次显示 Menu 之前都会调用,只要按一次 Menu 键,onPrepareOptionsMenu 就会调用一次,所以可以在这里动态改变 Menu。

上述代码的运行效果如图 3-49 和图 3-50 所示。点击 Menu 键,弹出图 3-49 所示的菜单,点击"分享"菜单时,因为是一个子菜单,所以弹出图 3-50 所示的子菜单。完整程序可参考本书附带的源代码。

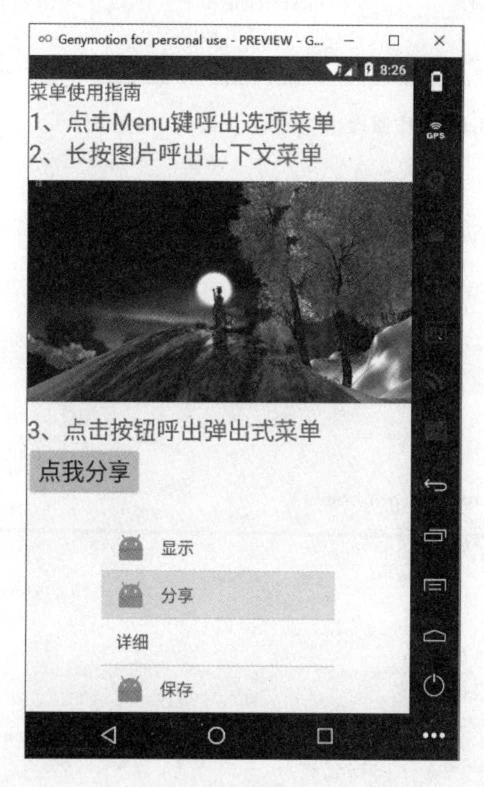

图 3-49 选项菜单运行效果

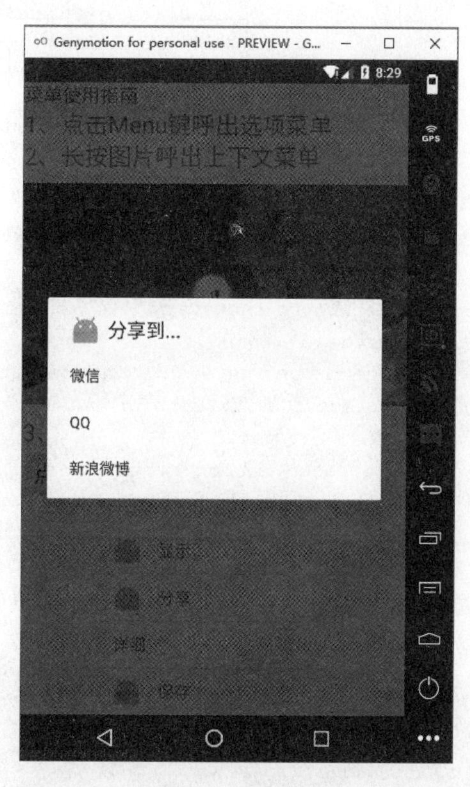

图 3-50 子菜单效果

3.5.2 上下文菜单

当用户长时间按住(超过 2s)某一个组件时,该组件关联的 ContextMenu(上下文菜单)就会显示出来。上下文菜单继承了 android. view. Menu,其用法如图 3-51 所示,可以像操作 OptionMenu 那样给 ContextMenu 增加菜单项。

创建 ContextMenu 后,将菜单注册到任意的 View 对象中(如基本控件、布局文件、ListView 的某一项等),当用户长按界面元素超过 2s 后,将自动出现上下文菜单,通过 onContextItemSelected()方法来响应用户的操作。

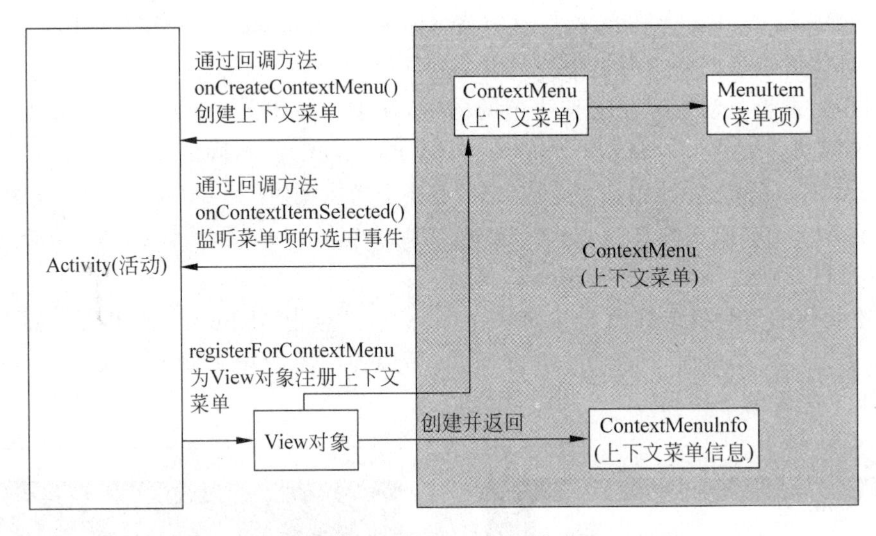

图 3-51 ContextMenu 工作原理

开发上下文菜单的具体步骤如下：

(1) 重写 Activity 的 onCreateContextMenu()方法。

```
//在 MenuContentActivity 中重写 onCreateContextMenu()
@Override
public void onCreateContextMenu(ContextMenu menu, View v,
ContextMenu.ContextMenuInfo menuInfo) {
    switch (v.getId()) {
        case R.id.img_menu:
            menu.setHeaderIcon(R.mipmap.ic_launcher);
            menu.setHeaderTitle("分享到......");
            menu.add(0, 200, 0, "QQ空间");
            menu.add(0, 201, 0, "朋友圈");
            SubMenu subMenu =menu.addSubMenu(0, 202, 0, "微博...");
            subMenu.add(0, 203, 0, "腾讯微博");
            subMenu.add(0, 204, 0, "新浪微博");
            break;

        default:
            break;
    }
    super.onCreateContextMenu(menu, v, menuInfo);
}
```

(2) 调用 Activity 的 registerForContextMenu(View view)方法为 View 组件注册上下文菜单，在本例中为一张图片注册上下文菜单：

```
imageView = (ImageView) findViewById(R.id.img_menu);
registerForContextMenu(imageView);
```

(3) 重写 onContextItemSelected(MenuItem m)方法响应菜单项的选择操作，调用 item.getItemId()方法得到被选择菜单项的 ID：

```
//在 MenuContentActivity 中重写 onContextItemSelected()
@Override
public boolean onContextItemSelected(MenuItem item) {
      switch (item.getItemId()) {
          case 200:
              Toast.makeText(MenuContentActivity.this, "你打算分享到 QQ 空间哇",
Toast.LENGTH_SHORT).show();
              break;
          case 203:
              Toast.makeText(MenuContentActivity.this, "你打算分享到腾讯微博哇",
Toast.LENGTH_SHORT).show();
              break;
                                               //其他菜单项的监听请自行添加 case 语句
        default:
              break;
      }
      return super.onContextItemSelected(item);
}
```

代码的运行效果如图 3-52 和图 3-53 所示。长按图片，弹出图 3-52 所示的菜单，点击"微博"菜单项时，因为是一个子菜单，所以弹出图 3-53 所示的子菜单。完整程序可参考本书附带的源代码。

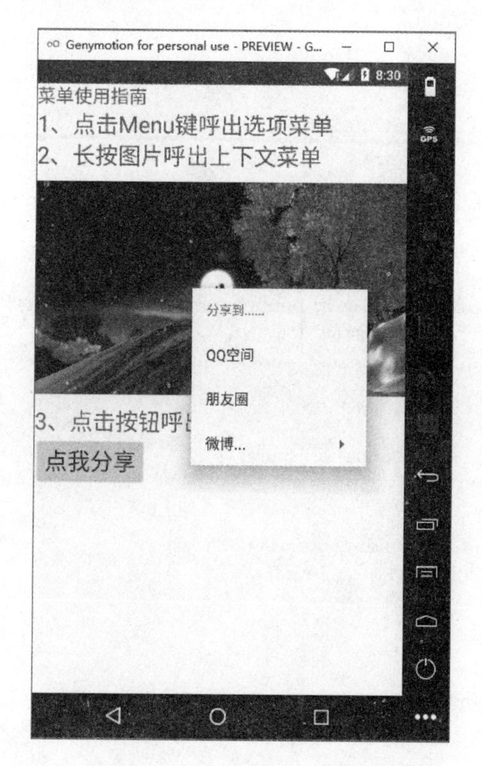

图 3-52 上下文菜单运行效果

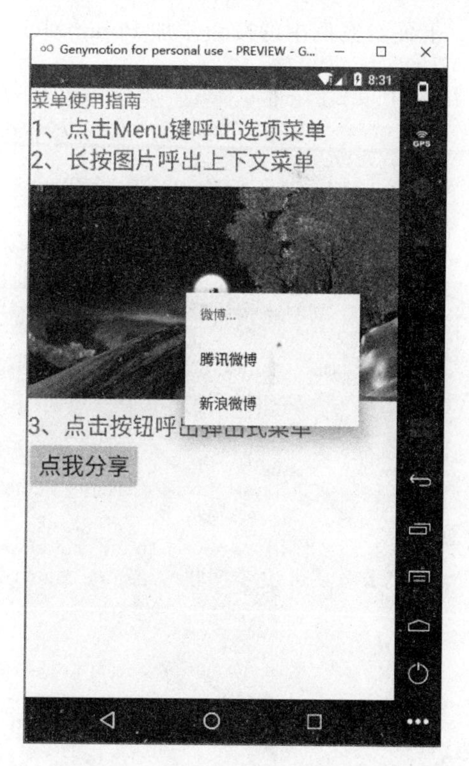

图 3-53 上下文子菜单运行效果

上下文菜单(ContextMenu)和选项菜单(OptionsMenu)的区别：

(1) 上下文没有快捷键，不能显示菜单项图标，但是可以通过 setHeaderIcon、setHeaderTitle、setHeaderView 来设置头。

(2) 每个 Activity 有且只有一个 Options Menu，它为整个 Activity 服务。

(3) 上下文菜单的拥有者是 Activity 中的 View，显式地通过 registerForContextMenu (View view)来为 View 指定是否拥有上下文菜单，多个 View 都可拥有 ContextMenu。

(4) onCreateOptionsMenu 只在第一次按 Menu 键时被调用，而 onCreate-ContextMenu 会在每一次长按 View 时被调用。

3.5.3 弹出式菜单

PopupMenu(弹出式菜单)也称为下拉菜单，它会在指定组件上弹出 PopupMenu，可以增加多个菜单项，并可以为菜单项增加子菜单，需要在 API 11 以上的版本中才能使用。

PopupMenu 可以增加多个菜单项，并可以为菜单项增加子菜单。

创建 PopupMenu 步骤如下：

(1) 调用 new PopupMenu(Context context，View anchor)创建下拉菜单，anchor 代表要激发该弹出菜单的组件。

(2) 调用 MenuInflater 的 inflate()方法将菜单资源填充到 PopupMenu 中。

(3) 调用 PopupMenu 的 show()方法显示弹出式菜单。

注意：PopupMenu 的事件监听 OnMenuItemClickListener。

下面给案例中的按钮添加 PopupMenu。

```
//监听按钮的点击事件
button.setOnClickListener(new View.OnClickListener() {
      @Override
      public void onClick(View v) {
                                           //创建 PopupMenu
        PopupMenu popupMenu =new PopupMenu(MenuContentActivity.this,button);
                                           //加载菜单资源
        popupMenu.getMenuInflater().inflate(R.menu.pop_menu, popupMenu.getMenu());
                                           //菜单时间监听
        popupMenu.setOnMenuItemClickListener(new PopupMenu.
            OnMenuItemClickListener() {
            @Override
              public boolean onMenuItemClick(MenuItem item) {
                switch (item.getItemId()) {
                  case R.id.mi_qq:
                        Toast.makeText(MenuContentActivity.this,
"你打算分享到 QQ 空间哇", Toast.LENGTH_SHORT).show();
                        break;

                    case R.id.mi_friend:
                        Toast.makeText(MenuContentActivity.this,
"你打算分享到朋友圈哇", Toast.LENGTH_SHORT).show();
                          break;
```

```
                            //其他菜单项的监听请自行添加 case 语句
                    default:
                        break;
                    }
                    return true;
                }
            });
            //使用反射,强制显示菜单图标
            try {
                Field field =popupMenu.getClass().getDeclaredField("mPopup");
                field.setAccessible(true);
                MenuPopupHelper mHelper = (MenuPopupHelper) field.get(popupMenu);
                mHelper.setForceShowIcon(true);
            } catch (IllegalAccessException | NoSuchFieldException e) {
                e.printStackTrace();
            }
                            //显示下拉菜单
            popupMenu.show();
        }
    });
```

代码的运行效果如图 3-54 所示。点击"点我分享"按钮,弹出图 3-54 所示的菜单,完整程序可参考本书附带的源代码。

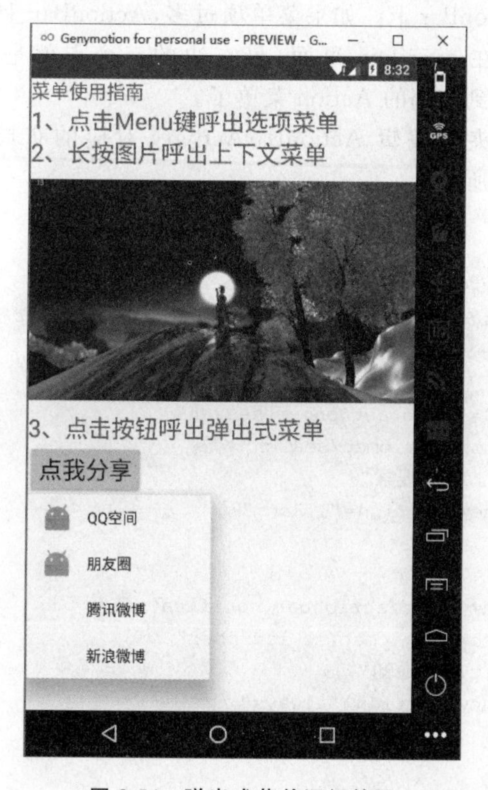

图 3-54 弹出式菜单运行效果

3.5.4 ActionBar 的使用

ActionBar 是在 Android 3.0 版本之后新增的一种导航栏功能，是 Android 3.0 版本以后菜单导航栏功能的主要风格。ActionBar 可以为应用程序提供一种全局统一的 UI 导航界面，标识当前操作界面的位置，并提供额外的用户动作、界面导航等功能；同时，ActionBar 也能自动适应屏幕分辨率的变化，提高用户体验。

1. ActionBar 的使用

在 Android 3.0 以后的版本中使用 ActionBar 非常简单，只需要在配置文件 AndroidManifest.xml 中添加 android:theme="@android:style/ Theme.Holo"，指定 Application 或者 Activity 的 theme 为 Theme.Holo 或其子类就可以了。

ActionBar 常用主题如下：

```
Theme.Holo
Theme.Holo.Light
Theme.Holo.Light.DarkActionBar
```

除了使用系统自带的主题外，也可以在系统主题的基础上自定义主题。自定义主题通常需要实现 theme.xml、styles.xml 等。

在 ActionBar 上可以摆放一些其他的 Action 菜单项，这些菜单项都会以图标或文字的形式直接显示在 ActionBar 上。如果菜单项过多，ActionBar 上不能全部显示，多出的一些菜单项将会隐藏在 overflow 里面（最右边的三个点就是 overflow 按钮），点击 overflow 按钮就可以看到全部的 Action 菜单了。

在 res/menu 文件夹下编辑 ActionBarActivity 对应的布局文件 actionbar_menu.xml，如文件清单 3-20 所示。

文件清单 3-20 actionbar_menu.xml

```
<?xml version="1.0" encoding="utf-8"?>
<menu xmlns:android="http://schemas.android.com/apk/res/android">

    <item
        android:id="@+id/actionbar_search"
        android:icon="@mipmap/search"
        android:title="搜索"
        android:showAsAction="ifRoom" />

    <item
        android:id="@+id/actionbar_notification"
        android:icon="@mipmap/ic_launcher"
        android:title="通知"
        android:showAsAction="always" />

    <item
```

```
        android:id="@+id/actionbar_settings"
        android:orderInCategory="100"
        android:title="设置"
        android:showAsAction="never" />

    <item
        android:id="@+id/actionbar_about"
        android:orderInCategory="101"
        android:title="关于"
        android:showAsAction="never" />

</menu>
```

这里通过 4 个<item>标签定义了 4 个 Action 菜单项。<item>标签中又有一些属性，其中 id 是该 Action 菜单项的唯一标识符，在事件监听中将通过 id 来区分用户操作的菜单项；icon 用于指定该菜单项的图标；title 用于指定该菜单项可能显示的文字(在图标能显示的情况下，通常不会显示文字)；orderInCategory 用于指定 Action 菜单项的显示顺序；showAsAction 则指定了该菜单项在 ActionBar 上的显示的位置，主要有以下几种值可选：

(1) always 表示菜单项永远显示在 ActionBar 中，如果屏幕空间不够则无法显示；

(2) ifRoom 表示在屏幕空间够的情况下，菜单项将显示在 ActionBar 中，如空间不够就显示在 overflow 溢出菜单中；

(3) never 表示菜单项永远显示在 overflow 中；

(4) withText 表示菜单的图标和菜单文本一起显示，通常在竖屏中不显示菜单文本。

当 Activity 启动时，系统会调用 Activity 的 onCreateOptionsMenu()方法来取出所有的 Action 按钮，我们只需要在这个方法中加载一个 menu 资源，并把所有的 Action 按钮都定义在资源文件里面就可以了。

```
//重写 ActionBarActivity 中的 onCreateOptionsMenu()
@Override
    public boolean onCreateOptionsMenu(Menu menu) {
        try {
            Class<?>menuClass =
Class.forName("com.android.internal.view.menu.MenuBuilder");
            Method menuMethod =menuClass
.getDeclaredMethod("setOptionalIconsVisible", boolean.class);
            menuMethod.setAccessible(true);
            menuMethod.invoke(menu, true);
        }catch (Exception e) {
            e.printStackTrace();
        }
        MenuInflater inflater =getMenuInflater();
        inflater.inflate(R.menu.actionbar_menu,menu);
```

```
        return super.onCreateOptionsMenu(menu);
    }
```

同样,为了能够显示 Menu 图标,依然使用 Java 反射机制进行修改。

2. ActionBar 的点击事件

当点击 ActionBar 中的按钮时,系统会调用 Activity 的 onOptionsItemSelected (MenuItem item)方法,我们可以调用 item 的 getItemId()方法和 menu 资源中的 ID 进行比较,从而辨别出用户点击的是哪一个按钮。

```
//重写 ActionBarActivity 中的 onOptionsItemSelected()
public boolean onOptionsItemSelected(MenuItem item) {
    switch (item.getItemId()) {
        case android.R.id.home:
            finish();
            return true;
        case R.id.actionbar_about:
            Toast.makeText(ActionBarActivity.this,"点击了 关于",
Toast.LENGTH_SHORT).show();
            return true;
        case R.id.actionbar_notification:
            Toast.makeText(ActionBarActivity.this,"点击了 通知",
Toast.LENGTH_SHORT).show();
            return true;
        case R.id.actionbar_search:
            Toast.makeText(ActionBarActivity.this,"点击了 搜索",
Toast.LENGTH_SHORT).show();
            return true;
        case R.id.actionbar_settings:
            Toast.makeText(ActionBarActivity.this,"点击了 设置",
Toast.LENGTH_SHORT).show();
            return true;
        default:
            return super.onOptionsItemSelected(item);
    }
}
```

在使用 ActionBar 时,通常也启用 ActionBar 图标导航的功能,通过导航可以允许用户根据当前应用的位置在不同界面之间切换。可以通过调用 setDisplayHomeAsUpEnabled()方法启用 ActionBar 图标导航功能。

在 Activity 的 onCreate()中添加如下代码:

```
ActionBar actionBar =getActionBar();
actionBar.setDisplayHomeAsUpEnabled(true);
```

可以看到,在 ActionBar 图标的左侧出现了一个向左的箭头,如图 3-55 所示。通常情况下，向左箭头表示返回到上一页的意思，当点击 ActionBar 图标时，系统同样会调用 onOptionsItemSelected()方法，并且此时的 itemId 是 android. R. id. home，因此最简单的实现就是在它的点击事件里面加入 finish()方法，可参考前面的代码。

图 3-55 ActionBar 显示效果

3. ActionBar 的使用要点

默认情况下，系统会使用< application >或< activity >中 icon 属性指定的图片来作为 ActionBar 的图标，但是也可以在< application >或者< activity >中通过 logo 属性来指定一张图片来作为 ActionBar 的图标，在 AndroidManifest. xml 中为 ActionBarActivity 设置 ActionBarActivity 图标的代码如下：

```
<activity
      android:name=".ActionBarActivity"
      android:logo="@mipmap/wh"
      android:theme="@android:style/Theme.Holo.Light.DarkActionBar"
</activity>
```

如果不想在应用中使用 ActionBar，可以通过以下两种方法移除 ActionBar：

(1) 将 theme 指定成 Theme. Holo. NoActionBar，表示使用一个不包含 ActionBar 的主题；

(2) 在 Activity 中调用以下方法将 ActionBar 隐藏：

```
ActionBar actionBar =getActionBar();
actionBar.hide();
```

overflow 按钮是否显示和手机的硬件有关，通常情况下，如果手机有物理 Menu 键，overflow 按钮一般不显示；当手机没有物理 Menu 键时，overflow 按钮就可以显示出来。如果想让 overflow 按钮能够一直显示，可以借助 Java 反射机制进行修改。Android 系统就是根据 ViewConfiguration 类中的 sHasPermanentMenuKey 静态变量的值来判断手机是否有物理 Menu 键的。通过反射的方式修改 sHasPermanentMenuKey 的值，让它取值为 false 就可以使 overflow 按钮一直显示，代码如下：

```
    try {
      ViewConfiguration config =ViewConfiguration.get(this);
Field menuKeyField =ViewConfiguration.class
.getDeclaredField("sHasPermanentMenuKey");
        menuKeyField.setAccessible(true);
        menuKeyField.setBoolean(config, false);
} catch (Exception e) {
      e.printStackTrace();
    }
```

3.5.5 ToolBar的使用

Toolbar是在Android 5.0版本开始时推出的一个Material Design风格的导航控件,Google推荐使用Toolbar作为Android客户端的导航栏,以此取代之前的Actionbar。与Actionbar相比,Toolbar要灵活很多。它不像Actionbar,必须固定在Activity的顶部,而是可以放到界面的任意位置。在设计ToolBar时,Google也留给了开发者很多可定制修改的余地,这些可定制修改的属性在API文档中都有详细介绍,有兴趣的读者可以去查阅。

接下来用ToolBar实现前面同样的导航效果。在res/menu文件夹下编辑ToolBarActivity对应的菜单文件toolbar_menu.xml,设置4个菜单项,文件写法与前面ActionBar菜单差不多,如文件清单3-21所示。

文件清单3-21 toolbar_menu.xml

```
<?xml version="1.0" encoding="utf-8"?>
<menu xmlns:android="http://schemas.android.com/apk/res/android"
    xmlns:app="http://schemas.android.com/apk/res-auto">

    <item
        android:id="@+id/toolbar_search"
        android:icon="@mipmap/search"
        android:title="搜索"
        app:showAsAction="ifRoom" />

    <item
        android:id="@+id/toolbar_notification"
        android:icon="@mipmap/ic_launcher"
        android:title="通知"
        app:showAsAction="ifRoom" />

    <item
        android:id="@+id/toolbar_settings"
        android:orderInCategory="100"
        android:title="设置"
        app:showAsAction="never" />

    <item
        android:id="@+id/toolbar_about"
        android:orderInCategory="101"
        android:title="关于"
        app:showAsAction="never" />
</menu>
```

在res/layout文件夹下创建布局编辑文件toolbar_layout.xml,使用Toolbar控件,注意要使用Toolbar的完整包名,如文件清单3-22所示。

文件清单 3-22 toolbar_layout.xml

```
<?xml version="1.0" encoding="utf-8"?>
<LinearLayout xmlns:android="http://schemas.android.com/apk/res/android"
    android:layout_width="match_parent"
    android:layout_height="match_parent"
    android:orientation="vertical">
    <android.support.v7.widget.Toolbar
        android:layout_width="match_parent"
        android:layout_height="wrap_content"
        android:background="?attr/colorPrimary"
        android:theme="?attr/actionBarTheme"
        android:minHeight="?attr/actionBarSize"
        android:id="@+id/toolbar" />
    <RelativeLayout
        android:layout_width="match_parent"
        android:layout_height="match_parent">
    </RelativeLayout>
</LinearLayout>
```

ToolBarActivity 核心代码如文件清单 3-23 所示。

文件清单 3-23 ToolBarActivity.java

```
public class ToolBarActivity extends Activity {
    @Override
    protected void onCreate(Bundle savedInstanceState) {
        super.onCreate(savedInstanceState);
        setContentView(R.layout.toolbar_layout);
        Toolbar toolbar = (Toolbar) findViewById(R.id.toolbar);
        // Toolbar 加载 menu 资源
        toolbar.inflateMenu(R.menu.toolbar_menu);
        //设置导航图标
        toolbar.setNavigationIcon(R.mipmap.ic_launcher);
        toolbar.setTitle("首页");
        toolbar.setTitleTextColor(getResources().getColor(android.R.color.white));
        //设置导航事件
        toolbar.setNavigationOnClickListener(new View.OnClickListener() {
            @Override
            public void onClick(View v) {
                finish();
            }
        });

        //toolbar 的 menu 点击事件的监听
        toolbar.setOnMenuItemClickListener(new Toolbar.OnMenuItemClickListener() {
            @Override
            public boolean onMenuItemClick(MenuItem item) {
                switch (item.getItemId()){
                    case R.id.toolbar_about:
                        Toast.makeText(ToolBarActivity.this,"点击了 关于",
```

```
Toast.LENGTH_SHORT).show();
                            return true;
                        case R.id.toolbar_notification:
                            Toast.makeText(ToolBarActivity.this,"点击了 通知",
Toast.LENGTH_SHORT).show();
                            return true;
                        case R.id.toolbar_search:
                            Toast.makeText(ToolBarActivity.this,"点击了 搜索",
Toast.LENGTH_SHORT).show();
                            return true;
                        case R.id.toolbar_settings:
                            Toast.makeText(ToolBarActivity.this,"点击了 设置",
Toast.LENGTH_SHORT).show();
                            return  true;
                        default:
                            return false;
                    }
                }
            });
        }
}
```

运行效果如图 3-56 所示,采用 ToolBar 的方式实现顶部导航效果。

图 3-56　ToolBar 运行效果

3.6 导航栏的使用

除了使用 ActionBar 和 ToolBar 作为顶部导航外，底部导航栏也是一种常见的导航方式，手机中一些常见的 APP，例如 QQ、微信、京东、淘宝等都有底部导航栏，用户可以随时切换界面，查看不同的内容。底部导航栏的实现方式很多，以前大多使用 TabHost 来实现，现在有了很多更好的选择，例如使用 ViewPager 和 Fragment 等。

下面分别通过 TabHost、ViewPager 和 Fragment 三种方式实现页面切换。在 Chapter03UI 项目中创建 NavigationActivity、MyTabActivity、ViewPagerActivity 和 MyFragmentActivity，其中 NavigationActivity 的页面展示如图 3-57 所示，在该页面点击不同按钮，分别跳转至 MyTabActivity、ViewPagerActivity 和 MyFragmentActivity。由于 NavigationActivity 的页面布局文件比较简单，在此就不给出详细内容。

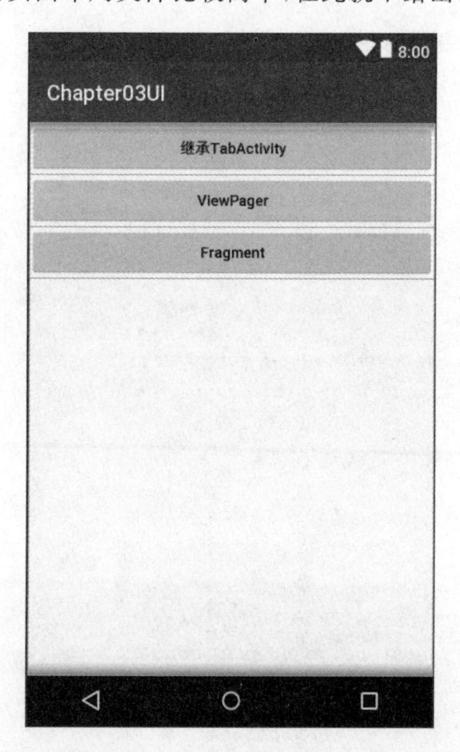

图 3-57 NavigationActivity 页面

3.6.1 TabHost 导航

TabHost 是整个 Tab 的容器，包含 TabWidget 和 FrameLayout 两个部分，TabWidget 是每个 Tab 的标签，FrameLayout 是 Tab 的内容。使用 TabHost 实现导航栏有两种方式。

(1) 继承 TabActivity。从 TabActivity 中用 getTabHost()方法获取 TabHost，各个 Tab 中的内容在布局文件中定义即可。TabHost 必须设置 android:id 为@android:id/

tabhost，TabWidget 必须设置 android:id 为@android:id/tabs，FrameLayout 必须设置 android:id 为@android:id/tabcontent。

（2）不继承 TabActivity。在布局文件中定义 TabHost 即可，但是 TabWidget 的 id 必须是@android:id/tabs，FrameLayout 的 id 必须是@android:id/tabcontent。

鉴于使用 TabHost 实现导航栏已经过时，关于 TabHost 的使用就不多做讲解，对 TabHost 感兴趣的读者可以去查阅其他资料以了解 TabHost 的使用。

以下编辑 MyTabActivity 对应的布局文件 activity_my_tab. xml。

（1）用 TabHost 和 RadioGroup 搭建基本布局，以 RadioGroup 代替 TabWidget，布局文件 activity_my_tab. xml 详细内容如文件清单 3-24 所示。

文件清单 3-24　activity_my_tab. xml

```
<?xml version="1.0" encoding="utf-8"?>
<TabHost xmlns:android="http://schemas.android.com/apk/res/android"
    android:id="@android:id/tabhost"
    android:layout_width="match_parent"
    android:layout_height="match_parent">

    <LinearLayout
        android:layout_width="match_parent"
        android:layout_height="match_parent"
        android:orientation="vertical" >

        <FrameLayout
            android:id="@android:id/tabcontent"
            android:layout_width="match_parent"
            android:layout_height="0.0dip"
            android:layout_weight="1.0" >
        </FrameLayout>

        <TabWidget
            android:id="@android:id/tabs"
            android:layout_width="match_parent"
            android:layout_height="wrap_content"
            android:visibility="gone" >
        </TabWidget>

        <RadioGroup
            android:layout_width="match_parent"
            android:layout_height="wrap_content"
            android:layout_gravity="bottom"
            android:background="@android:color/black"
            android:orientation="horizontal" >

            <RadioButton
                android:id="@+id/home_tab"
```

```
            style="@style/bottom_tab"
            android:checked="true"
            android:drawableTop="@mipmap/icon_home"
            android:text="首页" />

        <RadioButton
            android:id="@+id/near_tab"
            style="@style/bottom_tab"
            android:drawableTop="@mipmap/icon_nearby"
            android:text="附近" />

        <RadioButton
            android:id="@+id/order_tab"
            style="@style/bottom_tab"
            android:drawableTop="@mipmap/icon_order"
            android:text="预约" />

        <RadioButton
            android:id="@+id/mine_tab"
            style="@style/bottom_tab"
            android:drawableTop="@mipmap/icon_mine"
            android:text="我的" />
    </RadioGroup>
  </LinearLayout>
</TabHost>
```

(2) 编辑 res/values/styles.xml 文件,设置导航栏背景的样式 bottom_tab。

```
<-- 导航背景样式 -->
<style name="bottom_tab">
        <item name="android:textSize">@dimen/bottom_tab_font_size</item>
        <item name="android:textColor">@color/whiteColor</item>
        <item name="android:ellipsize">marquee</item>
        <item name="android:gravity">center_horizontal</item>
        <item name="android:background">@drawable/bottom_tab_bg</item>
        <item name="android:paddingTop">@dimen/bottom_tab_padding_up</item>
        <item name="android:layout_width">fill_parent</item>
        <item name="android:layout_height">wrap_content</item>
        <item name="android:button">@null</item>
        <item name="android:singleLine">true</item>
        <item name="android:drawablePadding">2dp</item>
        <item name="android:layout_weight">1.0</item>
</style>
```

在 res/drawable/文件夹下新建 bottom_tab_bg.xml 文件,在该文件中编辑按钮背景 selector。bottom_tab_bg.xml 文件如文件清单 3-25 所示。

文件清单 3-25 bottom_tab_bg.xml

```
<?xml version="1.0" encoding="utf-8"?>
<selector xmlns:android="http://schemas.android.com/apk/res/android">
    <item
        android:state_focused="true"
        android:state_enabled="true"
        android:state_pressed="false"
        android:drawable="@mipmap/bottom_tab_bg_s" />

    <item
        android:state_enabled="true"
        android:state_pressed="true"
        android:drawable="@mipmap/bottom_tab_bg_s" />

    <item
        android:state_enabled="true"
        android:state_checked="true"
        android:drawable="@mipmap/bottom_tab_bg_s" />
</selector>
```

(3) 针对导航的4个按钮,创建对应启动的Activity页面。页面文件分别为home_activity.xml,near_activity.xml,order_activity.xml,mine_activity.xml,页面设计分别如图3-58～图3-61所示。针对4个布局文件,创建对应的Activity,此步骤比较简单,在此就不再赘述。

图3-58 "首页"布局设计

图3-59 "附近"布局设计

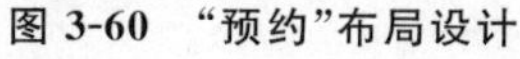
图 3-60 “预约”布局设计

图 3-61 “我的”布局设计

(4) 在 MyTabActivity 中实现按钮和内容的切换，核心代码如文件清单 3-26 所示。

文件清单 3-26 MyTabActivity. java

```
public class MyTabActivity extends TabActivity implements
CompoundButton.OnCheckedChangeListener{

    private TabHost tabhost;
    private RadioButton homeBtn;
    private RadioButton nearBtn;
    private RadioButton orderBtn;
    private RadioButton mineBtn;
    private Intent homeIntent;
    private Intent nearIntent;
    private Intent orderIntent;
    private Intent mineIntent;

    @Override
    protected void onCreate(Bundle savedInstanceState) {
        super.onCreate(savedInstanceState);
        setContentView(R.layout.tab_layout1);

        homeBtn = (RadioButton) findViewById(R.id.home_tab);
        nearBtn = (RadioButton) findViewById(R.id.near_tab);
        orderBtn = (RadioButton) findViewById(R.id.order_tab);
        mineBtn = (RadioButton) findViewById(R.id.mine_tab);

        homeIntent =new Intent(TableActivity1.this, HomeActivity.class);
        nearIntent =new Intent(TableActivity1.this, NearActivity.class);
```

```
        orderIntent =new Intent(TableActivity1.this, OrderActivity.class);
        mineIntent =new Intent(TableActivity1.this, MineActivity.class);

        tabhost =getTabHost();
        tabhost.addTab(tabhost.newTabSpec("home").setIndicator("0")
.setContent(homeIntent));
        tabhost.addTab(tabhost.newTabSpec("near").setIndicator("1")
.setContent(nearIntent));
        tabhost.addTab(tabhost.newTabSpec("order").setIndicator("2")
.setContent(orderIntent));
        tabhost.addTab(tabhost.newTabSpec("mine").setIndicator("3")
.setContent(mineIntent));

        homeBtn.setOnCheckedChangeListener(this);
        nearBtn.setOnCheckedChangeListener(this);
        orderBtn.setOnCheckedChangeListener(this);
        mineBtn.setOnCheckedChangeListener(this);
    }

    @Override
    public void onCheckedChanged(CompoundButton buttonView, boolean isChecked) {
        if (isChecked){
            switch (buttonView.getId()){
                case R.id.home_tab:
                    tabhost.setCurrentTab(0);
                    break;
                case  R.id.near_tab:
                    tabhost.setCurrentTab(1);
                    break;
                case R.id.order_tab:
                    tabhost.setCurrentTab(2);
                    break;
                case R.id.mine_tab:
                    tabhost.setCurrentTab(3);
                    break;
            }
        }
    }
}
```

最终运行效果如图 3-62 所示，点击底部按钮，导航到对应的页面。

3.6.2 ViewPager 的使用

早期 Android 应用通常使用 TabHost 实现页面之间的切换，现在更多地选择 ViewPager 与 Fragment 相结合的方式实现页面切换。Android 提供了一些专门的适配器——FragmentPagerAdapter 与 FragmentStatePagerAdapter 让 ViewPager 与 Fragment

图 3-62　底部导航

一起工作。ViewPager 与 Fragment 一起使用，提供了一种很好的方法来管理各个页面的生命周期。稍后将使用这种方式来重现前面的 TabHost 导航。在此之前，需要对 ViewPager 有充分的认识。

ViewPager 控件继承自 ViewGroup，即 ViewPager 是一个容器类，可以包含其他的 View 类，ViewPager 是 Android 扩展包 v4 包中的类，其继承结构如图 3-63 所示。

ViewPager 是一个允许使用者左右滑动页面的布局管理器，可通过一个适配器(PagerAdapter)管理要显示的页面。

在 Activity 里实例化 ViewPager 组件，并设置它的 Adapter，即可完成页面之间的滑动切换。

接下来对 ViewPagerActivity 进行编辑，让大家理解 ViewPager 的用法。

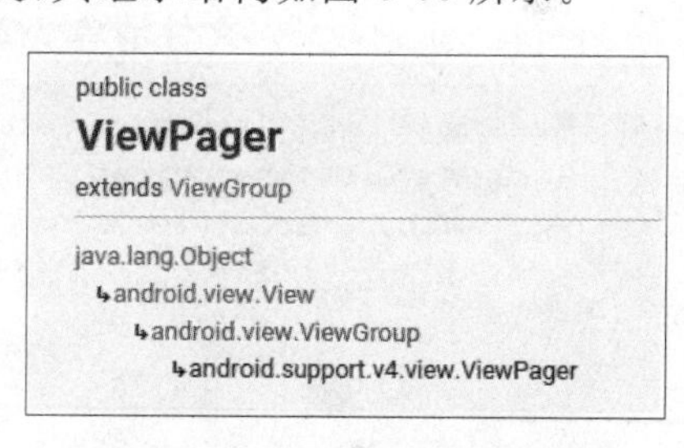

图 3-63　ViewPager 继承结构

(1) 编辑 ViewPagerActivity 对应的主页面布局 viewpager_layout. xml，在布局文件中只放置一个 ViewPager 控件。viewpager_layout. xml 内容如文件清单 3-27 所示。

文件清单 3-27　viewpager_layout. xml

```
<?xml version="1.0" encoding="utf-8"?>
<LinearLayout xmlns:android="http://schemas.android.com/apk/res/android"
    android:layout_width="match_parent"
    android:layout_height="match_parent">

    <android.support.v4.view.ViewPager
        android:id="@+id/viewpager"
        android:layout_width="match_parent"
        android:layout_height="match_parent">

    </android.support.v4.view.ViewPager>
</LinearLayout>
```

注意：ViewPager 引入时必须写完整 android. support. v4. view. ViewPager，如果按照一般控件那样直接写成 ViewPager，编译器会报错。

(2) 准备几个要切换的视图。这里直接选用在 TabHost 案例里的 4 个布局。

(3) 实现一个 PagerAdapter。

ViewPager 是通过适配器管理的，前面已经有了视图，这里的数据源就是一个包含 4 个 View 的列表 List。

PagerAdapter 作为"将多个页面填充到 ViewPager"的适配器的一个基类，多数情况下，可能更倾向于使用 FragmentPagerAdapter 或者 FragmentStatePagerAdapter 等实现 PagerAdapter 并且更加具体的适配器。

当实现一个 PagerAdapter 时，至少需要重写以下几个方法：

① instantiateItem(ViewGroup, int)负责初始化指定位置的页面，并且需要返回当前页面本身；

② destroyItem(ViewGroup, int, Object)负责移除指定位置的页面；

③ getCount()返回要展示的页面数量；

④ isViewFromObject(View, Object)里直接写"return view == object;"即可。

在本例中，自定义 MyPagerAdapter 代码如文件清单 3-28 所示。

文件清单 3-28 MyPagerAdapter. java

```
public class MyPagerAdapter extends PagerAdapter{
    private List<View>pageList;
    public MyPagerAdapter(List<View>pageList) {
        this.pageList =pageList;
    }

    @Override
    public int getCount() {
        return pageList.size();                //返回要展示的页面数量
    }

    @Override
    public Object instantiateItem(ViewGroup container, int position) {
        container.addView(pageList.get(position));
        return pageList.get(position);
    }

    @Override
    public void destroyItem(ViewGroup container, int position, Object object) {
        container.removeView(pageList.get(position));
    }

    @Override
    public boolean isViewFromObject(View view, Object object) {
        return view==object;
    }
}
```

(4) 在对应的 ViewPagerActivity 文件中声明 ViewPager、适配器及相关的页面，然后初始化 ViewPager 和适配器，绑定适配器。运行程序，这 4 个页面就可以滑动切换了。运行效果如图 3-64 所示。

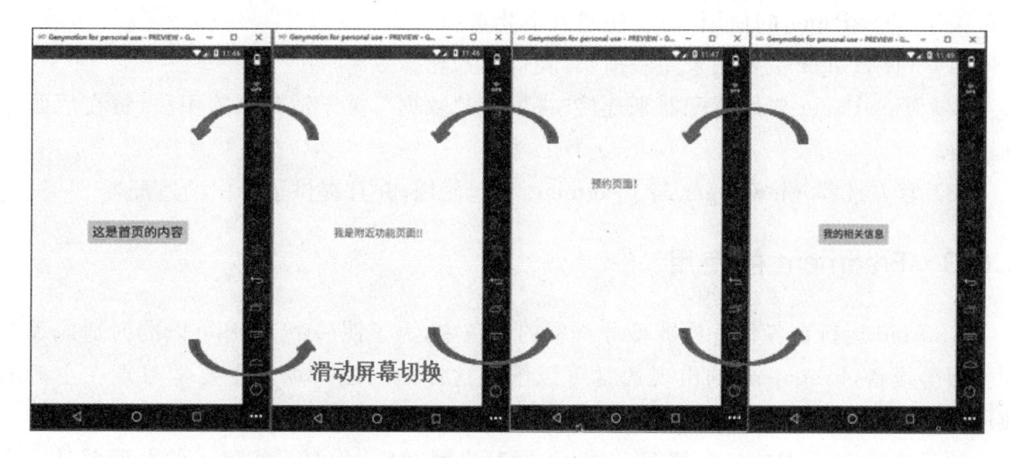

图 3-64 ViewPager 页面滑动效果

核心代码如文件清单 3-29 所示。

文件清单 3-29 ViewPagerActivity. java

```
public class ViewPagerActivity extends Activity {

    private ViewPager myViewPager;                              // 要使用的 ViewPager
    private View pageHome, pageNear, pageOrder, pageMine; // ViewPager 包含的页面
    private List<View> pageList; // ViewPager 包含的页面，一般给 Adapter 传的是一
个 List
    private MyPagerAdapter myPagerAdapter;                      // 适配器

    @Override
    protected void onCreate(Bundle savedInstanceState) {
        super.onCreate(savedInstanceState);
        setContentView(R.layout.viewpager_layout);
        myViewPager = (ViewPager) findViewById(R.id.viewpager);
        pageList =new ArrayList<View>();

        LayoutInflater inflater =getLayoutInflater();
        pageHome =inflater.inflate(R.layout.home_activity, null);
        pageNear =inflater.inflate(R.layout.near_activity, null);
        pageOrder =inflater.inflate(R.layout.order_activity, null);
        pageMine =inflater.inflate(R.layout.mine_activity, null);
        pageList.add(pageHome);
        pageList.add(pageNear);
        pageList.add(pageOrder);
        pageList.add(pageMine);
        myPagerAdapter =new MyPagerAdapter(pageList);
        myViewPager.setAdapter(myPagerAdapter);
```

```
        }
    }
```

关于 ViewPager 的使用,需要注意以下几点:

(1) ViewPager 主要用来左右滑动,类似图片轮播效果。

(2) ViewPager 要用适配器来连接"视图"和"数据",关于适配器的用法,将在后面详细讲解。

(3) 官方推荐 ViewPager 与 Fragment 一起使用,并且提供了专门的适配器。

3.6.3 Fragment 的使用

Android 运行在各种各样屏幕分辨率的设备中,为了使一个 APP 可以同时适应手机和平板等设备,Fragment 的出现通常可以解决这样的问题。本节主要学习现在主流的使用 ViewPager 和 Fragment 相结合的方式来实现的导航效果。

项目中的 ViewPager 会和 Fragment 同时出现,每一个 ViewPager 的页面就是一个 Fragment。接下来使用这种 ViewPager 和 Fragment 相结合的方式来实现之前 TabHost 的导航效果。

首先设计主界面。和 TabHost 导航页面的区别是:布局上面使用了 ViewPager,底部还是原先的 RadioButton。

编辑 MyFragmentActivity 的布局文件 fragment_layout.xml,编辑后的内容如文件清单 3-30 所示。

文件清单 3-30　fragment_layout.xml

```
<?xml version="1.0" encoding="utf-8"?>
<RelativeLayout xmlns:android="http://schemas.android.com/apk/res/android"
    android:layout_width="match_parent"
    android:layout_height="match_parent">

    <android.support.v4.view.ViewPager
        android:id="@+id/fragment_viewpager"
        android:layout_width="match_parent"
        android:layout_height="match_parent">
    </android.support.v4.view.ViewPager>

    <LinearLayout
        android:layout_width="match_parent"
        android:layout_height="wrap_content"
        android:layout_alignParentBottom="true"
        android:orientation="horizontal">

        <RadioGroup
            android:layout_width="match_parent"
            android:layout_height="wrap_content"
```

```
            android:layout_gravity="bottom"
            android:background="@android:color/black"
            android:orientation="horizontal" >

            <RadioButton
                android:id="@+id/home_tab"
                style="@style/bottom_tab"
                android:checked="true"
                android:drawableTop="@mipmap/icon_home"
                android:text="首页" />

            <RadioButton
                android:id="@+id/near_tab"
                style="@style/bottom_tab"
                android:drawableTop="@mipmap/icon_nearby"
                android:text="附近" />

            <RadioButton
                android:id="@+id/order_tab"
                style="@style/bottom_tab"
                android:drawableTop="@mipmap/icon_order"
                android:text="预约" />

            <RadioButton
                android:id="@+id/mine_tab"
                style="@style/bottom_tab"
                android:drawableTop="@mipmap/icon_mine"
                android:text="我的" />
        </RadioGroup>
    </LinearLayout>
</RelativeLayout>
```

接下来，针对前面已有的 home_activity. xml、near_activity. xml、order_activity. xml 和 mine_activity. xml 这 4 个 Layout 文件，准备 4 个 Fragment。在自定义 Fragment 时，需要继承 Fragment 父类，重写 onCreateView 等方法。

HomeFragment 文件如文件清单 3-31 所示。

文件清单 3-31　HomeFragment. java

```
public class HomeFragment extends Fragment {
    private Button btn_home;

    @Nullable
    @Override
    public View onCreateView(LayoutInflater inflater, @Nullable ViewGroup container,
@Nullable Bundle savedInstanceState) {
        return inflater.inflate(R.layout.home_activity, container, false);
    }
```

```
    @Override
    public void onActivityCreated(@Nullable Bundle savedInstanceState) {
        super.onActivityCreated(savedInstanceState);
        btn_home = (Button) getActivity().findViewById(R.id.btn_home);
        //点击跳转到另一个 Activity
        btn_home.setOnClickListener(new View.OnClickListener() {
            @Override
            public void onClick(View v) {
                Intent intent =new Intent();
                intent.setClass(getActivity(), MainActivity.class);
                startActivity(intent);
            }
        });
    }
}

MineFragment
public class MineFragment extends Fragment {
    @Nullable
    @Override
    public View onCreateView(LayoutInflater inflater, @Nullable ViewGroup container,
@Nullable Bundle savedInstanceState) {
        return inflater.inflate(R.layout.mine_activity,container,false);
    }
}

NearFragment
public class NearFragment extends Fragment {
    @Nullable
    @Override
     public  View  onCreateView (LayoutInflater  inflater, @ Nullable  ViewGroup
container,
@Nullable Bundle savedInstanceState) {
        return inflater.inflate(R.layout.near_activity,container,false);
    }
}

OrderFragment
public class OrderFragment extends Fragment {
    @Nullable
    @Override
    public View onCreateView(LayoutInflater inflater, @Nullable ViewGroup container,
@Nullable Bundle savedInstanceState) {
        return inflater.inflate(R.layout.order_activity,container,false);
    }
}
```

接下来需要对 ViewPager 设置适配器。Android 提供了一些专门的适配器让 ViewPager 与 Fragment 一起工作，即 FragmentPagerAdapter 与 FragmentStatePagerAdapter。

FragmentPagerAdapter 与 FragmentStatePagerAdapter 都继承自 PagerAdapter，二者存在许多相似之处，用法也差不多，它们之间最大的不同在于：访问过的页面不可见之后是否会保留在内存中，而这个区别也构成了它们使用场景的不同。FragmentPagerAdapter 最适用于那种少量且相对静态的页面，例如几个 Tab 页；而 FragmentStatePagerAdapter 更多用于大量页面，例如视图列表。

我们这里只有 4 个页面，选用 FragmentPagerAdapter。定义一个类 MyFragment-Adapter，继承自 FragmentPagerAdapter，并重写相关的方法，代码如文件清单 3-32 所示。

文件清单 3-32　MyFragmentAdapter. java

```
public class MyFragmentAdapter extends FragmentPagerAdapter {
    public final static int TAB_COUNT = 4;

    public MyFragmentAdapter(FragmentManager fm) {
        super(fm);
    }

    @Override
    public Fragment getItem(int position) {
        switch (position){
            case MyFragmentActivity.TAB_HOME:
                HomeFragment homeFragment = new HomeFragment();
                return homeFragment;

            case MyFragmentActivity.TAB_NEAR:
                NearFragment nearFragment = new NearFragment();
                return  nearFragment;

            case MyFragmentActivity.TAB_ORDER:
                OrderFragment orderFragment = new OrderFragment();
                return orderFragment;

            case MyFragmentActivity.TAB_MINE:
                MineFragment mineFragment = new MineFragment();
                return mineFragment;
        }
        return null;
    }

    @Override
    public int getCount() {
        return TAB_COUNT;
    }
```

```
}

public class MyFragmentActivity extends FragmentActivity  implements
View.OnClickListener{

    public static final int TAB_HOME = 0;
    public static final int TAB_NEAR = 1;
    public static final int TAB_ORDER = 2;
    public static final int TAB_MINE = 3;

    private ViewPager viewPager;
    private RadioButton homeBtn;
    private RadioButton nearBtn;
    private RadioButton orderBtn;
    private RadioButton mineBtn;

    @Override
    protected void onCreate(@Nullable Bundle savedInstanceState) {
        super.onCreate(savedInstanceState);
        setContentView(R.layout.fragment_layout);
        initView();
    }

    private void initView() {
        viewPager = (ViewPager) findViewById(R.id.fragment_viewpager);
        homeBtn = (RadioButton) findViewById(R.id.home_tab);
        nearBtn = (RadioButton) findViewById(R.id.near_tab);
        orderBtn = (RadioButton) findViewById(R.id.order_tab);
        mineBtn = (RadioButton) findViewById(R.id.mine_tab);
        homeBtn.setOnClickListener(this);
        nearBtn.setOnClickListener(this);
        orderBtn.setOnClickListener(this);
        mineBtn.setOnClickListener(this);

        MyFragmentAdapter adapter = new
MyFragmentAdapter(getSupportFragmentManager());
        viewPager.setAdapter(adapter);

        viewPager.addOnPageChangeListener(new ViewPager.OnPageChangeListener() {
            @Override
            public void onPageScrolled(int position, float positionOffset,
int positionOffsetPixels) {

            }

            @Override
            public void onPageSelected(int position) {
```

```
                switch (position) {
                    case TAB_HOME:
                        homeBtn.setChecked(true);
                        break;
                    case TAB_NEAR:
                        nearBtn.setChecked(true);
                        break;
                    case TAB_ORDER:
                        orderBtn.setChecked(true);
                        break;
                    case TAB_MINE:
                        mineBtn.setChecked(true);
                        break;
                    default:
                        break;
                }
            }

            @Override
            public void onPageScrollStateChanged(int state) {

            }
        });
    }

    @Override
    public void onClick(View v) {
        switch (v.getId()){
            case R.id.home_tab:
                viewPager.setCurrentItem(TAB_HOME);
                break;
            case  R.id.near_tab:
                viewPager.setCurrentItem(TAB_NEAR);
                break;
            case R.id.order_tab:
                viewPager.setCurrentItem(TAB_ORDER);
                break;
            case R.id.mine_tab:
                viewPager.setCurrentItem(TAB_MINE);
                break;
        }
    }
}
```

运行效果如图 3-65 所示。用户可以滑动切换 Fragment 页面，也可以通过点击底部导航按钮实现页面切换，这也是普遍使用的一种导航风格。

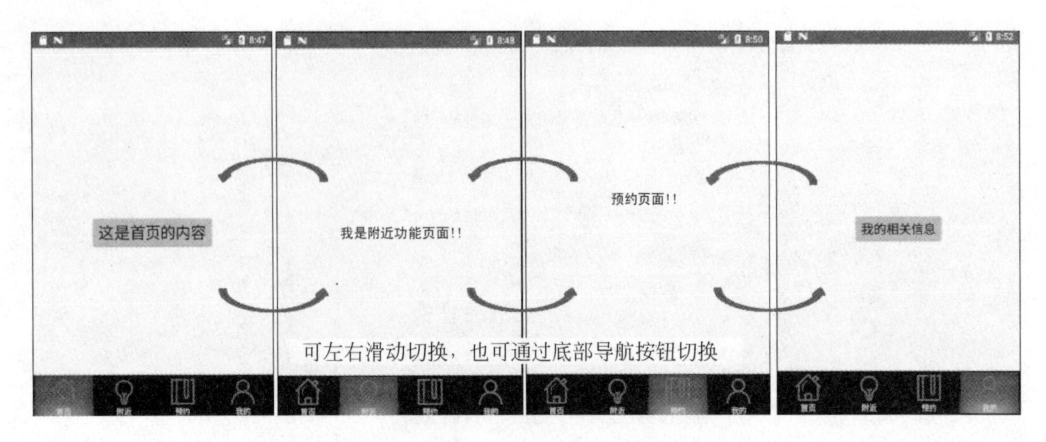

图 3-65 Fragment 运行效果

3.7 Adapter 及 AdapterView 的使用

在面向对象程序设计中接触过 MVC 模式，其实现原理：数据模型 M(Model)负责存放数据，通过控制器 C(Controller)将数据显示在相应的视图 V(View)上。在 Android 中也有这种类似于 MVC 框架的控件，它们不像前面介绍的那些控件一样拖曳到界面上就能用，而是需要通过适配器 Adapter 将数据添加到其上使用，这样的控件就是 AdapterView。AdapterView 相当于视图(V)，为了把数据(M)显示到视图上，需要通过适配器 Adapter 来充当控制器(C)。

3.7.1 常用 AdapterView

AdapterView 是一个抽象类，继承自 ViewGroup，其本质是个容器。AdapterView 派生出 AbsListView、AbsSpinner 和 AdapterViewAnimator 三个抽象类，在实际使用过程中，我们使用的是这三个抽象类的子类。AdapterView 组件作为一组非常重要的组件，主要以列表的形式展示多个具有相同格式的资源。常用的 AdapterView 有 AutoCompleteTextView(自动提示框)、Spinner(下拉列表)、ListView(列表)、GridView(网格图)等。

1. AutoCompleteTextView 控件

AutoCompleteTextView 类继承自 EditText 类，可以根据用户输入的文本弹出一个智能提示的下拉列表，这样便可以选择相应的选项。当输入字符串与事先为该控件定义的一组字符串集相关时，就会出现下拉选项供用户选择。

我们来看下面这个例子。

事先定义一组字符串集，这里用 String 数组表示。在实际应用中，数据可以来源于文件或数据库。

```
String[] contents = new String[]{"China","China1","china2","USA","USA1","USA2",
            "唱歌","china","cd","ch","chi","chin"};
```

页面布局很简单，无须过多说明，对应 Activity 的核心代码如下：

```
autoText = (AutoCompleteTextView) findViewById(R.id.autoCompleteTextView1);
ArrayAdapter<String> adapter1 = new ArrayAdapter<String> (this,
                    android.R.layout.simple_expandable_list_item_1, contents);
//将 adapter1 添加到 AutoCompleteTextView 中
autoText.setAdapter(adapter1);
autoText.setTextColor(Color.BLACK);
//设置输入 2 个字符后开始提示
autoText.setThreshold(2);
```

最终运行效果如图 3-66 所示，当输入 2 个字符后自动弹出输入提示框。

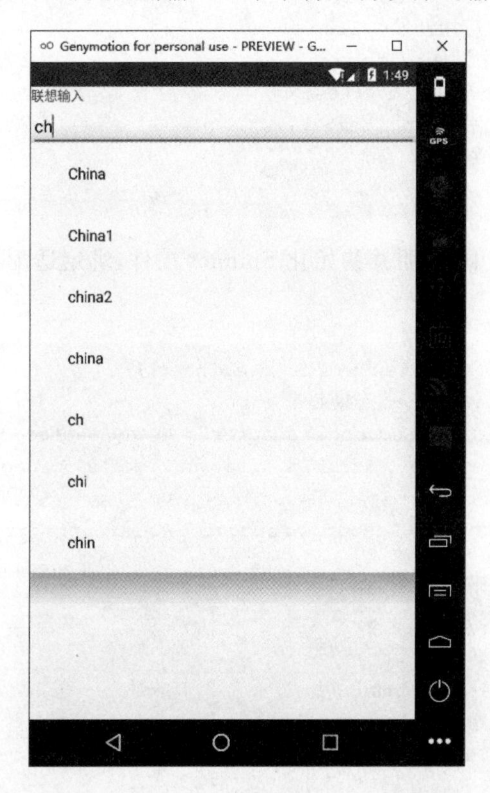

图 3-66 AutoCompleteTextView 案例效果

2. Spinner 控件

Spinner 是下拉列表控件，当用户点击控件时，下拉出选项列表供用户选择，Spinner 每次只显示用户选中的元素。

在 Spinner 的使用中，通常有以下两种方式为 Spinner 加载数据：

（1）在 res/values 文件夹下的 array.xml 文件中，事先定义好要加载的数据资源，然后使用 ArrayAdapter.createFromResource()把资源加载进来；

（2）直接在 Java 代码中使用 ArrayAdapter 对象，把 List<T>中的数据资源加载到 Spinner 中。

为了获取用户对 Spinner 下拉列表的选择，需要使用 setOnItemSelectedListener(onItemSelectedListener listener)方法为 Spinner 控件设置监听器，系统执行该方法时会传入一个实现了 Spinner.onItemSelectedListener 接口的匿名内部类对象，该匿名内部类实现了接口的 onItemSelected 方法，在该方法中可以根据传入的 position 参数匹配用户选择的下拉列表项。

通过下面这个案例学习 Spinner 控件的用法。

在 res/values/values.xml 文件中准备一个数组资源，内容如下：

```
<string-array name="corse_array">
    <item>Android 应用开发</item>
    <item>Java 程序设计</item>
    <item>Java Web 应用开发</item>
    <item>Html5 应用开发</item>
    <item>软件项目管理</item>
</string-array>
```

在对应的 Activity 中，声明并初始化 Spinner 控件，绑定适配器，实现事件监听，核心代码如下：

```
spinner = (Spinner) findViewById(R.id.spinner1);
//将可选内容与 ArrayAdapter 连接起来
final ArrayAdapter<CharSequence> adapter2 =ArrayAdapter.createFromResource(
this, R.array.corse_array, android.R.layout.simple_spinner_item);
//设置下拉列表的风格
adapter2.setDropDownViewResource (android.R.layout.simple_spinner_dropdown_
item);
//将 adapter2 添加到 spinner 中
spinner.setAdapter(adapter2);
//设置默认选中的是第一个
spinner.setSelection(0, true);
//添加事件 Spinner 事件监听
spinner.setOnItemSelectedListener(new AdapterView.OnItemSelectedListener() {
    @Override
    public void onItemSelected(AdapterView<?> parent, View view, int position,
long id) {
            String choice =getResources()
.getStringArray(R.array.corse_array)[position];
            Toast.makeText(AdapterViewActivity.this,
                            "你选的是"+choice, Toast.LENGTH_SHORT).show();
    }
```

```
    @Override
    public void onNothingSelected(AdapterView<?>parent) {
    }
});
```

最终运行效果如图3-67所示，用户点击Spinner控件，弹出下拉列表供用户选择。用户选择之后，事件监听如图3-68所示。

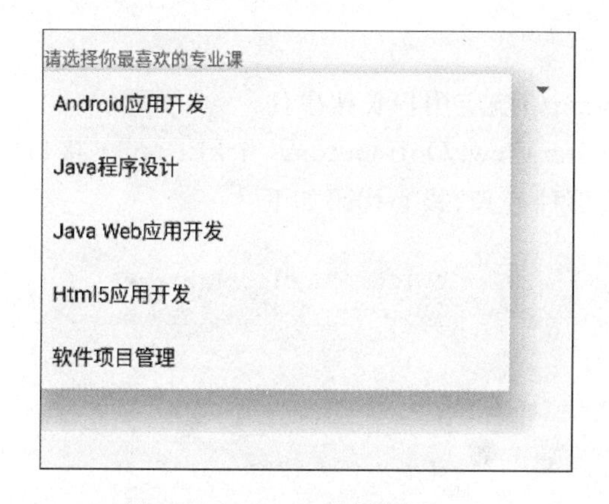

图3-67 Spinner案例效果(一)

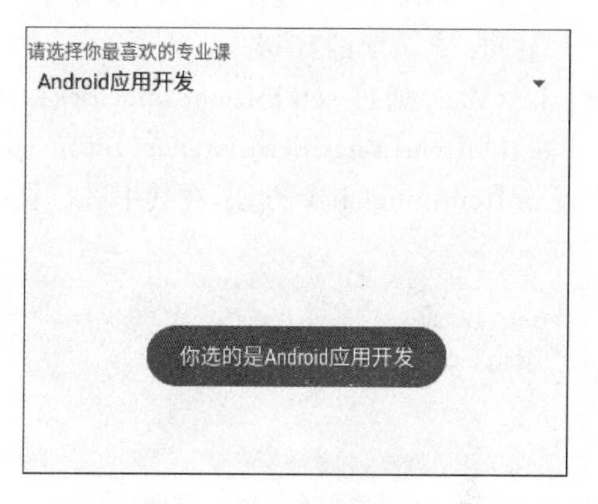

图3-68 Spinner案例效果(二)

3. ListView控件

ListView控件以列表形式展示内容，并根据数据的长度自适应显示内容。采用MVC模式将前端显示与后端数据分离，提供数据的List或数组相当于Model，ListView相当于视图View，Adapter对象相当于Control，将数据适配到ListView控件中。

在程序中使用ListView控件的步骤如下。

(1) 声明并初始化ListView控件。在程序中使用ListView显示数据时，可以在Activity布局中加入ListView控件，在Activity中进行声明并初始化；也可以直接让Activity继承ListActivity，继承ListActivity会把当前的整个Activity页面作为一个ListView。

(2) 构造Adapter对象，通过Adapter获取要显示的数据。

(3) 绑定Adapter，通过setAdapter()将ListView和Adapter绑定。

(4) 监听ListView列表的事件响应。在实际使用中，ListView一般常响应用户点击事件和长按事件。

通过调用ListView的setOnItemClickListener()方法，为ListView设置监听用户点击事件的监听器，该监听器需要实现AdapterView.OnItemClickListener接口，并且在onItemClick()方法中给出事件处理代码。

```
list.setOnItemClickListener(new AdapterView.OnItemClickListener(){…
public void onItemClick(AdapterView<?>parent, View view, int position, long id){
            .....
```

```
        }
    …})
```

相关参数说明：

① parent：发生点击动作的 ListView 对象；

② view：在 ListView 中被点击的 View；

③ position：点击项在 ListView 中的位置；

④ id：点击项的行 id。

ListView 通过 setOnItemLongClickListener()响应用户长按事件。

setOnItemLongClickListener()绑定 AdapterView. OnItemLongClickListener 接口，实现 onItemLongClick 方法，在其中进行长按事件处理，核心代码如下：

```
list.setOnLongItemClickListener(new AdapterView.OnItemLongClickListener(){…
public boolean onItemLongClick(AdapterView<?>parent, View view, int position,
long id){
            ….
    }
…}
)
```

相关参数说明：

① parent：发生点击动作的 ListView 对象；

② view：在 ListView 中被长按的 View；

③ position：被长按的列表项在 ListView 中的位置；

④ id：被长按的列表项的行 id。

3.7.2 Adapter

在使用 AdapterView 控件时，需要 Adapter 来填充数据。Adapter 对象在 AdapterView 控件和数据源之间扮演桥梁的角色，提供访问数据源的入口，将从数据源取出的数据项逐项加载到 AdapterView 控件中。在 Android 应用开发中主要使用 4 种 Adapter 对象：ArrayAdapter、SimpleAdapter、SimpleCursorAdapter 和自定义 Adapter (BaseAdapter)。

(1) ArrayAdapter：适用于列表项只含有文本信息的情况。

(2) SimpleAdapter：既可以处理列表项全是文本的情况，又可以处理列表项中包含其他控件(如图片、文本、按钮等)的情况。

(3) SimpleCursorAdapter：专门用来把一个 Cursor 中的数据映射到列表中，Cursor 中的每一条数据映射为列表中的一项，将在数据存储单元介绍它的具体用法。

(4) 自定义 Adapter：继承 BaseAdapter，根据 xml 文件中定义的样式进行列表项的填充，完全自定义数据适配方式，适用性最强。

1. ArrayAdapter

ArrayAdapter 类中定义有多个构造方法，这些构造方法如下：

```
public ArrayAdapter(Context context,int resource);
public ArrayAdapter(Context context,int resource,int textViewResourceId);
public ArrayAdapter(Context context,int resource,Object[] objects);
public ArrayAdapter(Context context,int resource,int textViewResourceId,Ojbect
[] objects);
public ArrayAdapter(Context context,int resource,List objects);
public ArrayAdapter(Context context, int resource, int textViewResourceId, List
objects);
```

针对 ArrayAdapter 构造方法中的参数说明如下：

(1) Context context：Context 上下文对象；

(2) int resource：对应列表项 item 的布局文件 ID；

(3) int textViewResourceId：列表项 item 布局中对应的 TextView 的 ID；

(4) Object[] object：需要显示数据的数组；

(5) List object：需要显示数据的集合。

接下来使用 ArrayAdapter 实现一个 ListView 页面，如文件清单 3-33 所示。

文件清单 3-33　ArrayAdapterListActivity.java

```
public class ArrayAdapterListActivity extends Activity {

    private ListView listView;
    private List<String>dataList;
    private ArrayAdapter<String>adapter;

    @Override
    protected void onCreate(Bundle savedInstanceState) {
        super.onCreate(savedInstanceState);
        setContentView(R.layout.array_listview_layout);
        listView = (ListView) findViewById(R.id.array_listview);
        dataList =new ArrayList<String> ();
        for (int i =0; i <=50; i++) {
            dataList.add("ListView 测试文本,第"+ (i+1) +"项");
        }
         adapter =new ArrayAdapter<String> (this, R.layout.array_adapter_item,
dataList);
        listView.setAdapter(adapter);

        listView.setOnItemClickListener(new AdapterView.OnItemClickListener() {
            @Override
            public void onItemClick(AdapterView<?>parent, View view,
int position, long id) {
```

```
                Toast.makeText(ArrayAdapterListActivity.this, "你点击的是"+
(position+1)+"项",Toast.LENGTH_SHORT).show();
            }
        });

        listView.setOnItemLongClickListener(new AdapterView.
OnItemLongClickListener() {
            @Override
            public boolean onItemLongClick(AdapterView<?>parent, View view,
int position, long id) {
                final int myPosition =position;
                new AlertDialog.Builder(ArrayAdapterListActivity.this)
                        .setIcon(R.mipmap.ic_launcher)
                        .setTitle("警告")
                        .setMessage("你确定要删除吗?")
                        .setPositiveButton("确定",
new DialogInterface.OnClickListener() {
                            @Override
                            public void onClick(DialogInterface dialog, int which) {
                                dataList.remove(myPosition);
                                adapter.notifyDataSetChanged();
                            }
                        })
                        .setNegativeButton("取消", null)
                        .create().show();
                return true;
            }
        });
    }
}
```

运行程序，当点击列表项时，触发 OnItemClickListener 事件，弹出 Toast 消息，如图 3-69 所示；当长按 ListView 列表项时，触发 OnItemLongClickListener 事件，弹出对话框，如图 3-70 所示。

2. SimpleAdapter

多数情况下，ListView 展示的内容不是单纯的文本，列表项中会有很多种组件，如 ImageView、Button、CheckBox 等。此时，ArrayAdapter 已不能满足我们的需求，可以通过 SimpleAdapter 来实现。

实现过程中，通过 SimpleAdapter 的构造器获取一个 Adapter 对象，其构造器如下：

```
SimpleAdapter(Context context, List<?extends Map<String, ?>>data, int resource,
String[] from, int[] to)
```

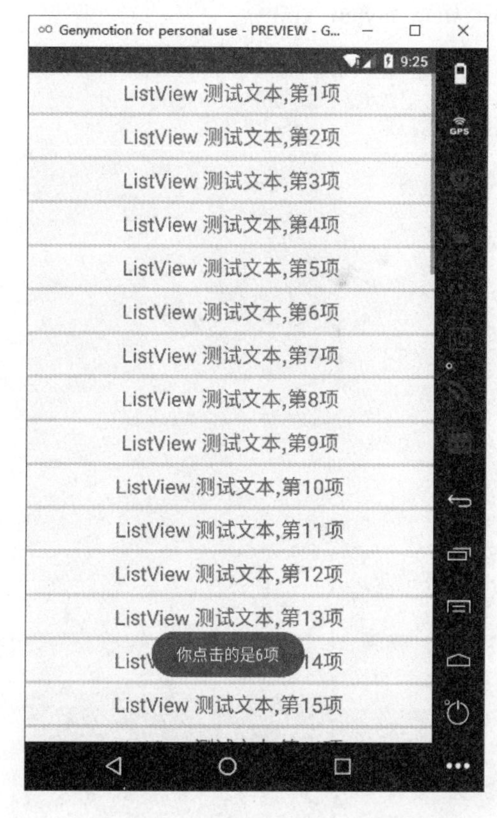

图 3-69 ListView 列表点击事件

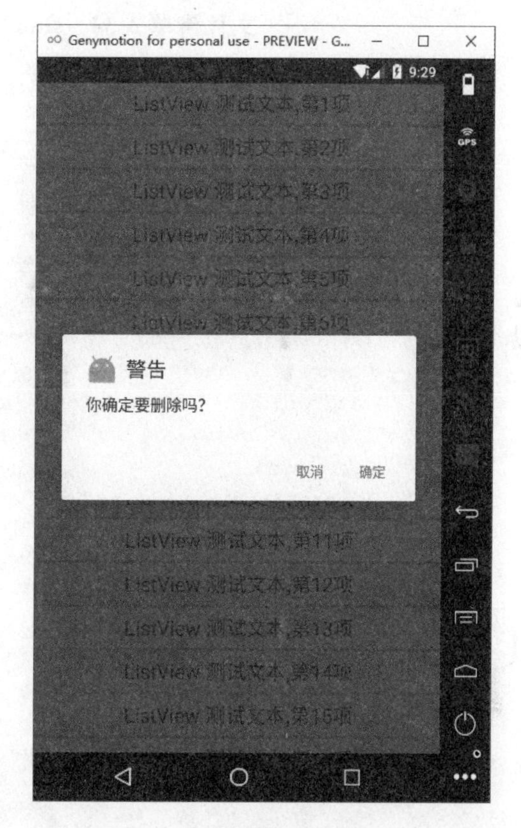

图 3-70 ListView 列表长按事件

相关参数说明：

(1) Contextcontext：Context 上下文对象，关联当前 SimpleAdapter 运行的上下文；

(2) List <? extends Map < String，? >>：data：数据集合，用于存储列表要显示的数据，data 中的每一项数据对应 ListView 中的每一项数据，每条项目的 key 要与 from 中指定内容一致；

(3) intresource：item 列表项布局文件的 id，这个布局控制列表项的显示，布局中必须包括 to 中定义的控件 id；

(4) 第 4 个参数 from：一个 String 数组，对应到 Map 上每一个< key，value >的 key 值；

(5) 第 5 个参数 to：是一个 int 数组，数组里面是 resource 自定义布局中各个控件的 id，需要与上面的 from 对应。

通过下面这个案例来讲解 SimpleAdapter 的具体用法。

与前面 ArrayAdapter 案例不同的是，我们需要设计一个如图 3-71 所示的列表项布局，列表项布局在/res/layout/中定义，名为 simple_list_item_layout. xml。

simple_list_item_layout. xml 文件内容如文件清单 3-34 所示。

图 3-71 ListView 列表项布局设计

文件清单 3-34 simple_list_item_layout. xml

```
<?xml version="1.0" encoding="utf-8"?>
<LinearLayout xmlns:android="http://schemas.android.com/apk/res/android"
    android:layout_width="match_parent"
    android:layout_height="match_parent"
    android:orientation="horizontal">

    <ImageView
        android:id="@+id/simple_image"
        android:layout_width="wrap_content"
        android:layout_height="wrap_content"
        android:src="@mipmap/ic_launcher" />

    <LinearLayout
        android:layout_width="wrap_content"
        android:layout_height="match_parent"
        android:layout_marginLeft="5dp"
        android:gravity="center_vertical"
        android:orientation="vertical" >

        <TextView
            android:id="@+id/simple_text"
            android:layout_width="wrap_content"
            android:layout_height="wrap_content"
            android:text="TextView"
            android:textSize="20sp" />
    </LinearLayout>

    <LinearLayout
        android:layout_width="match_parent"
        android:layout_height="match_parent"
        android:gravity="right" >

        <CheckBox
            android:id="@+id/simple_cbx"
            android:layout_width="wrap_content"
            android:layout_height="wrap_content"
            android:focusable="false"
            android:focusableInTouchMode="false" />

        <Button
            android:id="@+id/simple_btn"
            android:layout_width="wrap_content"
            android:layout_height="wrap_content"
            android:focusable="false"
            android:focusableInTouchMode="false"
            android:text="详情" />
    </LinearLayout>
</LinearLayout>
```

主类 Activity 的代码与前面类似，循环创建了一个 List < HashMap < String, Object >> 作为数据源，并初始化一个 SimpleAdapter 对象，绑定 ListView，监听 ListView 的点击事件，如文件清单 3-35 所示。

文件清单 3-35　SimpleAdapterListActivity. java

```
public class SimpleAdapterListActivity extends Activity {

    private List<HashMap<String, Object<<dataList;
    private SimpleAdapter adapter;
    private ListView simpleList;

    @Override
    protected void onCreate(Bundle savedInstanceState) {
        super.onCreate(savedInstanceState);
        setContentView(R.layout.array_listview_layout);

        dataList =new ArrayList<HashMap<String,Object<< ();
        HashMap<String, Object>map;
        for (int i =0; i <50; i++) {
            map =new HashMap<String, Object> ();
            map.put("img", R.mipmap.ic_launcher);
            map.put("text", "第"+ (i+1) +"个测试文本");
            map.put("cbx", "");
            map.put("btn", "详情");
            dataList.add(map);
        }
         adapter = new SimpleAdapter(this, dataList,R.layout.simple_list_item_
layout,
                new String[]{"img","text","cbx","btn"},new int[]{R.id.simple_
image,
R.id.simple_text,R.id.simple_cbx,R.id.simple_btn});

        simpleList = (ListView) findViewById(R.id.array_listview);
        simpleList.setAdapter(adapter);
        //列表点击事件
        simpleList.setOnItemClickListener(new AdapterView.OnItemClickListener() {
            @Override
            public void onItemClick(AdapterView<?>parent, View view,
int position, long id) {
                Toast.makeText(SimpleAdapterListActivity.this, "你点击的是"+
(position+1)+"项", Toast.LENGTH_SHORT).show();
            }
        });
    }
}
```

使用 SimpleAdapter 适配 ListView 的过程，实质就是使用 Map 的数据反复填充 XML 布局文件的各个控件的过程。上述代码运行效果如图 3-72 所示，完整程序可参考本书配套源代码。

3. BaseAdapter

BaseAdapter是一个抽象类,使用该类需要用户自己写一个适配器继承该类,并重载4个方法。这些方法分别是getView()、getCount()、getItem()和getItemId(),其中getView()最为重要,它是在每一次item从屏幕外滑进屏幕内时,或者程序刚开始创建第一屏item时,用来刷新它所在的ListView。与前面这些Adapter相比,BaseAdapter给开发人员提供了更大的自由。通过自定义Adapter的方式,我们可以自由控制列表的展示。

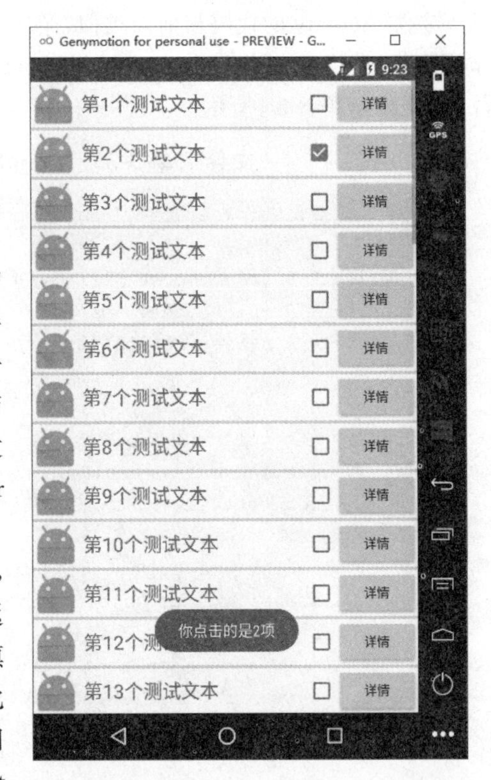

图3-72 SimpleAdapter案例运行效果

public abstract View getView (int position, View convertView, ViewGroup parent):返回列表项对应的视图,方法体中实例化视图填充器,用视图填充器,根据XML文件,实例化视图,根据布局找到控件,并设置属性,返回View视图。其中,position是指当前dataset的位置,通过getCount和getItem来使用。如果list向下滑动就是最底端的item的位置,如果向上滑动就是最上端的item的位置。convertView是指可以重用的视图,即刚刚出队的视图,parent则是显示数据的视图(如ListView、GridView等)。

int getCount():获得项目(Item)的数量。

Object getItem(int position):获得当前选项。

long getItemId(int position):获得当前选项的ID。

下面使用BaseAdapter实现一个类似QQ好友页面的列表。整体布局文件friend_list_layout.xml代码如文件清单3-36所示。

文件清单3-36 friend_list_layout.xml

```
<?xml version="1.0" encoding="utf-8"?>
<LinearLayout xmlns:android="http://schemas.android.com/apk/res/android"
    android:layout_width="match_parent"
    android:layout_height="match_parent"
    android:orientation="vertical">

    <LinearLayout
        android:layout_width="match_parent"
        android:layout_height="wrap_content"
        android:layout_marginTop="5dp"
```

```
        android:gravity="center_vertical|center_horizontal" >

        <ImageView
            android:id="@+id/imageView1"
            android:layout_width="wrap_content"
            android:layout_height="wrap_content"
            android:src="@mipmap/tangseng" />

        <TextView
            android:id="@+id/textView1"
            android:layout_width="wrap_content"
            android:layout_height="wrap_content"
            android:textSize="25sp"
            android:text="我的好友" />
    </LinearLayout>

    <ListView
        android:id="@+id/friendlist"
        android:layout_width="match_parent"
        android:layout_height="wrap_content"
        android:dividerHeight="3dp" >
    </ListView>
</LinearLayout>
```

页面效果如图 3-73 所示。

我们重点实现好友列表，好友列表中列表项的设计如图 3-74 所示。

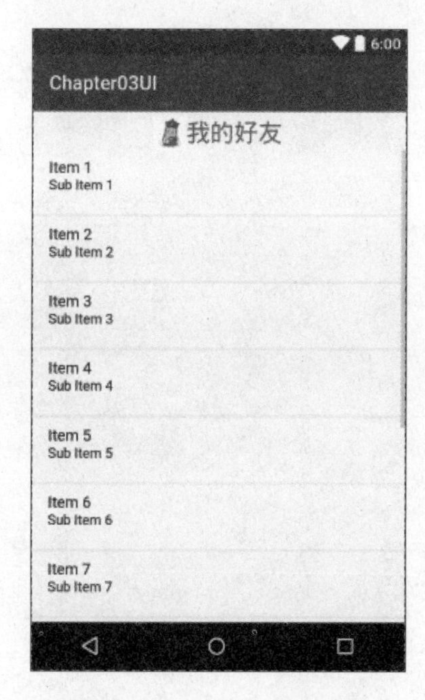

图 3-73 好友列表页面设计

图 3-74 列表项设计

对应的布局文件 friend_list_item_layout. xml 代码如文件清单 3-37 所示。

文件清单 3-37　friend_list_item_layout. xml

```
<?xml version="1.0" encoding="utf-8"?>
<LinearLayout xmlns:android="http://schemas.android.com/apk/res/android"
    android:layout_width="match_parent"
    android:layout_height="match_parent"
    android:orientation="horizontal">

    <ImageView
        android:id="@+id/iv_friend"
        android:layout_width="wrap_content"
        android:layout_height="match_parent"
        android:src="@mipmap/tangseng" />

    <LinearLayout
        android:layout_width="wrap_content"
        android:layout_height="match_parent"
        android:layout_marginLeft="5dp"
        android:orientation="vertical" >

        <TextView
            android:id="@+id/tv_friend_name"
            android:layout_width="wrap_content"
            android:layout_height="wrap_content"
            android:textSize="20sp"
            android:text="TextView" />

        <TextView
            android:id="@+id/tv_friend_msg"
            android:layout_width="wrap_content"
            android:layout_height="wrap_content"
            android:text="TextView" />
    </LinearLayout>

    <LinearLayout
        android:layout_width="match_parent"
        android:layout_height="match_parent"
        android:gravity="right" >

        <CheckBox
            android:id="@+id/cbx_friend"
            android:layout_width="wrap_content"
            android:layout_height="wrap_content"
            android:focusable="false"
            android:focusableInTouchMode="false"/>

        <Button
```

```
            android:id="@+id/btn_friend_detail"
            android:layout_width="wrap_content"
            android:layout_height="wrap_content"
            android:focusable="false"
            android:focusableInTouchMode="false"
            android:text="详情" />
    </LinearLayout>
</LinearLayout>
```

列表项中有按钮、复选框等组件，通常情况下这类组件会自动获取列表项的焦点，使得 ListView 列表项无法响应点击、长按等事件。可通过设置 android:focusable="false"，以及 android:focusableInTouchMode="false"来解决。

为了使 Button、CheckBox 等组件附带的事件能映射到 ListView 上，或者丰富每个 Item 的显示效果，如交替背景色等，我们需要在自定义适配器里操作。定义一个类，继承 BaseAdapter，重写 BaseAdapter 类的 4 个方法。

这里定义一个 MyAdapter 类，代码如文件清单 3-38 所示。

文件清单 3-38　MyAdapter.java

```
public class MyAdapter extends BaseAdapter{
    private Context context;                                //运行上下文
    private List<Map<String, Object<<listItems;             //好友信息集合
    private LayoutInflater listContainer;                   //视图容器
    //用于存储 CheckBox 选中状态
    public  Map<Integer,Boolean>cbxFlag =null;

//自定义控件集合
    public class ViewHolder{
        public ImageView image;
        public TextView name;
        public TextView msg;
        public CheckBox cbx;
        public Button detail;
    }

    public MyAdapter(Context context, List<Map<String, Object<<listItems) {
        this.context =context;
        listContainer =LayoutInflater.from(context);
        this.listItems =listItems;
        cbxFlag =new HashMap<Integer, Boolean>();
        init();
    }
//初始化所有 Checkbox 均为未选中状态
    private void init() {
        for (int i =0; i <listItems.size(); i++) {
            cbxFlag.put(i, false);
        }
```

```
    }

    @Override
    public int getCount() {
        return listItems.size();
    }

    @Override
    public Object getItem(int position) {
        return listItems.get(position);
    }

    @Override
    public long getItemId(int position) {
        return position;
    }

    public boolean hasChecked(int position) {
        return cbxFlag.get(position);
    }

    @Override
    public View getView(int position, View convertView, ViewGroup parent) {
        final int selectId =position;
        ViewHolder holder =null;

        if (convertView ==null) {
            holder =new ViewHolder();
            //获取 list_item 布局文件的视图
            convertView = listContainer. inflate (R. layout. friend _ list _ item _
layout, null);
            //获取控件对象
            holder.image = (ImageView) convertView.findViewById(R.id.iv_friend);
            holder.name = (TextView) convertView.findViewById(R.id.tv_friend_name);
            holder.msg = (TextView) convertView.findViewById(R.id.tv_friend_msg);
            holder.cbx = (CheckBox) convertView.findViewById(R.id.cbx_friend);
            holder.detail = (Button) convertView. findViewById (R. id. btn _ friend _
detail);
            //设置控件集到 convertView
            convertView.setTag(holder);
        }else {
            holder = (ViewHolder) convertView.getTag();
        }

        //设置颜色交替
        if (position% 2 ==0) {
            convertView.setBackgroundColor(Color.parseColor("#CAFFFF"));
        } else {
```

```
            convertView.setBackgroundColor(Color.parseColor("#B3FAFAFA"));
        }

        holder.image.setImageResource((Integer) listItems.get(position).get
("image"));
        holder.name.setText((String) listItems.get(position).get("name"));
        holder.msg.setText((String) listItems.get(position).get("msg"));
        //按钮点击事件
        holder.detail.setOnClickListener(new View.OnClickListener() {
            @Override
            public void onClick(View v) {
                new AlertDialog.Builder(context)
                        .setIcon((Integer) listItems.get(selectId).get("image"))
                        .setTitle((String) listItems.get(selectId).get("name"))
                        .setMessage((String) listItems.get(selectId).get("info"))
                        .setPositiveButton("确定", null)
                        .create()
                        .show();
            }
        });
        holder.cbx.setOnCheckedChangeListener(new
            CompoundButton.OnCheckedChangeListener() {
            @Override
            public void onCheckedChanged(CompoundButton buttonView,
boolean isChecked) {
                if (isChecked) {
                    cbxFlag.put(selectId, true);
                } else {
                    cbxFlag.put(selectId, false);
                }
            }
        });
        holder.cbx.setChecked(cbxFlag.get(selectId));
        return convertView;
    }
}
```

在 MyAdapter 中，定义了一个内部类 ViewHolder 来存放列表项里的 View 对象，使用 ViewHolder 和 convertView 来对列表进行优化。当第一次创建 convertView 对象时，把这些 ViewHolder 里的 view 找出来。然后用 convertView 的 setTag 方法将 viewHolder 设置到 Tag 中，当以后重用 convertView 时，只需从 convertView 中用 getTag 方法将 view 取出来就可以了。

接下来就是好友列表 Activity 的实现了。为了简化，我们提供一些静态数据，代码如文件清单 3-39 所示。

文件清单 3-39 FriendListActivity. java

```
public class FriendListActivity extends Activity{

    private ListView listView;
    private MyAdapter listaAdapter;
    private List<Map<String, Object<<listItems;

    private Integer[] imgeIDs ={R.mipmap.baigujing,R.mipmap.sunwukong,
            R.mipmap.zhubajie,R.mipmap.shaseng,R.mipmap.guanyin,
R.mipmap.baigujing,R.mipmap.baigujing,R.mipmap.baigujing,
R.mipmap.baigujing,R.mipmap.baigujing,R.mipmap.baigujing};

    private String[] friendNames ={"女儿国国王", "孙悟空","猪八戒", "沙僧", "观音姐姐",
"白骨精","东海龙女","龙套 1","龙套 2","龙套 3","龙套 4"};

    private String[] msgs ={"御弟哥哥,你好哇","师傅,小心妖怪","师傅,咱歇一歇",
"师傅,请喝水","你太墨迹了","吃你的肉可以长生?","你想干啥?",
"What are you 弄啥呢?","What are you 弄啥呢?",
"What are you 弄啥呢?","What are you 弄啥呢?"};

    private String[] infos ={"相见难别亦难,怎诉这胸中语万千","孙悟空是我的大徒弟",
"猪八戒就是一头懒猪!","沙悟净是个听话的好徒儿!",
"救苦救难观世音菩萨","白骨精,一直想吃我的肉,可惜你没后台!",
"小龙女,你在干啥?","你想咋滴?","你想咋滴?",
"你想咋滴?","你想咋滴?"};

    @Override
    protected void onCreate(Bundle savedInstanceState) {
        super.onCreate(savedInstanceState);
        setContentView(R.layout.friend_list_layout);

        listView =(ListView) findViewById(R.id.friendlist);
        listItems =getFriendItems();
        listaAdapter =new MyAdapter(this, listItems);
        listView.setAdapter(listaAdapter);
        //列表点击事件
        listView.setOnItemClickListener(new AdapterView.OnItemClickListener() {
            @Override
            public void onItemClick(AdapterView<?>parent, View view, int position,
long id) {
                if (listaAdapter.hasChecked(position)) {
                    Toast.makeText(FriendListActivity.this, "你想和"
                          +listItems.get(position).get("name")+"聊天哇!",
                          Toast.LENGTH_SHORT).show();
                }else {
                    Toast.makeText(FriendListActivity.this, "你点我干嘛?",
Toast.LENGTH_SHORT).show();
                }
```

```
            }
        });
    }

    private List<Map<String,Object<<getFriendItems() {
        List<Map<String, Object<<listItems =new ArrayList<Map<String, Object<<();
        for(int i =0; i <friendNames.length; i++) {
            Map<String, Object>map =new HashMap<String, Object>();
            map.put("image", imgeIDs[i]);                //图片资源
            map.put("name", friendNames[i]);             //好友名称
            map.put("msg", msgs[i]);                     //最新消息
            map.put("info", infos[i]);
            listItems.add(map);
        }
        return listItems;
    }
}
```

方法 getFriendItems 获取好友列表数据信息，返回的是一个 List < Map >数据，该数据作为 ListView 的数据源，通过 MyAdapter 将数据适配到 ListView 中。好友列表可响应用户的点击事件，如图 3-75 所示；也可以处理列表项里的复选框、按钮的相应事件，如图 3-76 所示。

图 3-75 好友列表点击事件

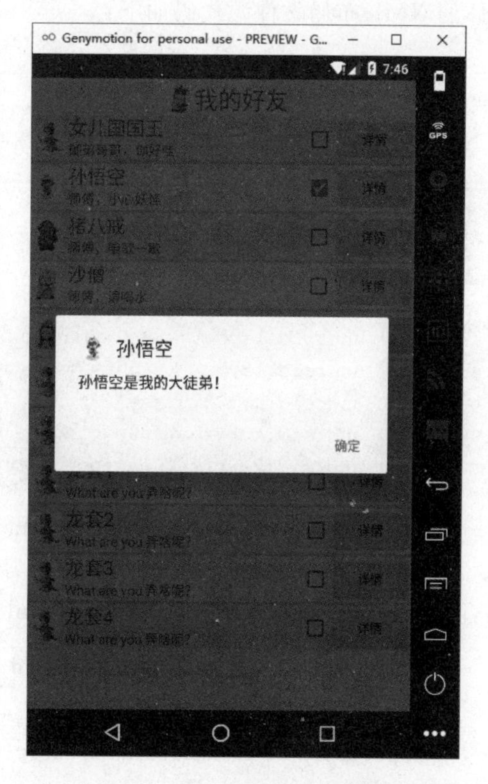

图 3-76 列表按钮点击事件

3.7.3 GridView 控件

ListView 适用列表是单列多行的形式，如果列表是多行多列的网状形式，则优先使用 GridView。GridView 以类似矩阵的方式来排列视图，其核心属性如表 3-6 所示。

表 3-6 GridView 属性说明

属 性	功 能 说 明
android:numColumns	设置列数，auto_fit 将列数设置为自动
android:columnWidth	设置每列的宽度，即 Item 的宽度
android:gravity	设置每个网格的比重位置。可选的值有 top、bottom、left、right、center_vertical、fill_vertical、center_horizontal、fill_horizontal、center、fill、clip_vertical，可以多选，用"|"分开
android:stretchMode	缩放模式，设置列应该以何种方式填充可用空间
android:horizontalSpacing	网格之间列的默认水平距离
android:verticalSpacing	设置网格之间行的默认垂直距离

接下来使用 GridView 实现一个人物卡牌的页面。在 XML 布局文件中，使用 LinearLayout 对整个界面进行垂直布局，然后在该布局中添加一个 GridView 控件即可。具体的 XML 布局文件源代码如下：

```
<?xml version="1.0" encoding="utf-8"?>
<LinearLayout xmlns:android="http://schemas.android.com/apk/res/android"
    android:layout_width="match_parent"
    android:layout_height="match_parent"
    android:orientation="vertical">

    <GridView
        android:id="@+id/grid_view"
        android:layout_width="match_parent"
        android:layout_height="wrap_content"
        android:numColumns="3"
        android:horizontalSpacing="10dp"
        android:verticalSpacing="10dp"/>
</LinearLayout>
```

人物卡牌页面中，每个人物卡牌的内容包括图像、名称和价格三部分，采用相对布局，网格项布局 grid_view_item_layout. xml 文件源代码如文件清单 3-40 所示。

文件清单 3-40 grid_view_item_layout. xml

```
<?xml version="1.0" encoding="utf-8"?>
<RelativeLayout xmlns:android="http://schemas.android.com/apk/res/android"
```

```
    android:layout_width="match_parent"
    android:layout_height="match_parent">

    <ImageView
        android:id="@+id/gridItemImage"
        android:layout_height="wrap_content"
        android:layout_width="wrap_content"
        android:layout_centerHorizontal="true">
    </ImageView>

    <TextView
        android:id="@+id/gridItemText"
        android:layout_width="wrap_content"
        android:layout_below="@+id/gridItemImage"
        android:layout_height="wrap_content"
        android:text="TextView01"
        android:layout_centerHorizontal="true">
    </TextView>

    <TextView
        android:id="@+id/gridItemPriceText"
        android:layout_width="wrap_content"
        android:layout_below="@+id/gridItemText"
        android:layout_height="wrap_content"
        android:text="TextView01"
        android:layout_centerHorizontal="true">
    </TextView>
</RelativeLayout>
```

在网格控件 GridView 中，常用的事件监听器有两个：OnItemSelectedListener 和 OnItemClickListener。其中，OnItemSelectedListener 用于 GridView 中项目选择事件监听，OnItemClickListener 用于项目点击事件监听。

要实现这两个事件监听很简单，实现 OnItemSelectedListener 和 OnItemClickListener 接口，并实现其抽象方法即可。其中，需要实现的 OnItemClickListener 接口的抽象方法如下：

```
public void onItemClick(AdapterView<?>parent, View view, int position, long id);
```

需要实现的 OnItemSelectedListener 接口的抽象方法有两个，分别如下：

```
public void onItemSelected(AdapterView<?>parent, View view, int position, long id);
public void onNothingSelected(AdapterView<?>parent);
```

人物卡牌页面 Activity 核心代码如文件清单 3-41 所示。

文件清单 3-41 GridViewActivity. java

```
public class GridViewActivity extends Activity {
    private GridView imageGridView;

    @Override
    protected void onCreate(Bundle savedInstanceState) {
        super.onCreate(savedInstanceState);
        setContentView(R.layout.grid_view_layout);

        imageGridView = (GridView) findViewById(R.id.grid_view);
        final List<HashMap<String, Object<<mapsList = new
ArrayList<HashMap<String, Object<< ();
        for (int i = 0; i < 60; i++) {
            HashMap<String, Object>hashMap = new HashMap<String, Object> ();
            hashMap.put("Image", R.mipmap.guanyin);
            hashMap.put("text", "人物卡牌"+i);
            hashMap.put("pro", "售价"+i);
            mapsList.add(hashMap);
        }
        SimpleAdapter adapter = new SimpleAdapter(getApplicationContext(),
            mapsList, R.layout.grid_view_item_layout,
new String[] { "Image", "text","pro" },
new int[] {R.id.gridItemImage , R.id.gridItemText , R.id.gridItemPriceText});
        imageGridView.setAdapter(adapter);

        imageGridView.setOnItemClickListener(new AdapterView.OnItemClickListener() {
            @Override
            public void onItemClick(AdapterView<?>parent, View view,
int position, long id) {
                Toast.makeText(GridViewActivity.this,"你选的是"+
mapsList.get(position).get("text"),Toast.LENGTH_SHORT)
.show();
            }
        });
    }
}
```

本例模拟了 60 张卡牌，采用 GridView 进行网格排列，并针对网格项添加了点击事件，页面运行效果如图 3-77 所示。

这里使用的是 SimpleAdapter，感兴趣的读者可用 BaseAdapter 实现同样的效果。

图 3-77 GridView 运行效果

本章小结

本章主要介绍了 Android UI 开发的相关知识。首先介绍了 Android UI 布局和常用控件，应重点掌握 View 与 ViewGroup 的功能、六大布局的特点，以及常规控件的用法。其次介绍了对话框与 Toast 等常用的 UI 交互，重点应掌握 AlertDialog 的用法、自定义 Dialog 的用法，以及常规的 Toast 用法。再次介绍了菜单与导航栏等界面交互的内容，重点应掌握菜单、ToolBar、ViewPager 与 Fragment 的用法，掌握这些主流的界面交互原理，有助于提高界面的友好与美观程度。最后讲解了 Adapter 与 AdapterView 的使用，重点应掌握自定义 Adapter 的 ListView 与 GridView 的用法特点，能做到灵活运用。

Android 应用界面作为手机 APP 的脸面，界面是否友好，是否美观，直接关系到应用程序能否获得用户的青睐与认可。通过本章的学习，希望大家能够对 Android 应用的 UI 界面开发有深入的理解，能够掌握常规的交互设计，设计并实现精美的 Android UI 界面交互功能。

习　题

1. 若想为输入框 EditText 添加提示文本信息,需要设置的属性是(　　)。

A. android:text　　B. android:hint

C. android:autoText　　D. android:freezesText

2. 以下属性中,可以定位"在指定控件的右边"的是(　　)。

A. android:layout_alignLeft　　B. android:layout_toRightOf

C. android:layout_left　　D. android:layout_alignRight

3. 将一个 TextView 的 android:layout_width 属性设置为 wrap_content 时,该组件将呈现的效果是(　　),设置为 match_parent 时,该组件将呈现的效果是(　　)。

A. 该文本域的宽度将填充父容器的宽度

B. 该文本域的高度将填充父容器的高度

C. 该文本域的宽度仅占据该组件的实际宽度

D. 该文本域的高度仅占据该组件的实际高度

4. 对于 AlertDialog 的描述不正确的是(　　)。

A. 使用 new 关键字创建 AlertDialog 的实例

B. 对话框的显示需要调用 show 方法

C. setPositiveButton 方法用来添加"确定"按钮

D. setNegativeButton 方法用来添加"取消"按钮

5. 下列关于 ListView 使用的描述中,错误的是(　　)。

A. 要使用 ListView,需要为该 ListView 使用 Adapter 方式传递数据

B. 要使用 ListView,该布局文件对应的 Activity 必须继承 ListActivity

C. ListView 中每一项的视图布局既可以使用内置的布局,也可以使用自定义的布局

D. ListView 中每一项被触摸时,将会触发 ListView 对象的 ItemClick 事件

6. 下列关于 BaseAdapter 适配器说法错误的是(　　)。

A. getView 方法只会调用一次

B. getCount 方法返回 ListView 的数据量

C. 可以利用 ConvertView 实现 ListView 优化

D. 可以在 getView 方法中完成控件的事件监听

7. ListView 中常用的适配器有________、________、________和________。

8. 线性布局主要有两种形式,当 android:orientation="horizontal"时,表示该布局是________,当 android:orientation="vertical"时,表示该布局是________。

9. 简要描述 Android 的常用布局及使用要点。

10. 如何优化 ListView?

11. 编写程序,仿照微信应用,模拟数据实现"微信"页面。

第4章 chapter 4

Android数据存储技术

主要内容：SharedPreferences，File操作，SQLite，ContentProvider
课　　时：10课时
知识目标：(1) 了解Android存储数据方式及其特点；
(2) 掌握SharedPreferences数据存储方式；
(3) 掌握File数据存储方式；
(4) 掌握SQLite数据存储方式；
(5) 掌握ContentProvider的使用。
能力目标：(1) 具备Android数据存储开发能力；
(2) 具备Android多线程开发的能力。

Android为开发者提供多种数据存储方式，开发者在实际开发过程中选择哪种方式依赖于特定需求，需要综合考虑数据存储的类型、需要空间大小、是否需要提供给其他应用程序使用等多方面因素。本章将对Android提供的几种数据存储方式进行详细讲解，帮助读者弄清这些存储方式的适用场景。

创建Android项目Chapter04Application，在该项目中完成本章的示例代码。创建的Chapter04Application项目结构如图4-1所示。

其中，MainActivity.java作为向导页面，以列表的形式展现当前应用的各个部分，当用户点击列表项时跳转至具体列表项代表的模块，MainActivity呈现样式如图4-2所示。

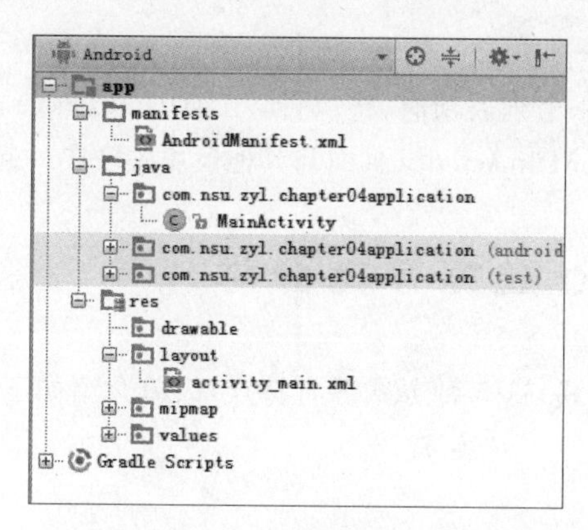

图4-1　Chapter04Application程序结构

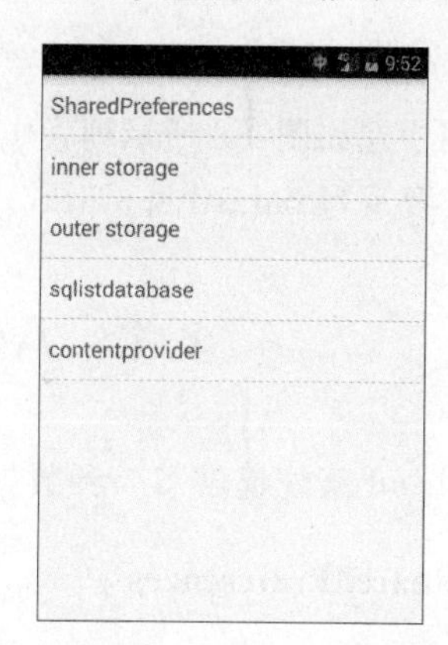

图4-2　MainActivity页面

我们让 MainActivity 继承 ListActivity。ListActivity 是一个专门显示列表的 Activity 类，它内置了 ListView 对象，只要设置了数据源，数据项就会自动地在列表中显示出来。ListActivity 和普通的 Activity 没有太大的差别，不同之处就是不需要使用 setConenetView()设置显示的布局样式，页面将以列表的形式显示，列表项内容通过调用 setAdapter 设置。

MainActivity.java 的源代码如文件清单 4-1 所示。

文件清单 4-1 MainActivity.java

```
package cn.edu.nsu.zyl.chapter04application;
import android.app.ListActivity;
import android.content.Intent;
import android.support.v7.app.AppCompatActivity;
import android.os.Bundle;
import android.view.View;
import android.widget.ArrayAdapter;
import android.widget.ListView;
import java.util.ArrayList;
public class MainActivity extends ListActivity {
    @Override
    protected void onCreate(Bundle savedInstanceState) {
        super.onCreate(savedInstanceState);
        ArrayList<String>indexList=new ArrayList<String>();
        indexList.add("SharedPreferences");
        indexList.add("inner storage");
        indexList.add("outer storage");
        indexList.add("sqlistdatabase");
        indexList.add("contentprovider")
        ArrayAdapter<String> adapter =new ArrayAdapter<String>(IndexActivity.
this, android.R.layout.simple_list_item_1, indexList);
        setListAdapter(adapter);
    }

    @Override
    protected void onListItemClick(ListView l, View v, int position, long id) {
        super.onListItemClick(l, v, position, id);
}
```

当用户在如图 4-2 所示列表中点击某一个列表项时，将执行 onListItemClick()方法，因此后续需要重写 onListItemClick()方法，将 MainActivity 页面与其他模块关联在一起。

4.1 Android 数据存储分类

Android 系统提供了 5 种数据存储方式，这 5 种方式各有特点，简单介绍如下。

1. SharedPreferences

SharedPreferences 以键值对的形式保存 int、String、boolean、float、long 等类型的数据。

在 Android 程序中，SharedPreferences 对象主要用于保存用户偏好设置、配置、状态等存储空间要求小的简单类型的数据。

2. File

Android 系统支持将数据以文件的形式保存在手机的内部和外部存储介质中。在程序中使用 I/O 流对文件进行读写操作。在 Android 程序中，文件系统主要用于存储图片、视频、音频等对存储空间要求大的数据。

3. SQLite

Android 系统自带支持基本的 SQL 语法的嵌入式数据库 SQLite，SQLite 是一种关系型数据库管理系统。在 Android 程序中，SQLite 主要用于存储批量的结构化数据。

4. 网络存储

Android 支持将数据通过网络存储在远程服务器上，在程序中可以通过网络在云端保存和获取数据。在 Android 程序中，可以使用网络存储方式将程序中的数据保存到远程云端。

5. ContentProvider

ContentProvider 是 Android 提供的一种常用的数据共享方式。由于数据在各个应用程序间通常是私密的，在程序中可以利用 ContentProvider 获取和保存其他应用程序暴露的数据。

4.2 SharedPreferences

SharedPreferences 是 Android 系统提供的一个轻量级数据存储方式，主要用于保存以简单数据类型（long、int、String、float、boolean）形式存在的状态、配置、用户偏好等信息。存储在 SharedPreferences 中的数据最终以键值对的形式保存在/data/data/<package-name>/shared_prefs 文件夹下的 XML 文件中。

4.2.1 获得 SharedPreferences 对象

由于 SharedPreferences 是一个接口，无法直接实例化，在实际开发过程中需要根据实际情况选择使用如下三种方式的某一种来获取对应的 SharedPreferences 对象。

（1）Context.getSharedPreferences(String name,int mode)：获取指定文件名对应的 SharedPreferences 对象，参数 name 是当前 SharedPreferences 对应的 XML 文件名，参数 mode 指定对该 SharedPreferences 的访问模式。

（2）Activity.getPreferences(int mode)：获取当前 Activity 对应的 SharedPreferences 对象，该对象对应的 XML 文件名称和 Activity 同名，参数 mode 指定对该

SharedPreferences 的访问模式。

(3) PreferenceManager. getDefaultSharedPreferences(Context):使用该方法可以获取每个应用都有的默认的配置文件 preferences. xml 对应的对象。

SharedPreferences 提供了一系列用于获取和存储数据的常用方法,如表 4-1 所示。

表 4-1　SharedPreferences 常用方法

方　法	描　述
boolean contains(String key)	判断是否包含某个配置信息
SharedPreferences. Editor edit()	获得对 Preferences 编辑的 Editor 对象
Map < String,? > getAll()	获取所有配置信息
boolean getBoolean(String key,Boolean defValue)	从配置文件中获取 boolean 类型的值
float getFloat(String key,float defValue)	从配置文件中获取 float 类型的值
Int getInt(String key,int defValue)	从配置文件中获取 int 类型的值
long geLong(String key,long defValue)	从配置文件中获取 long 类型的数据
String getString(String key,String defValue)	从配置文件中获取 String 类型的数据
Set < String > getStringSet (String key, Set < String > defValue)	从配置文件中获取 String 集合数据

SharedPreferences 以键值对的形式保存数据,因此可以利用 SharedPreferences 提供的 contains(String key)方法判断是否包括某个 key 对应的数据。方法 getXxx(String key,Xxx defalutValue)用于获取指定类型的数据,当不存在指定的 key 的数据时,返回第二个参数作为默认值。

4.2.2 SharedPreferences. Editor

SharedPreferences 只提供获取数据的方法,但不支持直接存储和修改数据。存储和修改数据需要使用 SharedPreferences. Editor 对象,SharedPreferences. Editor 对象通过调用 SharedPreferences 的 edit()方法得到。

SharedPreferences. Editor 提供了一系列用于向 SharedPreferences 存储和修改数据的方法,如表 4-2 所示。

表 4-2　SharedPreferences. Editor 常用方法

方　法	描　述
SharedPreferences. Editor clear()	清空数据
boolean commit()	提交编辑后的数据到 SharedPreferences 中
SharedPreferences. Editor putBoolean(String key, Boolean value)	向 SharedPreferences 存储指定 key 对应的 boolean 值
SharedPreferences. Editor putFloat (String key, float value)	向 SharedPreferences 存储指定 key 对应的 float 值

续表

方　法	描　述
SharedPreferences. Editor putInt(String key, int value)	向 SharedPreferences 存储指定 key 对应的 int 值
SharedPreferences. Editor putLong(String key, long value)	向 SharedPreferences 存储指定 key 对应的 long 值
SharedPreferences. Editor putString(String key, String value)	向 SharedPreferences 存储指定 key 对应的 String 值
SharedPreferences. Editor putStringSet(String key, Set<String> values)	向 SharedPreferences 存储指定 key 对应的 Set 类型的值
SharedPreferences. Editor remove(String key)	在 Editor 标记删除某个配置数据，当执行 commit()时，执行删除操作

SharedPreferences. Editor 提供 putXxx(String key, Xxx value)方法存储指定 key 对应的数据到 SharedPreferences 中；提供 remove(String key)方法删除 SharedPreferences 中保存的指定 key 对应的数据；提供 clear()方法用于删除 SharedPreferences 中保存的所有数据。存入数据和删除数据时，一定要在最后调用 commit()方法提交数据，否则存入和删除数据操作将不生效。

4.2.3 利用 SharedPreferences 读写数据

获取 SharedPreferences 中数据时，首先需要创建指定文件的 SharedPreferences 对象，然后根据 key 值获取存储的数据。具体步骤如下：

(1) 获取指定数据存储文件对应的 SharedPreferences 对象；

(2) 根据 key 值调用 getXxx(String key,defaultValue)获取存储的数据。

从 SharedPreferences 中获取数据的关键代码如下：

```
//获得私有类型的 SharedPreferences 对象
SharedPreferences sp=getSharedPreferences("data",MODE_PRIVATE);
//读取 SharedPreferences 中 key 为"name"对应的值,若值不存在返回""字符串
String name=sp.getString("name", "");
//读取 SharedPreferences 中 key 为"age"对应的值,若不存在返回 0
int age=sp.getInt("age", 0);
```

由于 SharedPreferences 对象本身不支持数据的存储和修改，当需要 SharedPreferences 存储数据时，首先需要创建指定文件的 SharedPreferences 对象，然后获取对应的 Editor 对象，接着调用 Editor 对象的 putXxx(Sting key, Xxx value)方法存储数据，最后调用 Editor 对象的 commit()方法提交数据。具体步骤如下：

(1) 获取指定数据存储文件对应的 SharedPreferences 对象；

(2) 调用 SharedPreferences 对象的 edit()方法获取 Editor 对象；

(3) 通过 Editor 对象存储 key-value 键值对数据；

(4) 通过 commit()方法提交数据。

利用 SharedPreferences 存储数据的关键代码如下：

```
SharedPreferences sp=getSharedPreferences("data",MODE_PRIVATE);
SharedPreferences.Editor editor=sp.edit();      //获取编辑器
editor.putString("name", "张三丰");              //存入 String 类型的数据
editor.putInt("age", 21);                       //存入 int 类型的数据
editor.commit();                                //提交修改
```

从 SharedPreferences 删除数据操作与存储数据操作类似，都需要先获取 Editor 对象，然后通过 Editor 对象删除/添加数据，最后提交。

删除 SharedPreferences 中存储数据的关键代码如下：

```
SharedPreferences sp=getSharedPreferences("data",MODE_PRIVATE);
SharedPreferences.Editor editor1=sp.edit();
editor.remove("name");
editor.clear();
editor.commit();
```

4.2.4 案例

下面通过程序展示在 Android 项目中使用 SharedPreferences 来存储和获取数据。在项目 Chapter04Application 中新建 SharedPreferencesSaveActivity 和 SharedPreferencesGetActivity，在 SharedPreferencesSaveActivity 中使用 SharedPreferences 保存用户输入的学号和姓名，并在 SharedPreferencesGetActivity 中获取保存在 SharedPreferences 中的数据并且显示。

编辑 SharedPreferencesSaveActivity 对应的布局文件 activity_shared_preferences_save.xml，在该布局文件中添加文本框和按钮等控件，并设置相关属性，完整的 activity_shared_preferences_save.xml 布局文件的代码如文件清单 4-2 所示。

文件清单 4-2 activity_shared_preferences_save.xml

```
<?xml version="1.0" encoding="utf-8"?>
<RelativeLayout xmlns:android="http://schemas.android.com/apk/res/android"
    xmlns:tools="http://schemas.android.com/tools"
    android:layout_width="match_parent"
    android:layout_height="match_parent"
    android:paddingBottom="@dimen/activity_vertical_margin"
    android:paddingLeft="@dimen/activity_horizontal_margin"
    android:paddingRight="@dimen/activity_horizontal_margin"
    android:paddingTop="@dimen/activity_vertical_margin"
    tools:context="com.nsu.zyl.
chapter04application.SharedPreferencesSaveActivity">
    <TextView
        android:layout_width="wrap_content"
```

```
        android:layout_height="wrap_content"
        android:text="学号："
        android:id="@+id/txtStuId"
        android:layout_alignParentTop="true"
        android:layout_alignParentStart="true"
        android:layout_marginTop="75dp" />

    <TextView
        android:layout_width="wrap_content"
        android:layout_height="wrap_content"
        android:text="姓名："
        android:id="@+id/txtStuName"
        android:layout_below="@+id/txtStuId"
        android:layout_alignParentStart="true"
        android:layout_marginTop="42dp" />

    <EditText
        android:layout_width="match_parent"
        android:layout_height="wrap_content"
        android:id="@+id/edtStuId"
        android:layout_alignTop="@+id/txtStuId"
        android:layout_toEndOf="@+id/txtStuId" />

    <EditText
        android:layout_width="match_parent"
        android:layout_height="wrap_content"
        android:id="@+id/edtStuName"
        android:layout_below="@+id/edtStuId"
        android:layout_alignStart="@+id/edtStuId" />

    <Button
        android:layout_width="match_parent"
        android:layout_height="wrap_content"
        android:text="登录"
        android:id="@+id/btnLogin"
        android:layout_centerVertical="true"
        android:layout_centerHorizontal="true" />
</RelativeLayout>
```

在 SharedPreferencesSaveActivity 类中重写 onCreate()方法，为“登录”按钮设置监听器，监听到用户的点击操作。在监听器处理方法中首先获得用户输入的用户名和密码，然后将获取的数据保存到 SharedPreferences 对象中，最后使用 Intent 跳转至 SharedPreferencesGetActivity。

SharedPreferencesSaveActivity 类的代码如文件清单 4-3 所示。

文件清单 4-3　SharedPreferencesSaveActivity. java

```
public class SharedPreferencesSaveActivity extends AppCompatActivity {
    private Button btnLogin;
    private EditText edtStuId,edtStuName;

    @Override
    protected void onCreate(Bundle savedInstanceState) {
        super.onCreate(savedInstanceState);
        setContentView(R.layout.activity_shared_preferences_save);
        btnLogin=(Button) findViewById(R.id.btnLogin);
        edtStuId=(EditText)findViewById(R.id.edtStuId);
        edtStuName=(EditText)findViewById(R.id.edtStuName);
        btnLogin.setOnClickListener(new View.OnClickListener() {
            @Override
            public void onClick(View v) {
                String strStuId=edtStuId.getText().toString();
                String strStuName=edtStuName.getText().toString();
                SharedPreferences sharedPreferences=getSharedPreferences("data",
Context.MODE_PRIVATE);
                SharedPreferences.Editor editor=sharedPreferences.edit();
                editor.putString("stuId",strStuId);
                editor.putString("stuName",strStuName);
                editor.commit();
                Intent intent = new Intent (SharedPreferencesSaveActivity.this,
SharedPreferencesGetActivity.class);
                startActivity(intent);
            }
        });
    }
}
```

SharedPreferencesGetActivity 对应的布局文件 activity_shared_preferences_get. xml 代码如文件清单 4-4 所示。

文件清单 4-4　activity_shared_preferences_get. xml

```
<?xml version="1.0" encoding="utf-8"?>
<RelativeLayout xmlns:android="http://schemas.android.com/apk/res/android"
    xmlns:tools="http://schemas.android.com/tools"
    android:layout_width="match_parent"
    android:layout_height="match_parent"
    android:paddingBottom="@dimen/activity_vertical_margin"
    android:paddingLeft="@dimen/activity_horizontal_margin"
    android:paddingRight="@dimen/activity_horizontal_margin"
    android:paddingTop="@dimen/activity_vertical_margin"
    tools:context="com.nsu.zyl. chapter04application.SharedPreferencesGetActivity">

    <TextView
```

```
            android:layout_width="wrap_content"
            android:layout_height="wrap_content"
            android:text="New Text"
            android:id="@+id/txtStuName"
            android:layout_below="@+id/txtStuId"
            android:layout_alignStart="@+id/txtStuId"
            android:layout_marginTop="77dp" />

        <TextView
            android:layout_width="wrap_content"
            android:layout_height="wrap_content"
            android:text="New Text"
            android:id="@+id/txtStuId"
            android:layout_alignParentTop="true"
            android:layout_centerHorizontal="true"
            android:layout_marginTop="69dp" />

        <TextView
            android:layout_width="wrap_content"
            android:layout_height="wrap_content"
            android:text="学号："
            android:id="@+id/textView4"
            android:layout_alignTop="@+id/txtStuId"
            android:layout_alignParentStart="true" />

        <TextView
            android:layout_width="wrap_content"
            android:layout_height="wrap_content"
            android:text="姓名："
            android:id="@+id/textView5"
            android:layout_alignTop="@+id/txtStuName"
            android:layout_alignParentStart="true" />
</RelativeLayout>
```

重写 SharedPreferencesGetActivity 中的 onCreate()方法，从 SharedPreferences 读取 SharedPreferencesSaveActivity 保存在 SharedPreferences 中的数据并在文本框中显示出来，代码如下：

```
public class SharedPreferencesGetActivity extends AppCompatActivity {
    private TextView txtStuId,txtStuName;
    @Override
    protected void onCreate(Bundle savedInstanceState) {
        super.onCreate(savedInstanceState);
        setContentView(R.layout.activity_shared_preferences_get);
        txtStuId=(TextView) findViewById(R.id.txtStuId);
        txtStuName=(TextView)findViewById(R.id.txtStuName);
        SharedPreferences sharedPreferences=getSharedPreferences("data",Context.
MODE_PRIVATE);
```

```
        String strStuId=sharedPreferences.getString("stuId","");
        String strStuName=sharedPreferences.getString("stuName","");
        txtStuId.setText(strStuId);
        txtStuName.setText(strStuName);
    }
}
```

重写 MainActivity 类中的 onListItemClick()方法，当用户点击列表项第一项时，启动 SharedPreferencesSaveActivity。

```
@Override
    protected void onListItemClick(ListView l, View v, int position, long id) {
        super.onListItemClick(l, v, position, id);
        Intent intent=new Intent();
        switch (position){
            case 0:
                intent.setClass(getApplicationContext(),
SharedPreferencesSaveActivity.class);
                break;
        }
        startActivity(intent);
    }
```

运行程序，在 MainActivity 页面点击“SharedPreferences”列表项，显示图 4-3 所示的界面，用户在该界面中输入学号“201710110”，姓名“张山”，点击“登录”按钮，跳转至图 4-4 所示的界面。

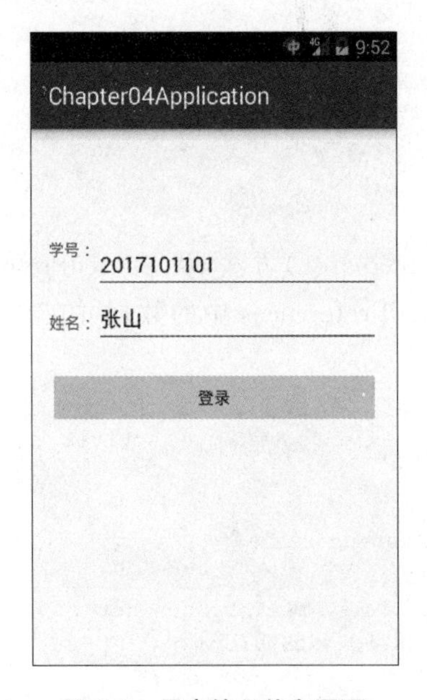

图 4-3　用户输入信息页面

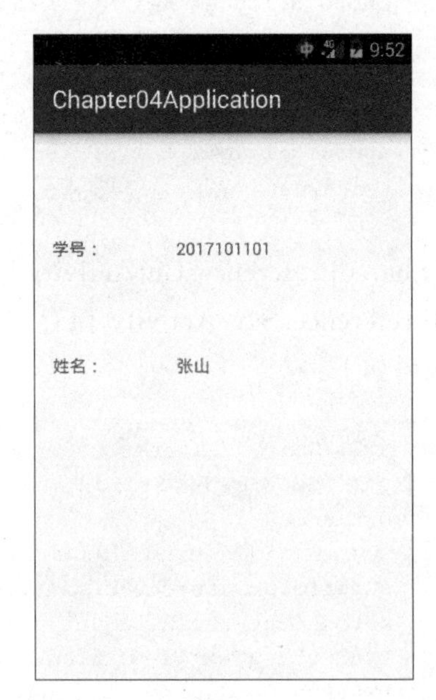

图 4-4　显示用户输入信息页面

保存到SharedPreferences中的数据，最终将以XML文件的形式保存数据。运行上述程序后，打开Android Device Monitor，在File Explorer窗口下，找到程序对应的文件夹，在shared_prefs文件夹下可以发现名为data.xml的文件，如图4-5所示，该文件以键值对的形式保存了存储在SharedPreferences中的数据。

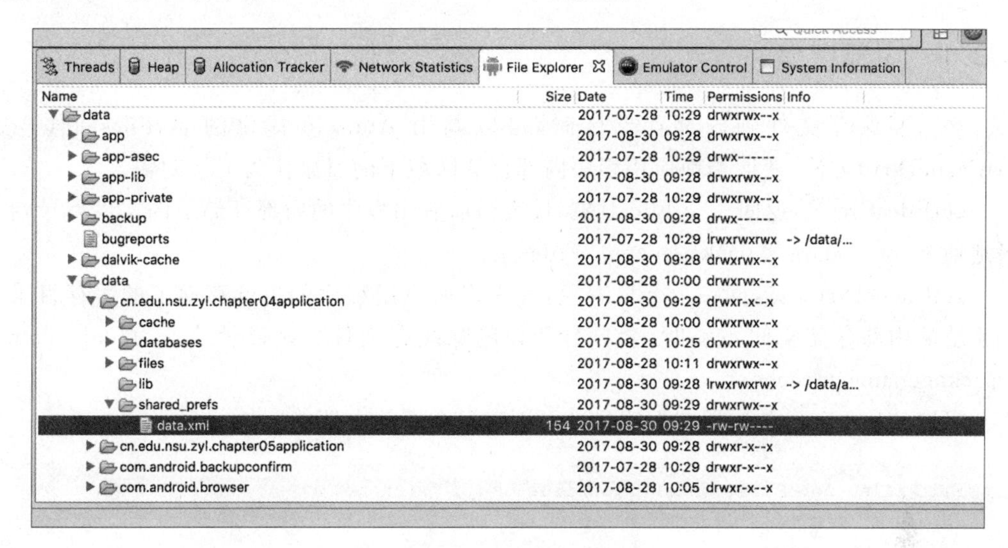

图4-5 data.xml

4.3 文件存储

文件存储是Android系统中一种基本的数据存储方式，与Java中的文件存储类似，Android支持以I/O流的形式对文件进行读取和写入操作。在Android中有两个文件存储区域：内部存储区和外部存储区。这两种区域的划分源自早期Android系统中内置的不可变的内存(internal storage)和可卸载的存储部件(external storage，类似SD Card)。虽然现在一些Android设备将内部存储区与外部存储区都做成了不可卸载的内置存储，但是这一整块还是从逻辑上被划分为内部存储区与外部存储区，只是现在不再以是否可卸载进行区分了。

内部存储区和外部存储区的主要区别如下：

(1) 内部存储区：

① 内部存储总是处于可用状态；

② 内部存储中的文件被创建它的应用程序私有，只能被创建它的应用程序访问；

③ 某个应用程序被卸载时，在内部存储中与该应用程序相关的文件都会被删除；

④ 在内部存储中存储文件可以确保文件不被用户或其他应用程序访问。

(2) 外部存储区：

① 外部存储并不总是处于可用状态，例如用户有时会通过USB存储模式挂载外部存储器，当取下挂载的这部分后，就无法对其进行访问了；

② 保存在外部存储区的文件可能被其他应用程序访问；

③ 当用户卸载某个应用程序时，系统仅仅会删除外部存储区域 external 根目录(getExternal-FilesDir())下的相关文件；

④ 外部存储区是在不需要严格的访问权限并且希望能够被其他应用程序所共享或者是允许用户通过计算机访问时的最佳存储区域。

4.3.1 内部存储

当需要保存文件到内部存储区时，可以调用 Android 提供的 getFilesDir()或 getCacheDir()方法，获取当前应用程序内部存储区域下的目录作为 File 对象。

getFilesDir()：返回一个 File 对象，代表当前应用程序的内部存储下的文件目录，目录地址为/data/data/< packagename >/files/。

getCacheDir()：返回一个 File 对象，代表当前应用程序的内部存储下的缓存目录，当系统的内部存储空间不够时，会自行选择删除缓存文件。该目录地址为/data/data/< packagename >/cache/。

可以使用 File()构造器在上述目录下创建一个新的文件：

```
File file =new File(context.getFilesDir(), filename);
```

为方便地对当前应用程序内部存储区域中的文件进行读和写操作，Android 提供 openFileOutput(String name, int mode)和 openFileInput(String name)两种方法，用于获取当前应用的内部存储目录下指定文件的输出流和输入流。

如下示例代码将"Hello world!"字符串保存到当前应用的内部存储目录下名字为 myfile 的文件中：

```
String filename ="myfile";
String string ="Hello world! ";
FileOutputStream outputStream;
try {
  outputStream =openFileOutput(filename, Context.MODE_PRIVATE);
  outputStream.write(string.getBytes());
  outputStream.close();
} catch (Exception e) {
  e.printStackTrace();
}
```

openFileOutput(Stringfilename,int mode)用于打开内部存储区指定文件的输出流，其中参数 name 表示文件名，mode 表示文件的操作模式。mode 的取值有 4 种：

(1) MODE_PRIVATE：该文件只能被当前程序读写，默认的操作方式。

(2) MODE_APPEND：该文件的内容可以追加。

(3) MODE_WORLD_READABLE：该文件的内容可以被其他文件读取。

(4) MODE_WORLD_WRITEABLE：该文件的内容可以被其他文件写入。

存储在内部储存区域的文件，被其创建的应用程序私有，保存在 Android 系统特定目录下(/data/data/< packagename >/files/),同一个应用创建的所有文件在该应用对应的包名(packagename)下,当应用程序被卸载,保存在其内部存储区中的文件也被删除。

4.3.2 案例(一)

下面通过实例展示在Android应用程序中如何对内部存储区中的文件进行读和写操作。在项目Chapter04Application中新建InnerStorageActivity,设计交互界面如图4-6所示。该案例实现将用户输入的信息保存在内部存储区域中指定文件中和从指定文件中读取数据的功能。

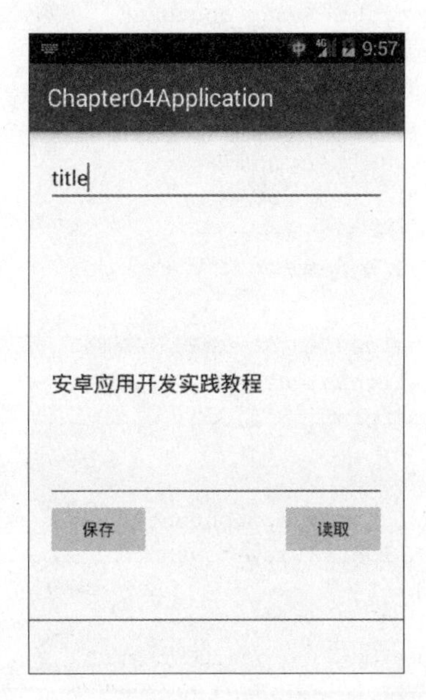

图4-6 InnerStorageActivity页面

InnerStorageActivity对应的布局文件activity_inner_storage.xml的代码如文件清单4-5所示。

文件清单4-5 activity_inner_storage.xml

```
<?xml version="1.0" encoding="utf-8"?>
<RelativeLayout xmlns:android="http://schemas.android.com/apk/res/android"
    xmlns:tools="http://schemas.android.com/tools"
    android:id="@+id/activity_inner_storage"
    android:layout_width="match_parent"
    android:layout_height="match_parent"
    android:paddingBottom="@dimen/activity_vertical_margin"
    android:paddingLeft="@dimen/activity_horizontal_margin"
    android:paddingRight="@dimen/activity_horizontal_margin"
    android:paddingTop="@dimen/activity_vertical_margin"

    tools:context="cn.edu.nsu.zyl.chapter04application.InternalStorageActivity">
    <EditText
```

```
        android:layout_width="match_parent"
        android:layout_height="wrap_content"
        android:id="@+id/edtFileName"
        android:layout_alignParentLeft="true"
        android:layout_alignParentRight="true"
        android:hint="请输入文件名"
          />

    <EditText
        android:layout_width="match_parent"
        android:layout_height="wrap_content"
        android:inputType="textMultiLine"
        android:minLines="8"
        android:maxLines="10"
        android:scrollbars="vertical"
        android:ems="10"
        android:id="@+id/edtContent"
        android:hint="输入文件内容"

        android:layout_below="@+id/edtFileName"
        android:layout_alignParentLeft="true"
        android:layout_marginTop="51dp" />

    <Button
        android:layout_width="wrap_content"
        android:layout_height="wrap_content"
        android:text="保存"
        android:id="@+id/btnSave"
        android:layout_below="@+id/edtContent"
        android:layout_alignParentLeft="true" />

    <Button
        android:layout_width="wrap_content"
        android:layout_height="wrap_content"
        android:text="读取"
        android:id="@+id/btnGet"
        android:layout_below="@+id/edtContent"
        android:layout_alignParentRight="true" />
</RelativeLayout>
```

上述布局文件中包含两个 EditText 控件，分别用于输入文件名和文件内容，两个 Button 控件用于与用户交互操作。该布局所对应的 InternalStorageActivity 源代码如文件清单 4-6 所示。

文件清单 4-6　InternalStorageActivity.java

```
package cn.edu.nsu.zyl.chapter04application;

import android.support.v7.app.AppCompatActivity;
```

```
import android.os.Bundle;
import android.text.TextUtils;
import android.view.View;
import android.widget.Button;
import android.widget.EditText;
import android.widget.Toast;

import java.io.FileInputStream;
import java.io.FileNotFoundException;
import java.io.FileOutputStream;
import java.io.IOException;

public class InternalStorageActivity extends AppCompatActivity {

    private Button btnGet, btnSave;
    private EditText edtFileName, edtContent;
    @Override
    protected void onCreate(Bundle savedInstanceState) {
        super.onCreate(savedInstanceState);
        setContentView(R.layout.activity_inner_storage);
        btnGet = (Button) findViewById(R.id.btnGet);
        btnSave = (Button) findViewById(R.id.btnSave);
        edtFileName = (EditText) findViewById(R.id.edtFileName);
        edtContent = (EditText) findViewById(R.id.edtContent);
        btnSave.setOnClickListener(new View.OnClickListener() {
            @Override
            public void onClick(View view) {
                String fileName =edtFileName.getText().toString();
                String fileContent =edtContent.getText().toString();
                if (!TextUtils.isEmpty(fileName)) {
                    try {
                        FileOutputStream fos=openFileOutput(fileName,MODE_APPEND);
                        fos.write(fileContent.getBytes());
                        fos.close();
                        edtContent.setText("");
                    } catch (FileNotFoundException e) {
                        e.printStackTrace();
                    } catch (IOException e) {
                        e.printStackTrace();
                    }
                }else{
                    Toast.makeText (InternalStorageActivity.this,"文件名不能为
空",Toast.LENGTH_LONG).show();
                }

            }
```

```
        });
        btnGet.setOnClickListener(new View.OnClickListener() {
            @Override
            public void onClick(View view) {
                String fileName=edtFileName.getText().toString();
                if(! TextUtils.isEmpty(fileName)){
                    try {
                        FileInputStream fis=openFileInput(fileName);
                        byte[] bytes=new byte[1024];
                        StringBuffer stringBuffer=new StringBuffer();
                        while(fis.read(bytes)! =-1){
                            stringBuffer.append(new String(bytes));
                        }
                        edtContent.setText(stringBuffer);
                    } catch (FileNotFoundException e) {
                        e.printStackTrace();
                    } catch (IOException e) {
                        e.printStackTrace();
                    }
                }else{
                    Toast.makeText (InternalStorageActivity.this,"文件名不能为
空",Toast.LENGTH_LONG).show();
                }
            }
        });

    }

}
```

在 onCreate()方法中,调用 findViewById()方法找到当前页面中的控件,然后分别为页面中的两个按钮控件设置监听器。在保存按钮的监听器处理方法中,首先获取用户输入的文件名和文件内容,判断文件名和文件内容是否为空,为空则给用户提示,不为空则调用 openFileOutput()获取指定文件的输出流,然后使用输出流对文件进行写操作。在读取按钮的监听器处理方法中,首先获取用户输入的文件名,判断文件名是否为空,为空则给用户提示,不为空则调用 openFileInput()获取指定文件的输入流,然后使用输入流读取文件内容,最后将读取的内容在文本框中显示出来。

同理,需要在 MainActivity 中的 onListItemClick()方法中关联 InternalStorageActivity,运行项目点击 inner storage 列表项跳转至 InternalStorageActivity,当用户点击“保存”按钮时,用户输入的内容将保存在指定文件中,打开 Android Device Monitor(如图 4-7 所示),在 File Explorer 窗口下,可以在/data/data/< packagename >/files/文件夹下找到保存的文件。点击“读取”按钮,将从指定名称的文件内容读出来并在文本框中显示出来。

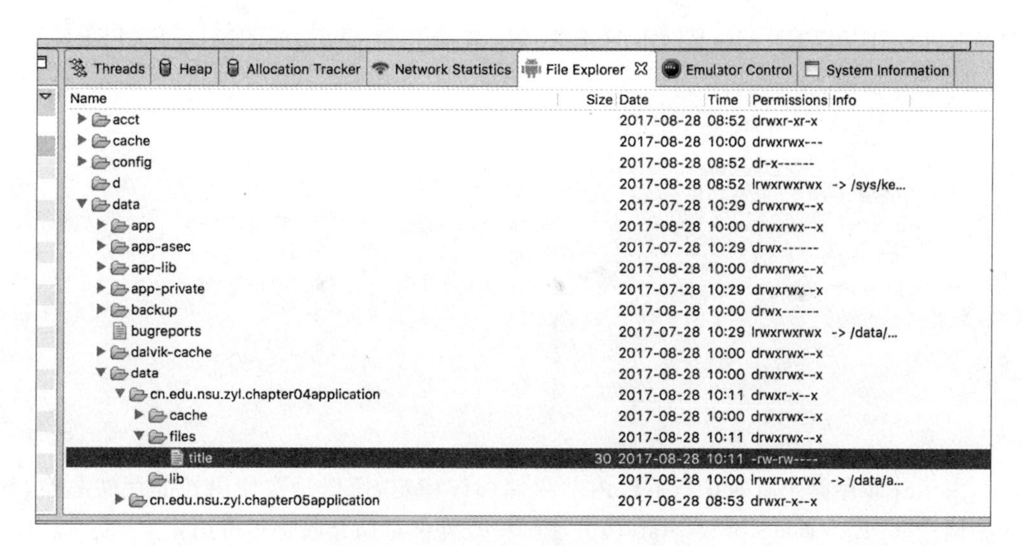

图 4-7 保存在当前应用内部的文件

4.3.3 外部存储

当把文件保存在外部存储区域(SD Card 或设备内嵌的存储卡等存储媒体)时,文件可以被其他应用程序共享,当把外部存储设备连接到计算机上时,这些文件可以在计算机端浏览、修改或删除。由于外部存储设备可能被移除、丢失或者处于其他状态,所以当对外部存储区的文件进行文件操作时,需要对其可用性进行检查。在程序中可以通过执行 getExternalStorageState()来查询外部设备的状态。若返回状态为 MEDIA_MOUNTED,则表示外部存储区域处于挂载状态,可以读写。

External Storage State 的状态列表如表 4-3 所示。

表 4-3 外部设备状态表

状 态	描 述
MEDIA_BAD_REMOVAL	在解除挂载前存储媒体已经被移除
MEDIA_CHECKING	存储媒体存在并在进行磁盘检查
MEDIA_EJECTING	存储媒体正在卸载过程中
MEDIA_MOUNTED	存储媒体已经挂载,并且挂载点可读写
MEDIA_MOUNTED_READ_ONLY	存储媒体已经挂载,挂载点只读
MEDIA_NOFS	存储媒体存在,但空白或使用了不支持的文件系统
MEDIA_REMOVED	存储媒体不存在,即被移除
MEDIA_SHARED	存储媒体正在通过 USB 共享
MEDIA_UNKNOWN	未知存储状态
MEDIA_UNMOUNTABLE	存储媒体无法挂载,一种典型状况是文件系统损坏
MEDIA_UNMOUNTED	存储媒体没有挂载

为了在外部存储区读写数据,必须在 AndroidManifest.xml 文件中添加相应权限。对文件增删操作需要添加 MOUNT_UNMOUNT_FILESYSTEMS 权限,写操作需要添

加 WRITE_EXTERNAL_STORAGE 权限，读操作需要添加 READ_EXTERNAL_STORAGE 权限。

在 AndroidManifest.xml 中添加权限代码如下：

```
<manifest ...>
    <uses-permission android:name="android.permission.WRITE_EXTERNAL_STORAGE" />
    <uses-permission android:name="android.permission.READ_EXTERNAL_STORAGE" />
    <uses-permission android:name="android.permission.MOUNT_UNMOUNT_
FILESYSTEMS" />

    ...
</manifest>
```

由于外部存储介质的状态不像内部存储那样稳定，所以每次使用外部存储之前，都应先调用 getExternalStorageState()方法来检查外部存储介质是否可用。

从外部储存区读取数据的示例代码如下：

```
String state =Environment.getExternalStorageState();
        if (Environment.MEDIA_MOUNTED.equals(state) ||
        Environment.MEDIA_MOUNTED_READ_ONLY.equals(state)) {
            File path =Environment.getExternalStorageDirectory();
            File file =new File(path, fileName);
            FileInputStream fis;
            try {
                fis =new FileInputStream(file);
                BufferedReader bufferedReader=new BufferedReader(new
InputStreamReader(fis));
                String data=bufferedReader.readLine();
                fis.close();
            } catch (Exception e) {
                e.printStackTrace();
            }
        }
```

上述代码首先获得当前外部存储区的状态，判断当前外部存储区状态是否处于可读写或者可读状态，如果可读写或者可读，调用 Environment.getExternalStorageDirectory()获得外部存储区根目录，最后利用 I/O 流对指定文件进行读操作。

同理，向外部存储区存储数据的示例代码如下：

```
String state =Environment.getExternalStorageState();
if (Environment.MEDIA_MOUNTED.equals(state)) {
    File path =Environment.getExternalStorageDirectory();
    File file =new File(path, "data.txt");
    FileOutputStream fos;
    try {
        fos =new FileOutputStream(file);
```

```
            fos.write(data.getBytes());
            fos.close();
        } catch (Exception e) {
            e.printStackTrace();
        }
    }
```

调用 Environment. getExternalStorageDirectory()方法得到外部存储根目录/mnt/sdcard/，使用该方法的好处是可以很灵活地指定文件存储路径。

此外，常用的外部文件存储路径可以通过如下几个方法获得：

(1) 调用 Environment. getDownloadCacheDirectory ()可以获得下载缓存区的根目录。

(2) 调用 Environment. getExternalStoragePublicDirectory (String type)可以获得用于存储特定类型文件的顶层共享外部存储器目录，参数 type 指定了特定的目录，这些目录都是以 DIRECTORY 开头，如 DIRECTORY_MUSIC、DIRECTORY_MOVIES、DIRECTORY_PICTURES 和 DIRECTORY_DOCUMENTS 等。获取外部存储区中存储图片的目录语句是：Environment. getExternalStoragePublicDirectory (Environment. DIRECTORY_PICTURES)；获取这些目录后，就可以在这些目录下进行文件的创建、读、写和删除操作。

通过上述方式创建的文件对用户和其他应用程序来说是公开的，可以被其他应用程序访问，当用户卸载创建这些文件的应用程序时，这些文件会被保留。

在外部存储区也可以创建应用程序私有的文件，当创建文件的应用程序被卸载时这些私有文件会被删除。想要将文件以私有形式保存在外部存储区中，可以通过执行 getExternalFilesDir()来获取相应的目录，并且传递一个指示文件类型的参数。每个以这种方式创建的目录都会被添加到外部存储区/mnt/sdcard/< packagename >/文件夹下。Android 4.4 版本开始，应用可以管理在它外部存储上的特定包名目录，而不用获取 WRITE_EXTERNAL_STORAGE 权限。例如，一个包名为 com. nsu. zyl. food 的应用，可以自由访问外存上的/Android/data/com. nsu. zyl. food/目录。

如下示例创建的文件会在用户卸载该应用程序时被系统删除：

```
public File getAlbumStorageDir(Context context, String albumName) {
    // Get the directory for the app's private pictures directory.
    File file =new File(context.getExternalFilesDir(
            Environment.DIRECTORY_PICTURES), albumName);
    if (!file.mkdirs()) {
        Log.e(LOG_TAG, "Directory not created");
    }
    return file;
}
```

如果想在应用程序被删除时文件仍然保留，可以使用 getExternalStoragePublicDirectory()来存储可以共享的文件。

4.3.4 案例(二)

在项目 Chapter04Application 中新建 ExternalStorageActivity,设计交互界面,如图 4-8 所示。该案例实现将用户输入的信息保存在外部存储区域中指定文件中和从指定的外部存储区域文件中读取数据的功能。

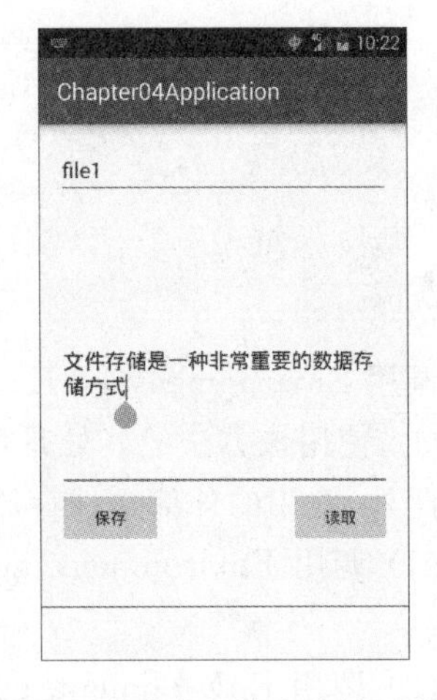

图 4-8 保存数据外包存储区文件

由图 4-8 可知,ExternalStorageActivity 具有和 InternalStorageActivity 相似的布局文件,因此我们不再给出 ExternalStorageActivity 的布局文件。ExternalStorageActivity 的源代码如文件清单 4-7 所示。

文件清单 4-7 ExternalStorageActivity.java

```
package cn.edu.nsu.zyl.chapter04application;

import android.Manifest;
import android.content.pm.PackageManager;
import android.os.Build;
import android.os.Environment;
import android.support.annotation.NonNull;
import android.support.v7.app.AppCompatActivity;
import android.os.Bundle;
import android.text.TextUtils;
import android.util.Log;
import android.view.View;
import android.widget.Button;
import android.widget.EditText;
import android.widget.Toast;

import java.io.File;
import java.io.FileInputStream;
import java.io.FileNotFoundException;
import java.io.FileOutputStream;
import java.io.IOException;

public class ExternalStorageActivity extends AppCompatActivity {
    private EditText edtFileName, edtContent;
    private Button btnSave, btnGet;
    private String fileName, fileContent;


    @Override
    protected void onCreate(Bundle savedInstanceState) {
        super.onCreate(savedInstanceState);
        setContentView(R.layout.activity_external_storage);
```

```
        edtFileName = (EditText) findViewById(R.id.edtFileName);
        edtContent = (EditText) findViewById(R.id.edtContent);
        btnSave = (Button) findViewById(R.id.btnSave);
        btnGet = (Button) findViewById(R.id.btnGet);
        btnSave.setOnClickListener(new View.OnClickListener() {
            @Override
            public void onClick(View view) {
                fileName =edtFileName.getText().toString();
                fileContent =edtContent.getText().toString();

                if (Build.VERSION.SDK_INT >=23) {
                    if (checkSelfPermission(Manifest.permission.WRITE_EXTERNAL_
STORAGE) !=PackageManager.PERMISSION_GRANTED) {
                        requestPermissions (new String[] {Manifest. permission.
WRITE_EXTERNAL_STORAGE}, 300);
                    } else {
                        saveContentToFile(fileName, fileContent);
                    }
                } else {
                    saveContentToFile(fileName, fileContent);
                }
            }
        });
        btnGet.setOnClickListener(new View.OnClickListener() {
            @Override
            public void onClick(View view) {
                fileName =edtFileName.getText().toString();

                if (Build.VERSION.SDK_INT >=23) {
                    if (checkSelfPermission(Manifest.permission.WRITE_EXTERNAL_
STORAGE) !=PackageManager.PERMISSION_GRANTED) {
                        requestPermissions(new String[]{Manifest.permission.
WRITE_EXTERNAL_STORAGE}, 100);
                    } else {

                        edtContent.setText(getContentFromFile(fileName));
                    }
                } else {
                    edtContent.setText(getContentFromFile(fileName));
                }
            }
        });
    }

    public void saveContentToFile(String fileName, String fileContent) {
        if (!TextUtils.isEmpty(fileName) && !TextUtils.isEmpty(fileContent)) {
            if (Environment.getExternalStorageState().equals(Environment.MEDIA_
MOUNTED) || !Environment.isExternalStorageRemovable()) {
                try {
                    String path =Environment.getExternalStoragePublicDirectory
(Environment.DIRECTORY_DOWNLOADS).getAbsolutePath();
                    Log.i("Path","save:"+path);
```

```
                    File file =new File(path, fileName);
                    FileOutputStream fos =new FileOutputStream(file);
                    fos.write(fileContent.getBytes());
                    fos.close();
                    edtContent.setText("");
                } catch (FileNotFoundException e) {
                    e.printStackTrace();
                } catch (IOException e) {
                    e.printStackTrace();
                }
            } else {
                 Toast.makeText(ExternalStorageActivity.this, "外部存储空间不可
用!", Toast.LENGTH_LONG).show();
            }
        } else {
             Toast.makeText(ExternalStorageActivity.this, "文件名或内容不能为
空!", Toast.LENGTH_LONG).show();
        }


    }

    @Override
    public void onRequestPermissionsResult(int requestCode, @NonNull String[]
permissions, @NonNull int[] grantResults) {
        super.onRequestPermissionsResult(requestCode, permissions, grantResults);
        if (grantResults[0] ==PackageManager.PERMISSION_GRANTED) {

            switch (requestCode) {
                case 100:
                    edtContent.setText(getContentFromFile(fileName));
                    break;
                case 300:
                    saveContentToFile(fileName, fileContent);
                    break;
            }
        } else {
            Toast.makeText(ExternalStorageActivity.this, "申请权限被拒绝", Toast.
LENGTH_LONG).show();

        }
    }

    public String getContentFromFile(String fileName) {
        if (!TextUtils.isEmpty(fileName)) {
            if (Environment.getExternalStorageState().equals(Environment.MEDIA_
MOUNTED) || !Environment.isExternalStorageRemovable()) {
                try {
```

```
                    String path =Environment.getExternalStoragePublicDirectory
(Environment.DIRECTORY_DOWNLOADS).getAbsolutePath();
                    Log.i("Path","path:"+path);
                    File file =new File(path, fileName);
                    FileInputStream fis =new FileInputStream(file);
                    byte[] bytes =new byte[512];
                    StringBuffer stringBuffer =new StringBuffer();
                    while (fis.read(bytes) !=-1) {
                        stringBuffer.append(new String(bytes, "utf-8"));
                    }
                    return stringBuffer.toString();
                } catch (FileNotFoundException e) {
                    e.printStackTrace();
                } catch (IOException e) {
                    e.printStackTrace();
                }
            } else {
                Toast.makeText(ExternalStorageActivity.this, "外部存储空间不可
用!", Toast.LENGTH_LONG).show();
            }
        } else {
            Toast.makeText(ExternalStorageActivity.this, "文件名不能为空!",
Toast.LENGTH_LONG).show();
        }
        return null;
    }
}
```

同理,需要在 MainActivity 的 onListItemClick()方法中关联 ExternalStorageActivity,运行项目点击 outer storage 跳转至 ExternalStorageActivity,当点击"保存"按钮时,用户输入的内容将保存在指定外部存储空间的 file1 文件中,如图 4-9 所示。

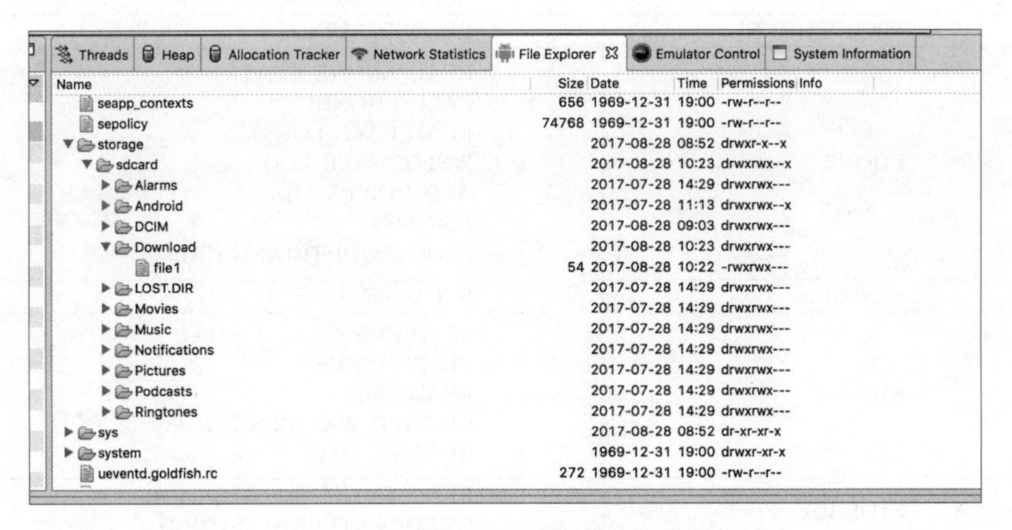

Name	Size	Date	Time	Permissions	Info
seapp_contexts	656	1969-12-31	19:00	-rw-r--r--	
sepolicy	74768	1969-12-31	19:00	-rw-r--r--	
storage		2017-08-28	08:52	drwxr-x--x	
sdcard		2017-08-28	10:23	drwxrwx--x	
Alarms		2017-07-28	14:29	drwxrwx---	
Android		2017-07-28	11:13	drwxrwx--x	
DCIM		2017-08-28	09:03	drwxrwx---	
Download		2017-08-28	10:23	drwxrwx---	
file1	54	2017-08-28	10:22	-rwxrwx---	
LOST.DIR		2017-07-28	14:29	drwxrwx---	
Movies		2017-07-28	14:29	drwxrwx---	
Music		2017-07-28	14:29	drwxrwx---	
Notifications		2017-07-28	14:29	drwxrwx---	
Pictures		2017-07-28	14:29	drwxrwx---	
Podcasts		2017-07-28	14:29	drwxrwx---	
Ringtones		2017-07-28	14:29	drwxrwx---	
sys		2017-08-28	08:52	dr-xr-xr-x	
system		1969-12-31	19:00	drwxr-xr-x	
ueventd.goldfish.rc	272	1969-12-31	19:00	-rw-r--r--	

图 4-9 file1 文件

4.3.5 权限管理

从 4.3.4 节对外部存储空间进行读写的例子可以发现，为了让程序正常运行，除了在 AndroidManifest. xml 中声明对外部存储设备的读写权限外，在代码中也对权限做了处理，为什么要这样呢？在 Android 6.0 版本之前，只需要在 AndroidManifest. xml 中声明当前应用所涉及的所有可能权限，在用户安装应用时，系统将展示所有声明的权限，用户安装即授予所有权限，取消则拒绝安装。一旦应用安装完成，在 AndroidManifest. xml 中申请的权限都会被系统默认授权，无法更改，因此存在安全隐患。

Android 6.0 版本开始，采用新的授权模型，用户可直接在运行时管理应用权限。在 Android 6.0 版本中将权限分为普通权限和危险权限。对于普通权限，在 AndroidManifest. xml 中声明即可；对于危险权限，需要开发者在代码中进行动态申请。也就是说，对于危险权限，在需要时首先判断用户是否授权，如果尚未授权将会提示用户是否授权，这样用户的自主性提高很多，这种模式让用户能够更好地了解和控制权限，同时为应用开发者精简了安装和自动更新过程。

Android 系统中的危险权限分为 9 组，取得一组中某一个权限的授权，则自动获取该组的所有授权。Android 危险权限组如表 4-4 所示。

表 4-4　危险权限组及权限

权限组	权限
CALENDAR	READ_CALENDAR WRITE_CALENDAR
CAMERA	CAMERA
CONTACTS	READ_CONTACTS WRITE_CONTACTS GET_ACCOUNTS
LOCATION	ACCESS_FINE_LOCATION ACCESS_COARSE_LOCATION
MICROPHONE	RECORD_AUDIO
PHONE	READ_PHONE_STATE CALL_PHONE READ_CALL_LOG WRITE_CALL_LOG ADD_VOICEMAIL USE_SIP PROCESS_OUTGOING_CALLS
SENSORS	BODY_SENSORS
SMS	SEND_SMS RECEIVE_SMS READ_SMS RECEIVE_WAP_PUSH RECEIVE_MMS
STORAGE	READ_EXTRENAL_STORAGE WRITE_EXTERNAL_STORAGE

因此,使用危险权限时,对于以 Android 6.0(API Level 23)或更高版本为目标平台的应用,需要在运行时检查和请求权限。具体流程是:首先调用 checkSelfPermission()方法,来判断应用是否已被授予权限,如果已经授予权限则直接执行业务操作,否则调用 shouldShowRequestPermissionRationale()判断是否向用户解释为何申请权限。如果需要,则弹出对话框提示用户申请权限原因,用户确认后调用 requestPermissions()申请权限,如果不需要,则直接调用 requestPermission 申请权限。

1. 检查权限

每次执行需要某一危险权限的操作时,都需要在代码中调用 ContextCompat.checkSelfPermission()方法检查是否具有该权限。

```
int permissionCheck =ContextCompat.checkSelfPermission(thisActivity,
        Manifest.permission.WRITE_EXTERNAL_STORAGE);
```

上述代码段检查 Activity 是否具有在日历中进行写入的权限。

如果应用具有此权限,方法将返回 PackageManager. PERMISSION_GRANTED,并且应用可以继续操作。如果应用不具有此权限,方法将返回 PERMISSION_DENIED,应用需要向用户请求权限。

2. 请求权限

如果应用没有所需权限的授权,需要调用 requestPermissions()方法获取权限。在应用运行需要获取相关权限授权时,会弹出一个请求对话框询问是否授予该程序相应权限,用户可以选择“允许”或者“拒绝”。

当第二次弹出请求对话框时,在对话框中增加一个“不再询问”的选项框,如果用户选择了该选项框,那么以后再次申请相关授权时将不再弹出对话框询问是否授予权限,也就无法获取权限了,只能到系统设置界面为应用授权。为了避免用户因不了解申请权限的原因,导致用户“拒绝授权”,Android 提供 shouldShowRequestPermissionRationale()方法,用于判断是否需要给用户解释。如果应用之前请求过此权限但用户拒绝了请求,此方法将返回 true。

以下代码可以检查应用是否具备对外部存储设备写权限,如不具备则请求该权限:

```
if (checkSelfPermission (Manifest. permission. WRITE _ EXTERNAL _ STORAGE)! =
PackageManager.PERMISSION_GRANTED)
{
  requestPermissions(new String[]
  {
     Manifest.permission.WRITE_EXTERNAL_STORAGE}, 300);
  }
}
```

3. 处理请求响应

当应用请求权限时,系统将向用户显示一个请求权限的对话框。当用户点击对话框中按钮响应时,系统将调用应用的 onRequestPermissionsResult()方法,向其传递用户响应。

```
@Override
    public void onRequestPermissionsResult(int requestCode, @NonNull String[]
permissions, @NonNull int[] grantResults) {
        super.onRequestPermissionsResult(requestCode, permissions, grantResults);
        if (grantResults[0] ==PackageManager.PERMISSION_GRANTED) {

            switch (requestCode) {
                case 100:
                    edtContent.setText(getContentFromFile(fileName));
                    break;
                case 300:
                    saveContentToFile(fileName, fileContent);
                    break;
            }
        } else {
            Toast.makeText(ExternalStorageActivity.this, "申请权限被拒绝", Toast.
LENGTH_LONG).show();

        }
    }
```

4.4 SQLite 数据库

对于结构更加复杂的数据,Android 提供内置的 SQLite 数据库存储数据。SQLite 是一款轻量级数据库,占用资源非常少,只需要几百 KB 的内存就够了,同时 SQLite 支持 SQL 语言和事务处理等功能。

SQLite 没有服务进程,它通过文件保存数据,该文件是跨平台的,可以放在其他平台中使用。SQLite 和其他数据库最大的不同就是对数据类型的支持,保存数据时,支持 NULL、INTEGER、REAL、TEXT 和 BLOB 5 种数据类型。创建一个表时,可以在 CREATE TABLE 语句中指定某列的数据类型,但是用户可以把任何数据类型放入任何列中。当某个值插入数据库时,SQLite 将检查它的类型。如果该类型与关联的列不匹配,则 SQLite 会尝试将该值转换成该列的类型。如果不能转换,则该值将作为其本身具有的类型存储,例如,可以把一个字符串(String)放入 INTEGER 列。

4.4.1 SQLite 数据库的使用

SQLite 使用 SQL 命令提供了完整关系型数据库能力,每个使用 SQLite 的应用程序都有一个该数据库的实例,并且在默认情况下仅限当前应用程序使用。数据存储在

Android 设置的/data/data/< packagename >/database 文件夹中。

Android SDK 提供了一系列对数据库进行操作的类和接口，便于对 SQLite 数据库的操作。

1. SQLiteDatabase 类

该类封装了一系列数据库操作的 API，可以对数据库进行增、删、改、查等操作。常用方法如表 4-5 所示。

表 4-5 SQLiteDatabase 方法表

状 态	描 述
create(SQLiteDatabase. CursorFactory factory)	创建数据库
openOrCreateDatabase (File file, SQLiteDatabase. CursorFactory factory)	创建或打开数据库
openOrCreateDatabase (String path, SQLiteDatabase. CursorFactory factory)	创建或打开数据库
public long insert (String table, String nullColumnHack, ContentValues values)	向指定的数据表中添加一条记录
public Cursor query (String table, String [ ] columns, String selection, String[] selectionArgs, String groupBy, String having, String orderBy)	该方法用于查询数据
public Cursor rawQuery(String sql, String[] selectionArgs)	执行带占位符的 SQL 查询
public int update (String table, ContentValues values, String whereClause, String[] whereArg)	修改特定数据
public int delete (String table, String whereClause, String [ ] whereArgs)	删除表中特定的记录
public void execSQL(String sql,Object[] bindArgs)	执行一条带有占位符的 SQL 语句
public void close()	关闭数据库

利用 openOrCreateDatabase()方法打开或者创建一个数据库时，它会自动检测是否存在这个数据库，如果存在则打开，不存在则创建一个数据库；创建成功则返回一个 SQLiteDatabase 对象，否则抛出异常 FileNotFoundException。

```
SQLiteDatabase
db= SQLiteDatabase.openOrCreateDatabase ("/data/data/com.nsu.db/databases/stu.
db",null);
```

直接调用 SQLiteDatabase 的 execSQL()方法执行创建表的 SQL 语句就可以完成表的创建。

```
String usersTable="create table users (_id integer primary key autoincrement,
username text, password  text, )";  //创建表的 SQL 语句
db.execSQL(usersTable);
```

创建的数据表必须有一个主键用于唯一标识表中的元素,上面创建表的 SQL 语句中的 primary key 指明_id 为主键,autoincrement 指明该主键自动增长。

对数据表进行增、删、改、查操作,有两种方法,一种方法是调用 SQLiteDatabase 提供的 insert()、delete()、update()和 query()方法分别实现插入、删除、修改和查询操作;另一种方法是定义增、删、改、查操作对应的 SQL 语句,直接调用 execSQL()方法执行 SQL 语句。

1) 插入操作

插入一条记录可以使用 SQLiteDatabase. insert(String table, String nullColumnHack, ContentValues values)方法实现,其中 table 代表要插入数据的表名,nullColumnHack 代表插入空行时该指定列的值为 Null,values 代表一行记录的数据。ContentValues 类似于 Map,它提供 put(String key, Xxx value)方法存入数据。

例如,向 users 表插入一条记录("Lucy","021475"),使用 SQLiteDatabase. insert()方法实现,代码如下:

```
ContentValues cValue =new ContentValues();
//添加用户名
cValue.put("username","Lucy");
//添加密码
cValue.put("password","021475");
//调用 insert()方法插入数据
db.insert("users",null,cValue);
```

如果使用 execSQL()方法,代码实现如下:

```
//插入数据 SQL 语句
String insertsql="insert into users (username,password)  values('Lucy','021475')";
//执行 SQL 语句
db.execSQL(sql);
```

2) 删除操作

删除一条记录可以使用 SQLiteDatabase. delete(String table, String whereClause, String[] whereArgs)方法实现,其中 table 代表删除数据的表名,满足该 whereClause 子句的记录将会被删除,whereArgs 用于为 whereClause 传入参数。例如,删除 users 表中_id=2 的记录,代码实现如下:

```
//删除条件
String whereClause ="_id=?";
//删除条件参数
String[] whereArgs ={2};
//执行删除
db.delete("users",whereClause,whereArgs);
```

如果使用 execSQL()方法,代码实现如下:

```
//插入数据 SQL 语句
String insertsql=" delete from users where _id =2 ";
//执行 SQL 语句
db.execSQL(sql);
```

3）修改操作

修改一条记录可以使用 SQLiteDatabase. update(String table, ContentValues values, String whereClause, String[] whereArgs)方法实现，其中table代表要修改记录的表，values代表要修改的数据，whereClause代表满足的相关条件，whereArgs代表为whereClause传入参数。例如，修改users表中_id＝10的记录将其更新为username＝Merry，代码实现如下：

```
//实例化内容值
ContentValues values =new ContentValues();
//在 values 中添加内容
values.put("username","Merry");
//修改条件
String whereClause ="_id=?";
//修改添加参数
String[] whereArgs={10};
//修改
db.update("usertable",values,whereClause,whereArgs);
```

如果使用execSQL()方法执行SQL语句实现，代码实现如下：

```
//修改 SQL 语句
String sql ="update stu_table set  snumber =654321 where id =1";
//执行 SQL
db.execSQL(sql);
```

4）查询操作

查询记录可以使用SQLiteDatabase. query(Boolean distinct, String table, String[] columns, String whereClause, String[] selectionArgs, String groupBy, String having, String orderBy, String limit)方法实现，其中distinct代表是否去除重复数据，table代表要查询数据的表，columns代表要查询的列名，whereClause代表查询子句，selectionArgs代表替换查询子句中占位符的参数值，groupBy代表分组方式，having代表分组方式，orderBy代表排序方式，limit用于限制由Select语句返回的数据数量。

2. SQLiteOpenHelper 类

SQLiteOpenHelper类是SQLiteDatabase一个辅助类。这个类的主要作用是生成一个数据库，并对数据库的版本进行管理。在实际开发过程中较少直接使用SQLiteDatabase的方法打开数据库，通常会继承SQLiteOpenHelper开发子类，并通过该类的getReadableDatabase()方法或getWritableDatabase()方法打开数据库。

SQLiteOpenHelper 提供的常用方法如表 4-6 所示。

表 4-6 SQLiteOpenHelper 类常用方法

状　态	描　述
public SQLiteOpenHelper(Context context, String name, CursorFactory factory, int version)	构造方法
public void onCreate(SQLiteDatabase db)	在数据库第一次生成时会调用这个方法
public void onUpgrade(SQLiteDatabase db, int oldVersion, int newVersion)	数据库版本更新时调用
pubic SQLiteDatabase getReadableDatabase()	创建或打开一个只读的数据库
public SQLiteDatabase getWritableDatabase()	创建或打开一个读写的数据库

SQLiteOpenHelper 类根据开发应用程序的需要，封装了创建和更新数据库使用的逻辑。定义类继承 SQLiteOpenHelper，至少需要在类中实现如下 3 个方法：

(1) 构造方法，需要在构造方法中调用父类 SQLiteOpenHelper 的构造方法。

(2) onCreate(SQLiteDatabase db)方法。应用第一次使用时会调用 onCreate(SQLiteDatabase db)方法生成数据库中的表。即在程序中调用 getWritableDatabase()或 getReadableDatabase()方法获取用于操作数据库的 SQLiteDatabase 实例时，如果数据库不存在，Android 系统会自动生成一个数据库，然后调用 onCreate()方法用于生成数据库的表。因为 onCreate()方法在初次生成数据库时才调用，重写 onCreate()方法时，可以生成数据表结构及添加数据到数据库中。

(3) onUpgrade(SQLiteDatabase db, int oldVersion, int newVersion)方法。用于升级软件时更新数据表结构，该方法在数据库的版本发生变化时调用，该方法中参数 oldVersion 代表数据库之前的版本号，参数 newVersion 代表数据库当前版本号。

在 SQLite 数据库中创建表的示例代码如下：

```
public class DBHelper extends SQLiteOpenHelper {
    private final static String CREATE_TABLE_SQL="create table users(_id integer
primary key username,address)";

    public DBHelper(Context context, String name, SQLiteDatabase.CursorFactory
factory, int version) {
        super(context, name, factory, version);
    }

    @Override
    public void onCreate(SQLiteDatabase db) {
        db.execSQL(CREATE_TABLE_SQL);
    }

    @Override
    public void onUpgrade(SQLiteDatabase db, int oldVersion, int newVersion) {
        sqLiteDatabase.execSQL("drop table if exists users");
        onCreate(sqLiteDatabase);
```

```
        }
    }
```

上述代码定义 SQLiteOpenHelper 的子类 DBHelper，在 onCreate()方法中完成表的创建工作。

3. Cursor 接口

Cursor 是一个游标接口，游标是系统为用户开设的一个数据缓冲区，用于存放 SQL 语句的执行结果。使用 SQLiteDatabase. query()方法时，会得到一个 Cursor 对象，Cursor 可以定位到结果集中的某一行，对数据读写。使用 Cursor 允许 Android 更有效地管理所需要的行和列，Cursor 中常用的方法如表 4-7 所示。

表 4-7 Cursor 常用方法

状态	描述
boolean moveToNext()	移动光标到下一行
int getInt(int columnIndex)	获取指定列的整数值
int getColumnIndex(String columnName)	返回指定列索引值，如果列不存在则返回－1
String getString(int columnIndex)	获取指定列的字符串
boolean moveToFirst()	移动光标到第一行
boolean moveToLast()	移动光标到最后一行
boolean moveToPrevious()	移动光标到上一行
boolean moveToPosition(int position)	移动光标到指定位置
int getCount()	返回 Cursor 中的行数
int getPosition()	返回当前 Cursor 的位置
String getColumnName(int columnIndex)	根据列的索引值获取列的名称
String[] getColumnNames()	获取 Cursor 所有列的名称的数组

4.4.2 SQLite 事务操作

SQLite 支持事务。事务是针对数据库的一组操作，可以由一条或者多条 SQL 语句组成。事务具有原子性，也就是说事务中的语句要么都执行，要么都不执行。

SQLiteDatabase 中包含两个方法用于控制事务。

(1) beginTransaction()：开始事务；

(2) endTransaction()：结束事务。

使用 SQLiteDatabase 的 beginTransaction()方法可以开启一个事务，程序执行到 endTransaction()方法时会检查事务的标志是否为成功，如果程序执行到 endTransaction()之前调用了 setTransactionSuccessful()方法设置事务的标志为成功则提交事务，如果没有调用 setTransactionSuccessful()方法则回滚事务。

```
//获取 SQLiteDatabase 对象
SQLiteDatabase db =dbOpenHelper.getWriteableDatabase();
//开启事务
db.beginTransaction();
try{
      //批量处理操作
      db.execSQL("SQL 语句 1", new Object[]{});
      db.execSQL("SQL 语句 2", new Object[]{});
      .....
      db.execSQL("SQL 语句 n",new Object[]{});
    //设置事务标志为成功,当结束事务时就会提交事务
    db.setTransactionSuccessful();
}catch(Exception e){
    Log.i("事务处理失败",e.toString());
}finally{
          //结束事务
         db.endTransaction();
db.close();
}
```

4.4.3 案例

下面开发一个员工信息管理程序,示范如何在 Android 应用中操作 SQLite 数据库。该程序提供对员工信息的增、删、改和查操作,并且员工信息以列表的形式展示。在 Chapter04Application 中新建 SQLiteOperateActivity,设计用户交互界面如图 4-10 所示。

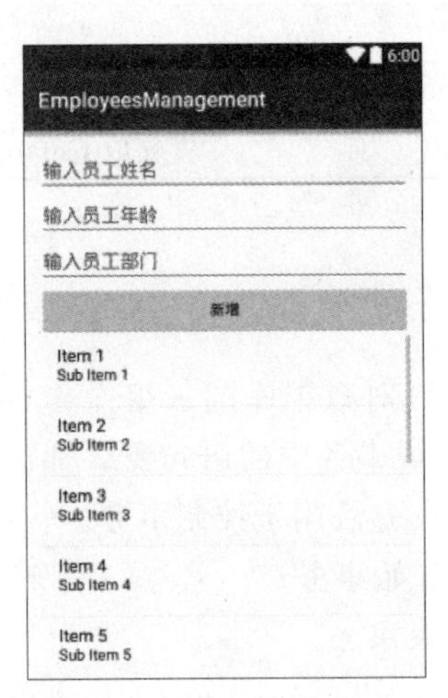

图 4-10 员工信息展示页面

(1) 根据图 4-10 所示页面,编辑 SQLiteOperateActivity 对应的布局文件 activity_sqlite_operate.xml,详细内容如文件清单 4-8 所示。

文件清单 4-8 activity_sqlite_operate.xml

```
<?xml version="1.0" encoding="utf-8"?>
<RelativeLayout xmlns:android="http://schemas.android.com/apk/res/android"
    xmlns:tools="http://schemas.android.com/tools"
    android:layout_width="match_parent"
    android:layout_height="match_parent"
    android:paddingBottom="@dimen/activity_vertical_margin"
    android:paddingLeft="@dimen/activity_horizontal_margin"
    android:paddingRight="@dimen/activity_horizontal_margin"
    android:paddingTop="@dimen/activity_vertical_margin"

tools:context="cn.edu.nsu.zyl.chapter04application.SQLiteOperateActivity">

    <EditText
        android:layout_width="match_parent"
        android:layout_height="wrap_content"
        android:id="@+id/edtName"
        android:hint="输入员工姓名"
        android:layout_alignParentTop="true" />

    <EditText
        android:layout_width="match_parent"
        android:layout_height="wrap_content"
        android:id="@+id/edtAge"
        android:layout_below="@+id/edtName"
        android:hint="输入员工年龄"
        android:layout_alignParentLeft="true"
        android:layout_alignParentStart="true"
        />

    <EditText
        android:layout_width="match_parent"
        android:layout_height="wrap_content"
        android:id="@+id/edtDepartment"
        android:layout_below="@+id/edtAge"
        android:hint="输入员工部门"
        android:layout_alignParentLeft="true"
        android:layout_alignParentStart="true" />

    <Button
        android:layout_width="match_parent"
        android:layout_height="wrap_content"
        android:text="新增"
        android:id="@+id/btnInsert"
        android:layout_below="@+id/edtDepartment"
        android:layout_alignParentLeft="true"
        android:layout_alignParentStart="true" />

    <ListView
```

```
        android:layout_width="wrap_content"
        android:layout_height="wrap_content"
        android:id="@+id/listView"
        android:layout_below="@+id/btnInsert"
        android:layout_alignParentLeft="true"
        android:layout_alignParentStart="true" />

</RelativeLayout>
```

页面中包含3个EditText控件用于用户输入新增员工的“姓名”“年龄”和“部门”信息，当用户点击“新增”按钮时，程序将3个EditText控件中输入的信息保存到数据库中。页面中的ListView用于显示数据库中存储的员工信息，支持对员工信息的删除和修改操作。

(2) 在res/layout文件夹下创建列表项布局文件itemt_employee.xml，列表项的布局如图4-11所示。

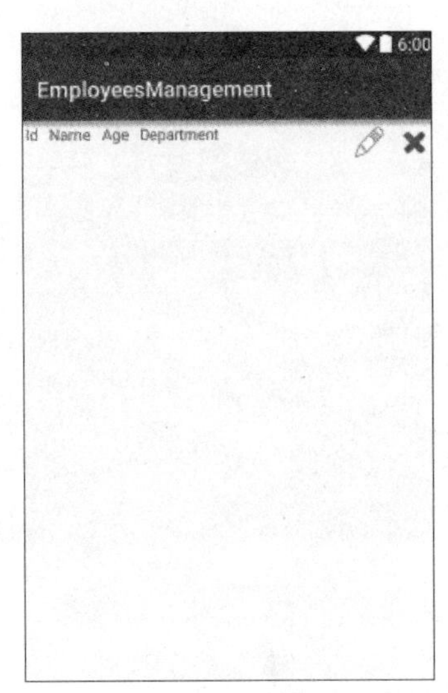

图4-11 列表项布局页面

列表项布局页面中包含4个TextView控件，分别用于显示员工ID、姓名、年龄和部门信息；两个ImageView控件，分别用于删除和编辑员工信息。item_employee.xml具体代码如文件清单4-9所示。

文件清单4-9 item_employee.xml

```
<?xml version="1.0" encoding="utf-8"?>
<RelativeLayout xmlns:android="http://schemas.android.com/apk/res/android"
    android:layout_width="match_parent"
    android:layout_height="match_parent">
    <TextView
        android:layout_width="wrap_content"
```

```
        android:layout_height="wrap_content"
        android:padding="5dp"
        android:text="Id"
        android:id="@+id/txtId"
        android:layout_alignParentTop="true"
        android:layout_alignParentLeft="true"
        android:layout_alignParentStart="true" />

    <TextView
        android:layout_width="wrap_content"
        android:layout_height="wrap_content"
        android:padding="5dp"
        android:text="Name"
        android:id="@+id/txtName"
        android:layout_alignParentTop="true"
        android:layout_toRightOf="@+id/txtId"
        android:layout_toEndOf="@+id/txtId" />

    <TextView
        android:layout_width="wrap_content"
        android:layout_height="wrap_content"
        android:padding="5dp"
        android:text="Age"
        android:id="@+id/txtAge"
        android:layout_alignParentTop="true"
        android:layout_toRightOf="@+id/txtName"
        android:layout_toEndOf="@+id/txtName" />

    <TextView
        android:layout_width="wrap_content"
        android:layout_height="wrap_content"
        android:padding="5dp"
        android:text="Department"
        android:id="@+id/txtDepartment"
        android:layout_alignParentTop="true"
        android:layout_toRightOf="@+id/txtAge"
        android:layout_toEndOf="@+id/txtAge" />

    <ImageView
        android:layout_width="wrap_content"
        android:layout_height="wrap_content"
        android:padding="5dp"
        android:id="@+id/imgDelete"
        android:layout_alignParentTop="true"
        android:layout_alignParentRight="true"
        android:layout_alignParentEnd="true"
        android:src="@android:drawable/ic_delete" />

    <ImageView
        android:layout_width="wrap_content"
        android:layout_height="wrap_content"
```

```
        android:padding="5dp"
        android:id="@+id/imgEdit"
        android:src="@android:drawable/ic_menu_edit"
        android:layout_alignParentTop="true"
        android:layout_toLeftOf="@+id/imgDelete"
        android:layout_toStartOf="@+id/imgDelete" />
</RelativeLayout>
```

(3) 创建数据库,定义 DBOpenHelper 类继承自 SQLiteOpenHelper,在 DBOpenHelper 中完成数据库的创建工作,DBOpenHelper 代码如文件清单 4-10 所示。

文件清单 4-10　DBOpenHelper. java

```
public class DBOpenHelper extends SQLiteOpenHelper {
    private final static String CREATE_TABLE_SQL="create table employees(_id
integer primary key autoincrement,name,age,department)";

    public DBOpenHelper(Context context,int version) {
        super(context,"zyl.db",null, version);
    }

    @Override
    public void onCreate(SQLiteDatabase sqLiteDatabase) {
        Log.i("DBOpenHelper","---onCreate called----");
        sqLiteDatabase.execSQL(CREATE_TABLE_SQL);
    }

    @Override
    public void onUpgrade(SQLiteDatabase sqLiteDatabase, int i, int i1) {
        Log.i("DBOpenHelper","----onUpgrade called----");
        sqLiteDatabase.execSQL("drop table if exists employees");
        onCreate(sqLiteDatabase);
    }
}
```

(4) 创建数据库业务操作类。在程序中需要对创建的数据库中表进行增、删、改、查操作,因此定义一个数据库业务操作类 EmployeesDAO 用于操作数据,具体代码如文件清单 4-11 所示。

文件清单 4-11　EmployeesDAO. java

```
public class EmployeesDAO {
    private DBOpenHelper dbOpenHelper;
    private static final String TABLENAME="employees";
    public EmployeesDAO(Context context){
        //创建 DBOpenHelper 对象
        dbOpenHelper=new DBOpenHelper(context,1);
    }
    //向数据库插入一条记录
```

```
    public void insert(String name,int age,String department){
        //获取数据库对象
        SQLiteDatabase sqLiteDatabase=dbOpenHelper.getReadableDatabase();
        ContentValues contentValues=new ContentValues();
        contentValues.put("name",name);
        contentValues.put("age",age);
        contentValues.put("department",department);
        //向表中插入记录
        sqLiteDatabase.insert(TABLENAME,null,contentValues);
        sqLiteDatabase.close();
    }
    //根据_id删除记录
    public int delete(int _id){
        SQLiteDatabase sqLiteDatabase=dbOpenHelper.getReadableDatabase();
        int count=sqLiteDatabase.delete(TABLENAME,"_id=?",new String[]{_id+""});
        sqLiteDatabase.close();
        return count;
    }
    //修改数据
    public int update(String name,int age,String department,int _id){
        SQLiteDatabase sqLiteDatabase=dbOpenHelper.getReadableDatabase();
        ContentValues contentValues=new ContentValues();
        contentValues.put("name",name);
        contentValues.put("age",age);
        contentValues.put("department",department);
        int count = sqLiteDatabase. update (TABLENAME, contentValues,"_id=?", new
String[]{_id+""});
        sqLiteDatabase.close();
        return  count;
    }
    //查询所有数据
    public ArrayList queryAll(){
        SQLiteDatabase sqLiteDatabase=dbOpenHelper.getReadableDatabase();
        Cursor cursor=sqLiteDatabase.query(TABLENAME,new String[]{"_id","name",
"age","department"},null,null,null,null,null);
        ArrayList list=new ArrayList();
        while(cursor.moveToNext()){
            HashMap map=new HashMap();

            map.put("_id",cursor.getInt(cursor.getColumnIndex("_id")));

            map.put("name",cursor.getString(cursor.getColumnIndex("name")));

            map.put("age",cursor.getInt(cursor.getColumnIndex("age")));

            map.put ( " department", cursor. getString ( cursor. getColumnIndex ( "
department")));
            list.add(map);
        }
```

```
        cursor.close();
        dbOpenHelper.close();
        return list;
    }

}
```

getWritableDatabase()方法以读写方式打开数据库，如果数据库只能读不能写，使用 getWritableDatabase()方法打开数据库将出错。

使用 getReadableDatabase()方法打开数据库，先以读写方式打开数据库，如果数据库只能读不能写，则以只读的方式打开数据库。因此，在程序中我们采用 getReadableDatabase()方法打开数据库。

(5) 创建自定义 Adapter。为了在 ListView 展示员工信息，并且实现对指定员工信息的修改和删除，自定义 EmployeesAdapter 类继承自 BaseAdapter 类。EmployeesAdapter 具体代码如文件清单 4-12 所示。

文件清单 4-12　EmployeesAdapter.java

```
public class EmployeesAdapter extends BaseAdapter {
    private Context context;
    private ArrayList list;

    public EmployeesAdapter(Context context,ArrayList list){
        this.context=context;
        this.list=list;
    }
    @Override
    public int getCount() {
        return list.size();
    }

    @Override
    public Object getItem(int i) {
        return list.get(i);
    }

    @Override
    public long getItemId(int i) {
        return (int)((HashMap)list.get(i)).get("_id");
    }

    @Override
    public View getView(final int i, View view, ViewGroup viewGroup) {
        View itemView=view;
        ViewHolder holder;
        if(itemView==null){
            LayoutInflater inflater= (LayoutInflater) context.getSystemService
(Context.LAYOUT_INFLATER_SERVICE);
```

```
            itemView=inflater.inflate(R.layout.item_employee,null);
            holder=new ViewHolder();
            holder.txtId=(TextView)itemView.findViewById(R.id.txtId);

            holder.txtName=(TextView)itemView.findViewById(R.id.txtName);

            holder.txtAge=(TextView)itemView.findViewById(R.id.txtAge);

            holder.txtDepartment=(TextView)itemView.findViewById(R.id.txtDepartment);

            holder.imgDelete=(ImageView)itemView.findViewById(R.id.imgDelete);

            holder.imgEdit=(ImageView)itemView.findViewById(R.id.imgEdit);
            itemView.setTag(holder);
        }else{
            holder=(ViewHolder) itemView.getTag();
        }

        holder.txtId.setText(((HashMap)list.get(i)).get("_id")+"");

        holder.txtName.setText(((HashMap)list.get(i)).get("name").toString());
holder.txtAge.setText(((HashMap)list.get(i)).get("age")+"");

        holder.txtDepartment.setText(((HashMap)list.get(i)).get("department").
toString());

        holder.imgEdit.setOnClickListener(new View.OnClickListener() {
            @Override
            public void onClick(View view) {
                AlertDialog.Builder builder=new AlertDialog.Builder(context);
                builder.setTitle("请输入更新后的值");
                View dialogView =LayoutInflater.from(context).inflate(R.layout.
dialog_edit,null);
                final EditText edtName = (EditText)  dialogView.findViewById(R.
id.edtName);
                final EditText edtAge = (EditText) dialogView.findViewById(R.id.
edtAge);
                final EditText edtDepartment = (EditText) dialogView.findViewById
(R.id.edtDepartment);

                edtAge.setText(((HashMap)list.get(i)).get("age").toString());

                edtName.setText(((HashMap)list.get(i)).get("name").toString());

                edtDepartment.setText (((HashMap) list.get (i)).get ("department").
toString());

                Button btnSubmit=(Button)dialogView.findViewById(R.id.btnSubmit);

                btnSubmit.setOnClickListener(new View.OnClickListener() {
                    @Override
                    public void onClick(View view) {
```

```
                        int _id=Integer.parseInt(((HashMap)list.get(i)).get("_
id").toString());
                        String name=edtName.getText().toString();
                        int age=Integer.parseInt(edtAge.getText().toString());
                        String department=edtDepartment.getText().toString();
                        HashMap map=(HashMap) list.get(i);
                        map.put("_id",_id);
                        map.put("name",name);
                        map.put("age",age);
                        map.put("department",department);

                        EmployeesDAO dao=new EmployeesDAO(context);
                        dao.update(name,age,department,_id);
                        notifyDataSetChanged();

                    }
                });
                builder.setView(dialogView);
                builder.show();

            }
        });

        holder.imgDelete.setOnClickListener(new View.OnClickListener() {
            @Override
            public void onClick(View view) {
                DialogInterface.OnClickListener listener= new DialogInterface.
OnClickListener() {
                    @Override
                    public void onClick(DialogInterface dialogInterface, int pi) {

                        EmployeesDAO dao=new EmployeesDAO(context);

dao.delete(Integer.parseInt(((HashMap)list.get(i)).get("_id").toString()));
                        list.remove(i);
                        notifyDataSetChanged();
                    }
                };
                AlertDialog.Builder builder=new AlertDialog.Builder(context);
                builder.setTitle("确定删除该记录?");
                builder.setNegativeButton("取消",null);

                builder.setPositiveButton("确定",listener);
                builder.show();
            }
        });

        return itemView;
    }
    private static class ViewHolder{
```

```
        public TextView txtId;
        public TextView txtName;
        public TextView txtAge;
        public TextView txtDepartment;
        public ImageView imgDelete;
        public ImageView imgEdit;
    }
}
```

(6) 编写界面交互代码 MainActivity。MainActivity 的具体代码如文件清单 4-13 所示。

文件清单 4-13　MainActivity. java

```
package cn.edu.nsu.zyl.chapter04application;

import android.database.sqlite.SQLiteOpenHelper;
import android.support.v7.app.AppCompatActivity;
import android.os.Bundle;
import android.view.View;
import android.widget.Button;
import android.widget.EditText;
import android.widget.ListView;

import java.util.ArrayList;

public class SQLiteOperateActivity extends AppCompatActivity {

    private EmployeesDAO employeesDAO;
    private EmployeesAdapter employeesAdapter;
    private ArrayList list;
    private EditText edtName,edtAge,edtDepartment;
    private ListView employeesListView;
    private Button btnInsert;

    @Override
    protected void onCreate(Bundle savedInstanceState) {
        super.onCreate(savedInstanceState);
        setContentView(R.layout.activity_sqlite_operate);

        edtName= (EditText) findViewById(R.id.edtName);
        edtAge= (EditText)findViewById(R.id.edtAge);
        edtDepartment= (EditText) findViewById(R.id.edtDepartment);
        employeesListView= (ListView)findViewById(R.id.listView);
        btnInsert= (Button)findViewById(R.id.btnInsert);


        employeesDAO=new EmployeesDAO(SQLiteOperateActivity.this);
```

```
        list=employeesDAO.queryAll();
        employeesAdapter=new EmployeesAdapter(SQLiteOperateActivity.this,list);
        employeesListView.setAdapter(employeesAdapter);

        btnInsert.setOnClickListener(new View.OnClickListener() {
            @Override
            public void onClick(View view) {
                String name=edtName.getText().toString();
                int age=Integer.parseInt(edtAge.getText().toString());
                String department=edtDepartment.getText().toString();
                employeesDAO.insert(name,age,department);
                list=employeesDAO.queryAll();
                employeesAdapter = new EmployeesAdapter (SQLiteOperateActivity.
this,list);
                employeesListView.setAdapter(employeesAdapter);
                edtName.setText("");
                edtAge.setText("");
                edtDepartment.setText("");
            }
        });
    }

}
```

运行程序执行,看到图 4-12 所示员工信息展示页面。

在 3 个 EditText 控件中分别输入员工姓名、年龄和部门,点击“新增”按钮,完成向数据库插入一条记录操作。效果如图 4-13 所示。

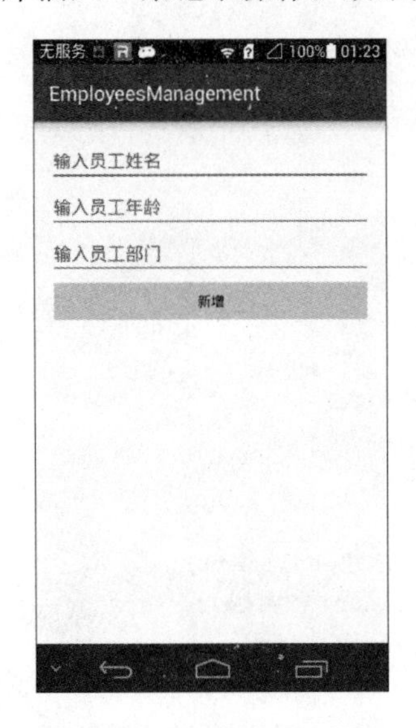

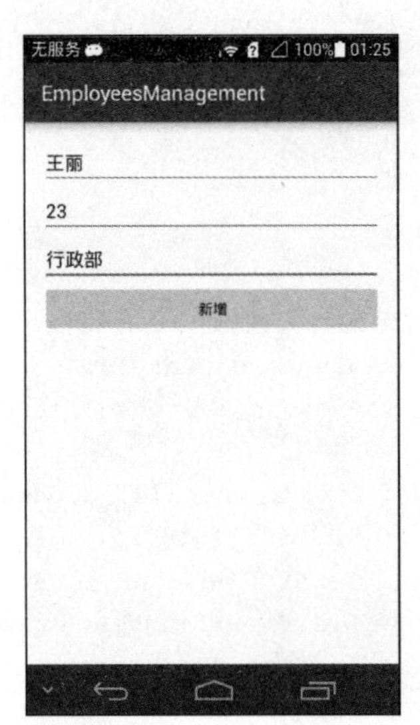

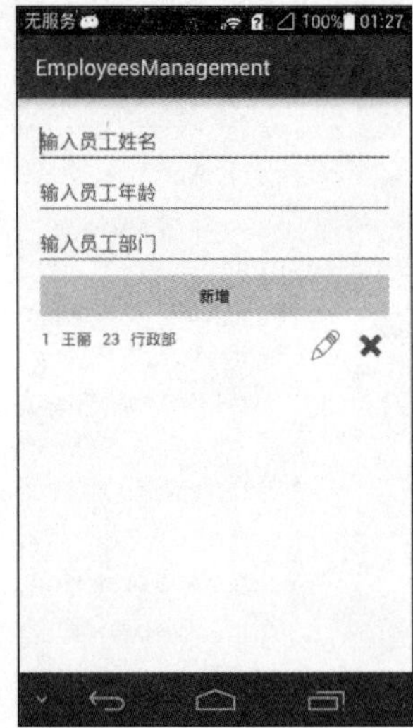

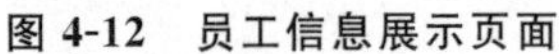
图 4-12 员工信息展示页面

图 4-13 插入一条记录

在ListView的列表项中，点击“删除”图标，执行删除操作，效果如图4-14所示。

图4-14 删除一条记录

在ListView列表项中点击“修改”图标，实现修改一条记录的操作，效果如图4-15所示。

图4-15 修改一条记录

4.5 ContentProvider

Android系统中不同的应用程序分别运行在各自的进程中，应用程序之间的数据不能相互访问。ContentProvider是Android系统提供的用于数据共享的组件，利用

ContentProvider 可以在隐藏实现细节基础上，向其他应用程序提供访问和操作数据的接口，其他应用程序通过 ContentResolver 来获取使用 ContentProvider 暴露的数据。

利用 ContentProvider 共享数据的工作原理如图 4-16 所示。

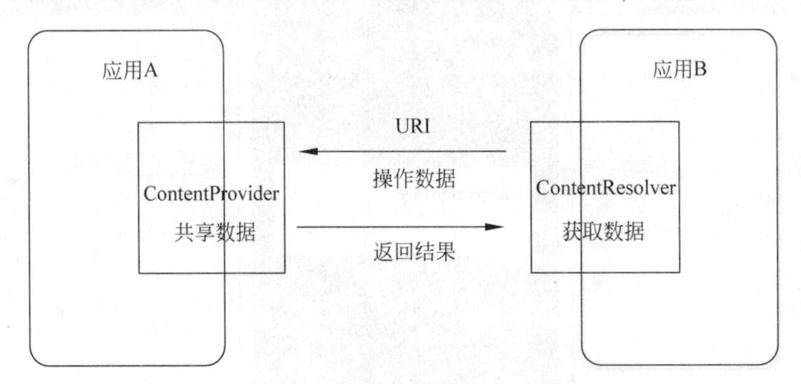

图 4-16 ContentProvider 工作原理

由图 4-16 所示 ContentProvider 工作原理可知，ContentProvider 通过 URI 形式向外提供数据，其他应用程序通过 URI 操作指定的数据。

ContentProvider 的 URI 字符串格式如下：

```
content://<authority>/<path>/id
```

URI 组成部分说明如下：

(1) content://：ContentProvider 的通用标准前缀，表示该 URI 用于访问 ContentProvider 资源。

(2) authority：是在 AndroidManifest.xml 文件中为 ContentProvider 指定的 autority，该值是唯一的，用于表示当前的 ContentProvider。Android 系统由这个部分找到对应的 ContentProvider。

(3) path：资源部分，当访问者需要访问不同资源时，这个部分可以动态改变。

(4) id：数据编号，用于唯一确定一条记录。

需要使用 URI 工具类的 parse()方法将字符串转化为 URI 如下：

```
Uri uri=Uri.parse("conent://cn.nsu.lab.contentprovider/data");
```

4.5.1 自定义 ContentProvider

Android 系统通过继承 ContentProvider 类来创建数据提供者，由于 ContentProvider 是抽象类，所以在创建的类中需要重写 ContentProvider 中的抽象方法。这些抽象方法如表 4-8 所示。

getType(Uri uri)用于返回 URI 路径指定数据的类型，如果指定数据的类型属于集合型(多条数据)，getType()方法返回的字符串应该以"vnd.android.cursor.dir/"开头；

如果属于非集合(单条数据),则返回的字符串以"vnd. android. curosr. item/"开头。

表 4-8 ContentProvider 的抽象方法

状　态	描　述
publicboolean onCreate()	创建 ContentProvider 时调用
Public int delete(Uri uri, String selection, String[] selectionArgs)	根据传入的 URI 删除指定条件下的数据
publilc Uri insert(Uri uri, ContentValues values)	根据传入的 URI 插入数据
public Cursor query(Uri uri, String[] projection, String selection, String[] selectionArgs, String sortOrder)	根据传入的 URI 查询指定条件下的数据
public int update(Uri uri, ContentValues values, String selection, String[] selectionArgs)	根据传入的 URI 更新指定条件的数据
public String getType(Uri uri)	返回指定 URI 代表数据的 MIME 类型

自定义 ContentProvider 的创建步骤如下:

(1) 创建内容提供者。

创建一个继承 ContentProvider 的 Java 类,根据实际业务逻辑,重写类中的 6 个方法。

(2) 注册内容提供者。定义的 ContentProvider 需要在 AndroidManifest. xml 中注册后才能够被其他应用程序访问。ContentProvider 对应< application/>元素的< provider >元素,注册 ContentProvider 时需要对应的< provider >元素中声明一系列的属性。注册 ContentProvider 的语法格式如下:

```
<provider
        android:name="string"
        android:authorities="list"
        android:enabled=["true" | "false"]
        android:exported=["true" | "false"]
        android:grantUriPermissions=["true" | "false"]
        android:icon="drawable resource"
        android:initOrder="integer"
        android:label="string resource"
        android:multiprocess=["true" | "false"]
        android:permission="string"
        android:process="string"
        android:readPermission="string"
        android:syncable=["true" | "false"]
        android:writePermission="string">
    ...
</provider>
```

常用属性的说明如下：

1. android:name

用于指定ContentProvider实现类的名称，这个属性应该使用完整的Java类名来设定（如com. nsu. zyl. UsersContentProvider），这个属性没有默认值，必须给这个属性设定一个值。

2. android:authorities

用于指定该ContentProvider对应的URI列表，当有多个URI时，用分号来分割。为了避免冲突，URI应该使用Java样式的命名规则（如com. nsu. zyl. contentprovider）。这个属性没有默认值，至少要指定一个URI。

3. android:enabled

用于指定该ContentProvider是否能够被系统安装。值为true，表示可以安装，否则不能安装。默认值是true。

4. android:exported

用于指定该ContentProvider是否能够被其他的应用程序组件使用。如果设置为true，则可以被使用，否则不能被使用。默认值是true。虽然能够使用这个属性来公开ContentProvider，但是依然可以用permission属性来限制对它的访问。

5. android:grantUriPermission

用于设定那些对ContentProvider的数据没有访问权限的访问者，是否能够被授予访问的权限，这个权限是临时性的，它会克服由readPermission、writePermission和permission属性的设置限制。如果这个属性设置为true，那么权限就可以授予访问者，否则不会授予访问者。

6. android:icon

用于定义一个代表ContentProvider的图标。如果这个属性没有设置，那么就会使用应用程序的<application>元素的icon属性值来代替。

7. android:initOrder

用于定义ContentProvider被实例化的顺序，这个顺序是相对于相同进程所拥有的其他内容提供器的。当内容提供器之间有相互依赖时，就需要设置这个属性，以确保它们能够按照其依赖的顺序被创建。这个属性值是一个简单的整数，大的数字要被优先初始化。

8. android:label

用于给ContentProvider定义一个用户可读的标签。如果这个属性没有设置，那么它会使用< application >元素的 label 属性值来代替。

9. android:multiprocess

用于设定是否能够在每个使用该内容提供器的客户端进程中都创建一个内容提供器的实例,如果设置为 true,就能够在其每个客户端进程中创建一个实例,否则不可以。默认值是 false。

10. android:permission

用于设定客户端在读写 ContentProvider 的数据时必须要有的权限的名称。这个属性为同时设置读写权限提供了一种便利的方法。但是 readPermission 和 writePermission 属性的优先级要比这个属性高。如果 readPermission 属性也被设置了,那么它就会控制对内容提供器的查询访问。如果 writePermission 属性被设置,它就会控制对内容提供器数据的修改访问。

11. android:process

用于定义 ContentProvider 运行所在的进程名称。通常,应用程序的所有组件都运行在给应用程序创建的默认进程中。它有与应用程序包相同的名称。< application >元素的 process 属性能够给其所有的组件设置一个不同的默认进程,但是每个组件都能够用它们自己的 process 属性来覆盖这个默认设置,从而允许把应用程序分离到不同的多个进程中。

如果这个属性值是用":"开头的,在需要这个 ContentProvider 时,系统就会给这个应用程序创建一个新的私有进程,并且对应的 Activity 也要运行在这个私有进程中。如果用小写字母开头,那么 Activity 则会运行在一个用这个属性值命名的全局进程中,它提供了对 ContentProvider 的访问权限。这样就允许不同应用程序的组件能够共享这个进程,从而减少对系统资源的使用。

12. android:readPermission

用于设置查询内容提供器的数据时客户端所必须要有的权限。

13. android:syncable

用于设定内容提供器控制下的数据是否要与服务器上的数据进行同步,如果设置为 true,则要同步,否则不需要同步。

14. android:writePermission

用于设置修改内容提供器的数据时客户端所必须要有的权限。

以上属性中,android:name 和 android:authrities 属性必须设置,其他属性根据需要进行配置。

4.5.2 访问 ContentProvider

对于使用 ContentProvider 共享的数据，其他应用程序可以使用 ContentResolver 访问和操作。ContentResolver 通过 URI 来查询 ContentProvider 中提供的数据。ContentProvider 中重写的方法需要由调用者 ContentResolver 来调用才能触发，ContentResolver 传入 URI 参数来调用这些方法。为了确定 ContentProvider 实际能处理的 URI 以及方法中 URI 参数能够操作的数据，Android 提供了一个辅助工具类 UriMatcher 用于匹配 URI。

UriMatcher 的常用方法如表 4-9 所示。

表 4-9　UriMatcher 常用方法

方法名称	功能描述
public UriMatcher(int code)	创建 UriMatcher 对象时调用，参数通常使用 UriMatcher.NO_MATCH，表示路径不满足条件返回－1
Public void addURI (String authority, String path, int code)	用于向 UriMatcher 对象注册 URI，其中 authority 和 path 组成一个 URI，code 代表该 URI 对应的标识码
publilc int match(Uri uri)	根据注册 URI 匹配获取对应的标识码，如果匹配不到，则返回－1

除此之外，Android 还提供一个用于操作 URI 字符串的工具类——ContentUris。该类提供如表 4-10 所示方法。

表 4-10　ContentUris 常用方法

方法名称	功能描述
public static Uri withAppendedId(Uri contentUri,long id)	用于为路径加上 ID
Public static long parseId(Uri contentUri)	从路径中获取 ID

withAppendedId()方法的使用：

```
Uri uri=Uri.parse("content://cn.nsu.lab.contentprovider/data");
Uri newUri=ConentUris.withAppendedId(uri,1);
```

此时，newUri 为 content://cn. nsu. lab. contentprovider/data/1。

parseId(uri)方法的使用：

```
Uri uri=Uri.parse("content://cn.nsu.lab.contentprovider/data/1");
long dataId=ContentUris.parseId(uri);     //dataid的值为 1
```

访问 ContentProvider 的步骤如下：

1. 获取 ContentResolver 对象

```
ContentResolver resolver=getContentResolver();
```

2. 操作数据

ContentResolver 类提供了与 ContentProvider 类相同签名的 4 个方法。

(1) public Uri insert(Uri uri, ContentValues values): 该方法用于往 ContentProvider 添加数据。

(2) public int delete(Uri uri, String selection, String[] selectionArgs): 该方法用于从 ContentProvider 删除数据。

(3) public int update(Uri uri, ContentValues values, String selection, String[] selectionArgs): 该方法用于更新 ContentProvider 中的数据。

(4) public Cursor query(Uri uri, String[] projection, String selection, String[] selectionArgs, String sortOrder): 该方法用于从 ContentProvider 中获取数据。

这些方法的第一个参数为 URI,代表要操作的 ContentProvider 和对其中的哪些数据进行操作,使用 ContentResolver 访问 ContentProvider 共享数据的代码如下:

```
Uri  uri=Uri.parse("com.nsu.zyl.contentprovider");
    ContentResolver resolver=getContentResolver();
    Cursor cursor=resolver.query(uri,new String[]{"id","isbn","bookname","
author", "price", "publisher", "address","date"}, null,null,null);
    while(cursor.moveToNext()){
          int id=cursor.getInt(0);
          String isbn=cursor.getString(1);
          String bookname=cursor.getString(2);
          String author=cursor.getString(3);
          double price=cursor.getDouble(4);
          String publisher=cursor.getString(5);
          String address=cursor.getString(6);
        }
        cursor.close();
```

上述代码使用 ContentResolver 对象的 query()方法实现对其他应用数据的查询功能。

4.5.3 案例

本节通过案例讲解如何通过自定义 ContentProvider 向外暴露数据,其他应用程序利用 ContentProvider 获取和操作被暴露的数据。

在应用程序中自定义 ContentProvider,需要完成如下几项工作:

(1) 建立数据存储系统,原则上可以使用任何方式存储,但大多数 ContentProvider 使用 Android 文件存储方式和 SQLite 数据库存储方式保存数据。

(2) 继承 ContentProvider 类来提供数据访问。

(3) 在当前应用程序中的 AndroidManifest.xml 中声明定义的 ContentProvider。

前面章节中,在项目 Chapter04Application 中定义了 SQLiteOpenHelper 的子类 DBHelper 创建 SQLite 数据库用于保存员工的信息,因此这部分将定义 ContentProvider 向其他应用程序提供访问 SQLite 数据库中的数据的接口。

定义类 EmployeesContentProvider 继承 ContentProvider,用于实现暴露数据库 employees 表中数据的功能。EmployeesContentProvider 代码如文件清单 4-14 所示。

文件清单 4-14 EmployeesContentProvider.java

```
package cn.edu.nsu.zyl.chapter04application;

import android.content.ContentProvider;
import android.content.ContentUris;
import android.content.ContentValues;
import android.database.Cursor;
import android.database.sqlite.SQLiteDatabase;
import android.net.Uri;
import android.util.Log;

public class EmployeesContentProvider extends ContentProvider {
    private DBOpenHelper dbOpenHelper;
    private static final String TABLENAME="employees";

    public EmployeesContentProvider() {

    }

    @Override
    public int delete(Uri uri, String selection, String[] selectionArgs) {
        SQLiteDatabase database=dbOpenHelper.getReadableDatabase();
        int count=database.delete(TABLENAME,selection,selectionArgs);
        database.close();
        return count;
    }

    @Override
    public String getType(Uri uri) {

        throw new UnsupportedOperationException("Not yet implemented");

    }

    @Override
    public Uri insert(Uri uri, ContentValues values) {

        SQLiteDatabase database=dbOpenHelper.getReadableDatabase();
```

```
            long rowId=database.insert(TABLENAME,null,values);
            database.close();
            if(rowId>0){
                Uri insertUri=ContentUris.withAppendedId(uri,rowId);
                return insertUri;
            }else{
                return null;
            }

        }

        @Override
        public boolean onCreate() {
            dbOpenHelper=new DBOpenHelper(getContext(),1);
            return true;
        }

        @Override
        public Cursor query(Uri uri, String[] projection, String selection,
                            String[] selectionArgs, String sortOrder) {
            SQLiteDatabase database=dbOpenHelper.getReadableDatabase();
            Cursor cursor=database.query(TABLENAME,projection,selection,selectionArgs,
    null,null,sortOrder);

            Log.i("ContentProvider","cursor.size="+cursor.getColumnCount());
            return cursor;
        }

        @Override
        public int update(Uri uri, ContentValues values, String selection,
                          String[] selectionArgs) {
            SQLiteDatabase database=dbOpenHelper.getReadableDatabase();
            int count=database.update(TABLENAME,values,selection,selectionArgs);
            database.close();
            return count;
        }
    }
```

上述代码中的EmployeesContentProvider类重写ContentProvider中的6个方法，实现了对数据库表employees的增、删、改、查等操作。

(4) 在当前应用的AndroidManifest.xml文件中注册EmployeesContentProvider。

EmployeesContentProvider在AndroidManifest.xml中的配置信息如下：

```
<provider
        android:name=".EmployeesContentProvider"
    android:authorities="cn.edu.nsu.zyl.employeescontentprovider"
        android:enabled="true"
        android:exported="true">
</provider>
```

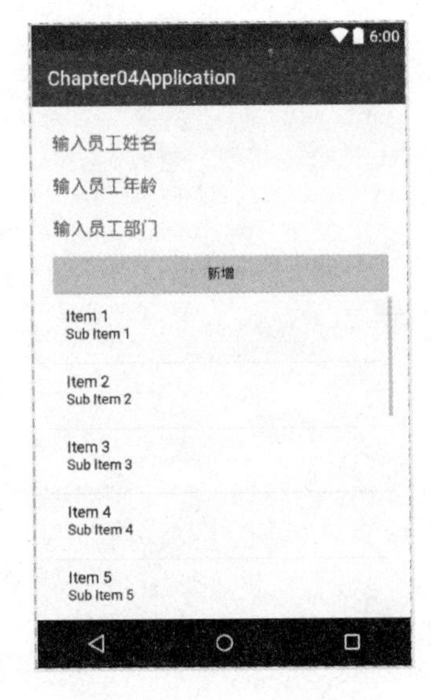

图 4-17 程序主界面

android:name 指定该 ContentProvider 的实现类为 EmployeesContentProvider，android: authorities 指定该 ContentProvider 对应的 URI 为 content:// cn. edu. nsu. zyl. employeescontentprovider，android: exported = true 指定该内容提供器能够被其他的应用程序组件使用。

接下来在 Chapter04Application 项目中创建 ContentProviderActivity，在程序中通过 ContentResolver 来访问 EmployeesContentProvider 提供的数据。程序交互界面如图 4-17 所示，在该界面中包含用于输入员工姓名、年龄和部门的文本框、用于点击的“新增”按钮以及用于显示所有员工信息的 ListView 控件。

ContentProviderActivity 对应界面的布局文件 activity_content_provider. xml 如文件清单 4-15 所示。

文件清单 4-15　activity_content_provider. xml

```
<?xml version="1.0" encoding="utf-8"?>
<RelativeLayout xmlns:android="http://schemas.android.com/apk/res/android"
    xmlns:tools="http://schemas.android.com/tools"
    android:layout_width="match_parent"
    android:layout_height="match_parent"
    android:paddingBottom="@dimen/activity_vertical_margin"
    android:paddingLeft="@dimen/activity_horizontal_margin"
    android:paddingRight="@dimen/activity_horizontal_margin"
    android:paddingTop="@dimen/activity_vertical_margin"

    tools:context="cn.edu.nsu.zyl.chapter04application.ContentProviderActivity">

    <EditText
        android:layout_width="match_parent"
        android:layout_height="wrap_content"
        android:id="@+id/edtName"
        android:hint="输入员工姓名"
        android:layout_alignParentTop="true" />

    <EditText
        android:layout_width="match_parent"
        android:layout_height="wrap_content"
        android:id="@+id/edtAge"
        android:layout_below="@+id/edtName"
        android:hint="输入员工年龄"
        android:layout_alignParentLeft="true"
        android:layout_alignParentStart="true"
        />
```

```
        <EditText
            android:layout_width="match_parent"
            android:layout_height="wrap_content"
            android:id="@+id/edtDepartment"
            android:layout_below="@+id/edtAge"
            android:hint="输入员工部门"
            android:layout_alignParentLeft="true"
            android:layout_alignParentStart="true" />

        <Button
            android:layout_width="match_parent"
            android:layout_height="wrap_content"
            android:text="新增"
            android:id="@+id/btnInsert"
            android:layout_below="@+id/edtDepartment"
            android:layout_alignParentLeft="true"
            android:layout_alignParentStart="true" />
        <ListView
            android:layout_width="wrap_content"
            android:layout_height="wrap_content"
            android:id="@+id/listView"
            android:layout_below="@+id/btnInsert"
            android:layout_alignParentLeft="true"
            android:layout_alignParentStart="true" />

    </RelativeLayout>
```

在主界面中使用 ListView 来展示获取的员工信息数据，因此需要在/res/layout/的目录下创建 item_employees_list.xml 文件，编写 item 的布局。item_employees_list.xml 的图形化界面如图 4-18 所示。

图 4-18 ListView item 布局

item_employees_list.xml 文件的代码如文件清单 4-16 所示。

文件清单 4-16 item_employees_list.xml

```
<?xml version="1.0" encoding="utf-8"?>
<RelativeLayout xmlns:android="http://schemas.android.com/apk/res/android"
    android:layout_width="match_parent"
    android:layout_height="match_parent">
    <TextView
        android:layout_width="wrap_content"
        android:layout_height="wrap_content"
        android:padding="5dp"
        android:text="Id"
        android:id="@+id/txtId"
        android:layout_alignParentTop="true"
        android:layout_alignParentLeft="true"
        android:layout_alignParentStart="true" />

    <TextView
        android:layout_width="wrap_content"
        android:layout_height="wrap_content"
        android:padding="5dp"
        android:text="Name"
        android:id="@+id/txtName"
        android:layout_alignParentTop="true"
        android:layout_toRightOf="@+id/txtId"
        android:layout_toEndOf="@+id/txtId" />

    <TextView
        android:layout_width="wrap_content"
        android:layout_height="wrap_content"
        android:padding="5dp"
        android:text="Age"
        android:id="@+id/txtAge"
        android:layout_alignParentTop="true"
        android:layout_toRightOf="@+id/txtName"
        android:layout_toEndOf="@+id/txtName" />

    <TextView
        android:layout_width="wrap_content"
        android:layout_height="wrap_content"
        android:padding="5dp"
        android:text="Department"
        android:id="@+id/txtDepartment"
        android:layout_alignParentTop="true"
        android:layout_toRightOf="@+id/txtAge"
        android:layout_toEndOf="@+id/txtAge" />

    <ImageView
        android:layout_width="wrap_content"
        android:layout_height="wrap_content"
        android:padding="5dp"
```

```
        android:id="@+id/imgDelete"
        android:layout_alignParentTop="true"
        android:layout_alignParentRight="true"
        android:layout_alignParentEnd="true"
        android:src="@android:drawable/ic_delete" />

    <ImageView
        android:layout_width="wrap_content"
        android:layout_height="wrap_content"
        android:padding="5dp"
        android:id="@+id/imgEdit"
        android:src="@android:drawable/ic_menu_edit"
        android:layout_alignParentTop="true"
        android:layout_toLeftOf="@+id/imgDelete"
        android:layout_toStartOf="@+id/imgDelete" />
</RelativeLayout>
```

为了实现在 ListView 显示从 ContentProvider 中获取的数据，自定义 EmployeesAdapter，EmployeesAdapter 代码如文件清单 4-17 所示。

文件清单 4-17　EmployeesAdapter.java

```
package cn.edu.nsu.zyl.chapter04application;

import android.app.AlertDialog;
import android.content.Context;
import android.content.DialogInterface;
import android.view.LayoutInflater;
import android.view.View;
import android.view.ViewGroup;
import android.widget.BaseAdapter;
import android.widget.Button;
import android.widget.EditText;
import android.widget.ImageView;
import android.widget.TextView;

import java.util.ArrayList;
import java.util.HashMap;

/**
 * Created by colinzhong on 2017/8/13.
 */

public class EmployeesAdapter extends BaseAdapter {
    private Context context;
    private ArrayList list;

    public EmployeesAdapter(Context context,ArrayList list){
        this.context=context;
```

```
        this.list=list;
    }
    @Override
    public int getCount() {
        return list.size();
    }

    @Override
    public Object getItem(int i) {
        return list.get(i);
    }

    @Override
    public long getItemId(int i) {
        return i;
    }

    @Override
    public View getView(final int i, View view, ViewGroup viewGroup) {
        View itemView=view;
        ViewHolder holder;
        if(itemView==null){
            LayoutInflater inflater= (LayoutInflater) context.getSystemService
(Context.LAYOUT_INFLATER_SERVICE);

itemView=inflater.inflate(R.layout.item_sqlite_operator,null);
            holder=new ViewHolder();
            holder.txtId= (TextView)itemView.findViewById(R.id.txtId);

holder.txtName= (TextView)itemView.findViewById(R.id.txtName);

holder.txtAge= (TextView)itemView.findViewById(R.id.txtAge);

holder.txtDepartment= (TextView)itemView.findViewById(R.id.txtDepartment);

holder.imgDelete= (ImageView)itemView.findViewById(R.id.imgDelete);

holder.imgEdit= (ImageView)itemView.findViewById(R.id.imgEdit);
            itemView.setTag(holder);
        }else{
            holder= (ViewHolder) itemView.getTag();
        }
        holder.txtId.setText(((HashMap)list.get(i)).get("_id")+"");

holder.txtName.setText(((HashMap)list.get(i)).get("name").toString());
```

```
        holder.txtAge.setText(((HashMap)list.get(i)).get("age")+"");

holder.txtDepartment.setText(((HashMap)list.get(i)).get("department").toString());

        holder.imgEdit.setOnClickListener(new View.OnClickListener() {
            @Override
            public void onClick(View view) {
                AlertDialog.Builder builder=new AlertDialog.Builder(context);
                builder.setTitle("请输入更新后的值");
                final View dialogView = LayoutInflater.from(context).inflate(R.
layout.dialog_edit,null);
                final EditText edtName = (EditText)  dialogView.findViewById(R.
id.edtName);
                final EditText edtAge = (EditText) dialogView.findViewById(R.id.
edtAge);
                final EditText edtDepartment = (EditText) dialogView.findViewById
(R.id.edtDepartment);
                builder.setView(dialogView);
                final AlertDialog dialog=builder.create();

edtAge.setText(((HashMap)list.get(i)).get("age").toString());

edtName.setText(((HashMap)list.get(i)).get("name").toString());

edtDepartment.setText(((HashMap)list.get(i)).get("department").toString());

                Button btnSubmit= (Button)dialogView.findViewById(R.id.btnSubmit);

                btnSubmit.setOnClickListener(new View.OnClickListener() {
                    @Override
                    public void onClick(View view) {
                        int _id=Integer.parseInt(((HashMap)list.get(i)).
get("_id").toString());
                        String name=edtName.getText().toString();
                        int age=Integer.parseInt(edtAge.getText().toString());
                        String department=edtDepartment.getText().toString();
                        HashMap map= (HashMap) list.get(i);
                        map.put("_id",_id);
                        map.put("name",name);
                        map.put("age",age);
                        map.put("department",department);
                        EmployeesDAO dao=new EmployeesDAO(context);
                        dao.update(name,age,department,_id);
                        notifyDataSetChanged();
                        dialog.dismiss();
                    }
```

```
                });
                dialog.show();




            }
        });

        holder.imgDelete.setOnClickListener(new View.OnClickListener() {
            @Override
            public void onClick(View view) {
                 DialogInterface.OnClickListener listener=new DialogInterface.
OnClickListener() {
                    @Override
                    public void onClick(DialogInterface dialogInterface, int pi) {

                        EmployeesDAO dao=new EmployeesDAO(context);

dao.delete(Integer.parseInt(((HashMap)list.get(i)).get("_id").toString()));
                        list.remove(i);
                        notifyDataSetChanged();
                    }
                };
                AlertDialog.Builder builder=new AlertDialog.Builder(context);
                builder.setTitle("确定删除该记录?");
                builder.setNegativeButton("取消",null);

                builder.setPositiveButton("确定",listener);
                builder.show();
            }
        });

        return itemView;
    }
    private static class ViewHolder{
        public TextView txtId;
        public TextView txtName;
        public TextView txtAge;
        public TextView txtDepartment;
        public ImageView imgDelete;
        public ImageView imgEdit;
    }

}
```

ContentProviderActivity 实现如下交互功能：

(1) 使用 ContentResolver 查询数据并将数据显示在 ListView 上；

(2) 点击"新增"按钮，利用 ContentResolver 插入数据到 SQLite 数据库；

(3) 点击列表某项中"编辑"按钮，弹出对话框完成编辑操作，利用 ContentResolver 实现对 SQLite 数据库中数据修改功能；

(4) 点击列表项实现删除 SQLite 数据库中数据的功能。

编辑 ContentProviderActivity，代码如文件清单 4-18 所示。

文件清单 4-18 ContentProviderActivity.java

```
package cn.edu.nsu.zyl.chapter04application;

import android.content.ContentResolver;
import android.content.ContentValues;
import android.database.Cursor;
import android.net.Uri;
import android.support.v7.app.AppCompatActivity;
import android.os.Bundle;
import android.view.View;
import android.widget.Button;
import android.widget.EditText;
import android.widget.ListView;

import java.util.ArrayList;
import java.util.HashMap;

public class ContentProviderActivity extends AppCompatActivity {

    private EmployeesAdapter employeesAdapter;
    private ArrayList list;
    private EditText edtName,edtAge,edtDepartment;
    private ListView employeesListView;
    private Button btnInsert;
    private Uri uri=Uri.parse("content://
        cn.edu.nsu.zyl.employeescontentprovider");
    private ContentResolver resolver;


    @Override
    protected void onCreate(Bundle savedInstanceState) {
        super.onCreate(savedInstanceState);
        setContentView(R.layout.activity_content_provider);
        edtName=(EditText) findViewById(R.id.edtName);
        edtAge=(EditText)findViewById(R.id.edtAge);
        edtDepartment=(EditText) findViewById(R.id.edtDepartment);
        employeesListView=(ListView)findViewById(R.id.listView);
```

```
        btnInsert= (Button)findViewById(R.id.btnInsert);

        list=initList();
        employeesAdapter=new EmployeesAdapter(ContentProviderActivity.this,list);
        employeesListView.setAdapter(employeesAdapter);

        btnInsert.setOnClickListener(new View.OnClickListener() {
            @Override
            public void onClick(View view) {
                String name=edtName.getText().toString();
                int age=Integer.parseInt(edtAge.getText().toString());
                String department=edtDepartment.getText().toString();
                ContentValues contentValues=new ContentValues();
                contentValues.put("name",name);
                contentValues.put("age",age);
                contentValues.put("department",department);
                resolver.insert(uri,contentValues);
                // list.clear();
                list=initList();
                employeesAdapter=new EmployeesAdapter(ContentProviderActivity.
this,list);
                employeesListView.setAdapter(employeesAdapter);
                edtName.setText("");
                edtAge.setText("");
                edtDepartment.setText("");
            }
        });
    }
    public ArrayList initList(){
        ArrayList list=new ArrayList();
        resolver=getContentResolver();
        Cursor cursor= resolver.query(uri,new String[]{"_id","name","age","
department"},null,null,null);
        while (cursor.moveToNext()){
            HashMap map=new HashMap();
            map.put("_id",cursor.getString(0));
            map.put("name",cursor.getString(1));
            map.put("age",cursor.getInt(2));
            map.put("department",cursor.getString(3));
            list.add(map);
        }
        return list;
    }
}
```

运行该程序，主界面显示如图4-19所示，页面列表中的数据是通过ContentResolver从EmployeesContentProvider获取。

在主页面的文本编辑框中输入需要插入到EmployeesContentProvider的员工信息，点击“新增”按钮，用户输入的数据就会作为一条记录插入到EmployeesContentProvider对应的数据库中，如图4-20所示。

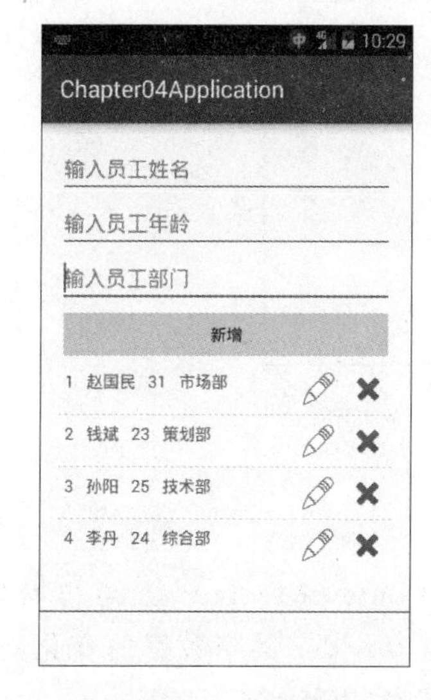

图 4-19 主页面

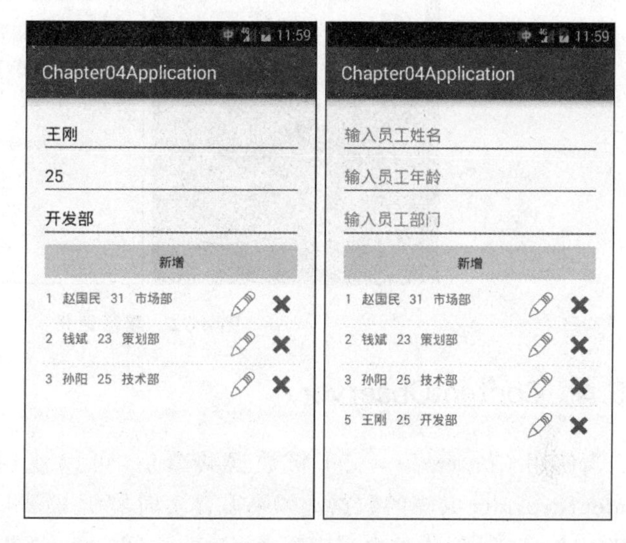

图 4-20 插入一条记录

点击列表项中的“编辑”按钮，可以完成对该列表对应记录的修改，如图4-21所示。

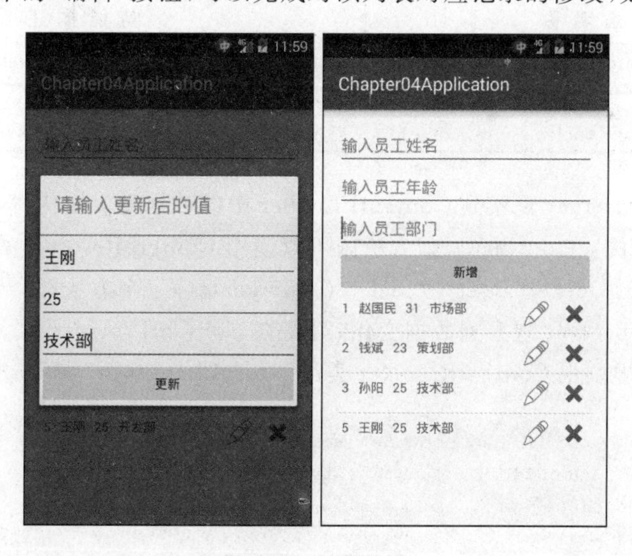

图 4-21 编辑信息

点击列表项，弹出“删除”确认对话框，确定后就可以完成删除操作，如图4-22所示。

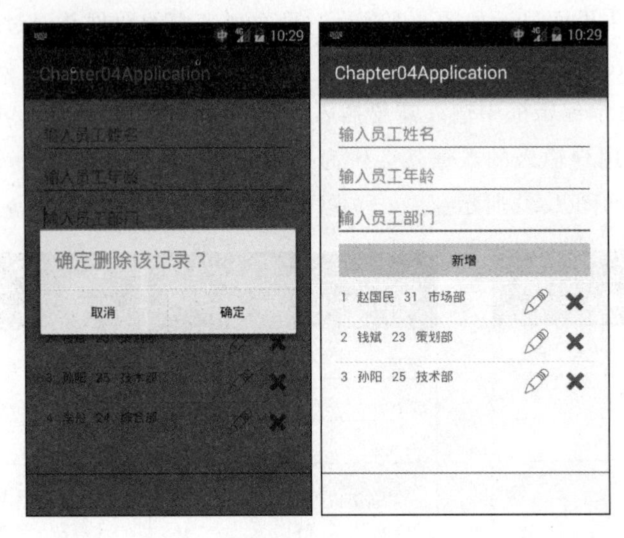

图 4-22 删除记录

4.5.4 ContentObserver

当使用 ContentProvider 将数据共享后，可以使用 ContentResolver 访问和操作 ContentProvider 共享的数据。如果需要实时监听 ContentProvider 共享数据的变化，可以使用 Android 提供的内容观察者(ContentObserver)来实现(表 4-11)。

表 4-11 ContentObserver 常用方法

方法名称	功能描述
public void ContentObserver(Handler handler)	ContentObserver 构造方法，ContentObserver 的所有子类都必须调用该构造方法
public void onChange(boolean selfChange)	当观察到 URI 代表的数据发生变化时，会触发该方法

当 ContentObserver 观察到 ContentProvider 中的数据发生变化时，会回调执行内部的 onChange()方法，在 onChange()方法内可以使用 ContentResolver 查询变化的数据。

在 ContentProvider 中 insert()、delete()和 update()三个方法都会引起数据的变化，为了让 ContentObserver 观察到数据变化，需要在 ContentProvider 的 insert()、delete()和 update()方法中调用 ContentResolver 类的 notifyChange(uri,null)方法。示例如下：

```
public class PersonContentProvider extends ContentProvider {
    public Uri insert(Uri uri, ContentValues values) {
      db.insert("person", "personId", values);
      getContext().getContentResolver().notifyChange(uri, null);
    }
}
```

为观察 ContentProvider 中数据的变化，需要在程序中注册 ContentObserver：

```
getContentResolver().registerContentObserver(Uri.parse("content:// cn.edu.nsu.
colin.bookcontentprovider/words"), true, new BookObserver(new Handler()));
```

4.5.5 系统 ContentProvider

Android 系统提供许多可以直接使用的系统 ContentProvider，例如：

(1) MediaProvider：用来查询磁盘上多媒体文件；

(2) ContactsProvider：用来查询联系人信息；

(3) CalendarProvider：用来提供日历相关信息的查询；

(4) BookmarkProvider：用来提供书签信息的查询。

这些 ContentProvider 的使用大同小异，使用它们对应的 URI 地址就可以进行增、删、改、查的操作。Android 官方文档中提供相应 ContentProvider 的 URI 和 ContentProvider 操作的数据列的列名，可以根据需要查阅 Android 官方文档。

本节通过一个访问 MediaProvider 程序为例，展示如何使用系统提供的 ContentProvider。MediaProvider 作为系统级别的应用程序在系统上运行，专门负责收集多媒体文件(音频、视频、文件)相关的信息。当需要获取这类文件相关的信息时，可以向 MediaProvider 发起查询的请求。MediaProvider 在开机启动后，会在后台“监听”磁盘上文件的变化，特定情况下，会自动更新媒体文件的信息，例如磁盘上是否增加、修改或删除媒体文件等。

首先，需要确定访问的 URI，Android 为多媒体提供了如下的 URI：

(1) MediaStore. Audio. Media. EXTERNAL_CONTENT_URI：存储在外部设备的音频文件；

(2) MediaStore. Audio. Media. INTERNAL_CONTENT_URI：存储在手机内部的音频文件；

(3) MediaStore. Images. Media. EXTERNAL_CONTENT_URI：存储在外部设备的图片文件；

(4) MediaStore. Images. Media. INTERNAL_CONTENT_URI：存储在内部设备的图片文件；

(5) MediaStore. Video. Media. EXTERNAL_CONTENT_URI：存储在外部设备的视频文件；

(6) MediaStore. Video. Media. INTERNAL_CONTENT_URI：存储在内部设备的视频文件。

我们创建的应用程序主要功能是查看当前手机外部存储中所有图片，因此查询请求的 URI 地址为：

```
Uri uri =MediaStore.Images.Media.EXTERNAL_CONTENT_URI;
```

接下来，确定要请求的图片文件信息对应的字段名。需要查询图片的名字、详细信息和所在地址，这些信息在 MediaProvider 中都有对应的字段名称。

```
String[] searchKey = new String[] {
MediaStore.Images.Media.DISPLAY_NAME        //对应图片的名字
MediaStore.Images.Media.DESCRIPTION,        //对应图片的详细信息
MediaStore.Images.Media.DATA                //对应图片保存位置
};
```

创建用于查看当前手机外部存储区域中图片的 Activity，该 Activity 对应的布局如图 4-23 所示。

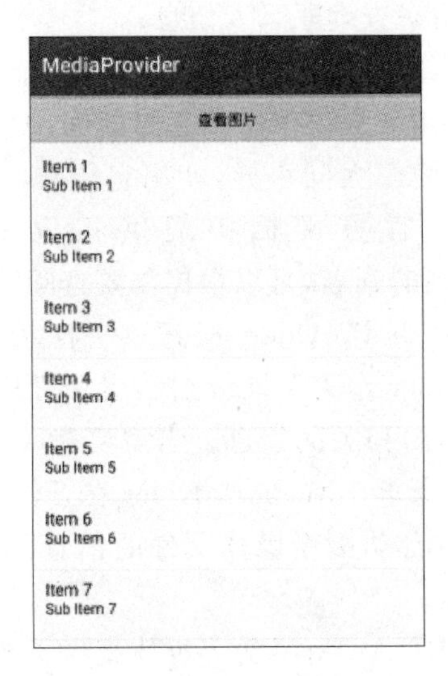

图 4-23 主页面布局

布局文件 main. xml 源代码如文件清单 4-19 所示。

文件清单 4-19 main. xml

```
<LinearLayout xmlns:android="http://schemas.android.com/apk/res/android"
    xmlns:tools="http://schemas.android.com/tools"
    android:layout_width="match_parent"
    android:layout_height="match_parent"
    android:orientation="vertical"
    tools:context=".MainActivity" >

    <Button
        android:id="@+id/btnSearch"
        android:layout_width="match_parent"
        android:layout_height="wrap_content"
        android:text="查看图片"/>

    <ListView
        android:id="@+id/imgList"
        android:layout_width="match_parent"
        android:layout_height="wrap_content"></ListView>
</LinearLayout>
```

在主页面中使用 ListView 展示所有的图片信息，对于 ListView 中列表项的布局为 items. xml，具体代码如文件清单 4-20 所示。

文件清单 4-20　items. xml

```
<?xml version="1.0" encoding="utf-8"?>
<LinearLayout xmlns:android="http://schemas.android.com/apk/res/android"
    android:layout_width="match_parent"
    android:layout_height="match_parent"
    android:orientation="vertical" >
    <TextView
        android:id="@+id/txt1"
        android:layout_width="match_parent"
        android:layout_height="wrap_content"/>
    <TextView
        android:id="@+id/txt2"
        android:layout_width="match_parent"
        android:layout_height="wrap_content"/>
</LinearLayout>
```

为列表设置点击事件监听器，当用户点击列表项时，弹出显示图片的对话框，对话框对应的布局文件为 view. xml，该布局文件具体代码如文件清单 4-21 所示。

文件清单 4-21　view. xml

```
<?xml version="1.0" encoding="utf-8"?>
<LinearLayout xmlns:android="http://schemas.android.com/apk/res/android"
    android:layout_width="match_parent"
    android:layout_height="match_parent"
    android:orientation="vertical" >

    <ImageView
        android:id="@+id/image1"
        android:layout_width="match_parent"
        android:layout_height="match_parent" />

</LinearLayout>
```

在程序中首先获取 ContentResolver 对象，让它使用前面的参数向 MediaProvider 发起查询请求，查询的结果存放在 Cursor 中。

```
//通过 ContentResolver 查询所有图片信息
Cursor curos =getContentResolver().query(
        MediaStore.Images.Media.EXTERNAL_CONTENT_URI, null, null, null, null);
```

遍历 Cursor，得到它指向的每一条查询到的信息，当 Cursor 指向某条数据时获取它携带每个字段的值。

```
Cursor curos =getContentResolver().query(
                        MediaStore.Images.Media.EXTERNAL_CONTENT_URI, null, null,
                        null, null);
                while (curos.moveToNext()) {
                    // 获取图片显示的名字
                    String name =curos.getString(curos
                        .getColumnIndex(MediaStore.Images.Media.DISPLAY_NAME));
                    // 获取图片的详细信息、
                    String desc =curos.getString(curos
                        .getColumnIndex(MediaStore.Images.Media.DESCRIPTION));
                    // 保存图片名的位置数据
                    byte[] data =curos.getBlob(curos
                        .getColumnIndex(MediaStore.Images.Media.DATA));
                    // 将图片名添加到 names 集合中
                    names.add(name);
                    // 将图片描述添加到 desc 集合中
                    descs.add(desc);
                    // 将图片保存路径添加到 filenames 集合中
                    filenames.add(new String(data, 0, data.length -1));
                }
```

Cursor 使用后要把它关闭。

```
cursor.close();
```

MainActivity 的具体代码如文件清单 4-22 所示。

文件清单 4-22　MainActivity. java

```
public class MainActivity extends AppCompatActivity {

    private Button btnSearch;
    private ListView imgList;

    ArrayList<String>names =new ArrayList<String>();
    ArrayList<String>descs =new ArrayList<String>();
    ArrayList<String>filenames =new ArrayList<String>();

    @Override
    protected void onCreate(Bundle savedInstanceState) {
        super.onCreate(savedInstanceState);
        setContentView(R.layout.activity_main);
        btnSearch = (Button) findViewById(R.id.btnSearch);

        imgList = (ListView) findViewById(R.id.imgList);

        btnSearch.setOnClickListener(new View.OnClickListener() {
```

```
            @Override
            public void onClick(View v) {
                // 清空 names、desc、fileName 集合里原有的数据
                names.clear();
                descs.clear();
                filenames.clear();
                // 通过 ContentResolver 查询所有图片信息
                Cursor curos =getContentResolver().query(
                        MediaStore.Images.Media.EXTERNAL_CONTENT_URI, null, null,
                        null, null);
                while (curos.moveToNext()) {
                    // 获取图片显示的名字
                    String name =curos.getString(curos
                            .getColumnIndex(MediaStore.Images.Media.DISPLAY_NAME));
                    // 获取图片的详细信息、
                    String desc =curos.getString(curos
                            .getColumnIndex(MediaStore.Images.Media.DESCRIPTION));
                    // 保存图片名的位置数据
                    byte[] data =curos.getBlob(curos
                            .getColumnIndex(MediaStore.Images.Media.DATA));
                    // 将图片名添加到 names 集合中
                    names.add(name);
                    // 将图片描述添加到 desc 集合中
                    descs.add(desc);
                    // 将图片保存路径添加到 filenames 集合中
                    filenames.add(new String(data, 0, data.length -1));
                }
                // 创建一个 List 集合的元素是 Map
                List<Map<String, Object<<listitems =new ArrayList<Map<String,
Object<<();
                // 将 names、descs 两个集合对象的数据转换到 Map 集合
                for (int i =0; i <names.size(); i++) {
                    Map<String, Object>listitem =new HashMap<String, Object>();
                    listitem.put("name", names.get(i));
                    listitem.put("desc", descs.get(i));
                    listitems.add(listitem);
                }
                SimpleAdapter simple =new SimpleAdapter(MainActivity.this,
                        listitems, R.layout.items, new String[]{"name",
                        "desc"}, new int[]{R.id.txt1, R.id.txt2});
                imgList.setAdapter(simple);

            }
        });
        imgList.setOnItemClickListener(new AdapterView.OnItemClickListener() {

            @Override
            public void onItemClick(AdapterView<?>arg0, View arg1, int arg2,
```

```
                                long arg3) {
                //加载 view.xml 界面布局代表视图
                View view =getLayoutInflater().inflate(R.layout.view, null);
                //获取 viewDialog 中 ImageView 组件
                ImageView image1 = (ImageView) view.findViewById(R.id.image1);
                //设置 image 显示指定的图片

                image1.setImageBitmap(BitmapFactory.decodeFile(filenames
                        .get(arg2)));
                //使用对话框显示用户点击的图片
                new AlertDialog.Builder(MainActivity.this).setView(view)
                        .setPositiveButton("确定", null).show();

            }
        });

    }
}
```

运行效果如图 4-24 所示。

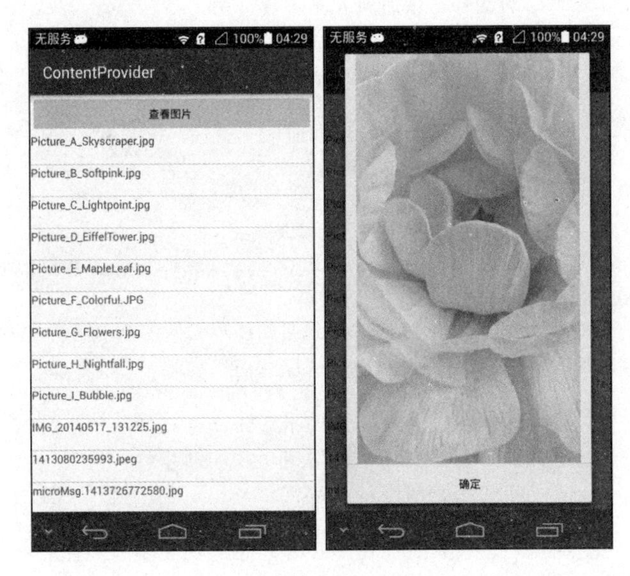

图 4-24　运行效果

本章小结

本章主要讲述 Android 数据存储技术，首先介绍 Android 中常见的数据存储方式，然后依次介绍使用 SharedPreferences、文件、SQLite 数据库存储和获取数据，最后介绍利用 ContentProvider 对外提供数据操作的接口。ContentProvider 为存储和获取数据提供了统一的接口，从而可以在不同应用程序之间共享数据。ContentProvider 对数据进行封装，不用关心数据存储的细节。Android 为常见的一些数据提供了默认的 ContentProvider（包括音频、视频、图片和通讯录等）。

习　　题

1. Android 中可用的数据存储方式有________、________、________、________和________。

2. SharedPreferences 存储的是________的数据。

3. 使用外部存储区之前需要调用________方法来检查存储介质是否可用。

4. 在 Android 中使用________唯一标识 ContentProvider 提供的内容。

5. 使用 SQLite 数据库时，创建数据库以及数据库版本更新需要继承________。

6. 下面关于 SharedPreferences 说法错误的是（　　）。

A. SharedPreferences 以键值对的形式保存并取回数据

B. 使用 getSharedPreferences 或 getPreferences()方法得到 SharedPreferences

C. SharedPreferences 可以保存任意类型的数据

D. 调用 SharedPreferences.Editor 对象保存数据时，需要调用 commit()方法提交修改

7. 下面关于 SQLite 数据库说明正确的是（　　）。

A. SQLite 是一个开源的嵌入式非关系数据库

B. SQLite 保存数据时必须指定数据类型

C. SQLite 数据库单独占用一个进程

D. SQLiteDatabase 对象的 execSQL(String sql)方法可以执行各种 SQL 语句

8. 下列关于 ContentResolver 描述错误的是（　　）。

A. 可以直接操作数据库数据

B. 操作其他应用数据必须知道包名

C. 只能操作 ContentProvider 暴露的数据

D. 可以操作 ContentProvider 的任意数据

9. 简述在 Android 系统中如何使用文件保存和获取数据。

10. 简述使用 ContentProvider 对外提供数据的步骤。

第5章 chapter 5

服务与广播

主要内容：Service的创建与注册，Service的两种启动方式，常用系统Service的使用，BroadcastReceiver的创建与使用，普通广播与有序广播，监听系统广播

课　　时：8课时

知识目标：(1) 掌握Service的创建与注册；
(2) 掌握Service的启动；
(3) 了解常用系统Service的使用；
(4) 掌握BroadcastReceiver的创建与注册；
(5) 掌握普通广播和有序广播的区别；
(6) 了解如何监听系统广播。

能力目标：(1) 具备Service开发能力；
(2) 具备BroadcastReceiver开发能力。

为了更好地通过示例讲解Service和BroadcastReceiver在Android中的使用，我们首先在AndroidStudio中新建Android项目Chapter05Application，并在该项目中完成本章相关章节示例代码。新建的Chapter05Application项目文件结构如图5-1所示。

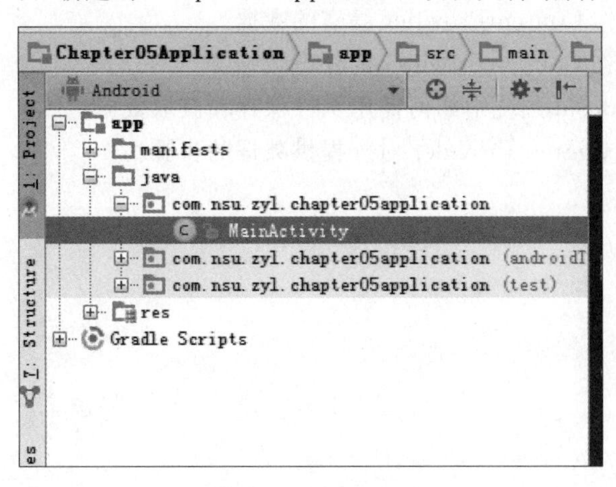

图5-1　Chapter05Application项目文件结构

其中，MainActivity 是 Chapter05Application 项目的入口页面，也是本章相关示例的导航页面。MainActivity 以列表的形式列出本章涉及的相关示例，当用户点击 MainActivity 页面中的列表项时，应用将跳转至某一具体示例。由于 MainActivity 页面以列表的形式展示，所以我们没有为 MainActivity 设置布局页面，直接使 MainActivity 继承 ListActivity。最初 MainActivity 的内容如文件清单 5-1 所示。

文件清单 5-1　MainActivity. java

```
package com.nsu.zyl.chapter05application;
import android.app.ListActivity;
import android.content.Intent;
import android.os.Bundle;
import android.view.View;
import android.widget.ArrayAdapter;
import android.widget.ListAdapter;
import android.widget.ListView;

public class MainActivity extends ListActivity {

    @Override
    protected void onCreate(Bundle savedInstanceState) {
        super.onCreate(savedInstanceState);
        setContentView(R.layout.activity_main);
        String[] items=new String[]{"startService 启动 Service","bindService 启动
Service","IntentService 的使用","通知服务","短信服务","广播接收器的创建与注册",
"接收有序广播"};
        ListAdapter adapter = new ArrayAdapter (MainActivity. this, android. R.
layout.simple_list_item_1,  items);
        setListAdapter(adapter);
    }

    @Override
    protected void onListItemClick(ListView l, View v, int position, long id) {
        super.onListItemClick(l, v, position, id);

    }
  }
}
```

运行 Chapter05Application 将显示图 5-2 所示页面。

为了实现点击 MainActivity 页面列表项跳转至某一具体页面的操作，在后面的案例中我们将逐步更新 onListItemClick()方法，从而将各个小节相关案例关联起来。

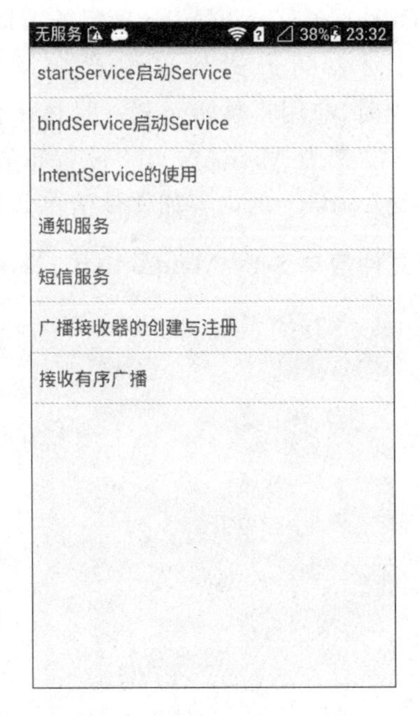

图 5-2 MainActivity 页面

5.1 Service

Activity 是一种具有用户界面的组件，当 Activity 的界面不在前台显示时，Activity 将由运行状态转换为暂停或者停止状态，当系统空间资源紧张时，处于暂停或停止状态的 Activity 将被销毁。对于那些需要在后台长时间运行的业务使用 Activity 就不合适了，这时需要借助 Android 提供的另一种组件——Service。Service（服务）是一种能够在后台长期运行且不提供用户界面的应用程序组件。很多应用使用 Service 组件实现在后台长时间执行的功能，例如在后台执行音乐播放或文件下载等功能。Service 启动后，即使用户切换到另一应用程序，Service 仍然可以在后台运行。

5.1.1 Service 的创建与注册

在程序中开发 Service 与开发 Activity 类似，需要经过创建和注册两个步骤。创建自定义 Service 类时，需要继承 Service 类或它的子类，创建好的 Service 类需要在 AndroidManifest.xml 中注册后才能被其他应用程序组件启动。

1. 创建 Service

自定义的 Service 类需要继承 Service 类或它的子类，与 Activity 类似，在 Service 生命周期中也定义了一系列的回调方法，可以根据实际需要对这些生命周期方法进行重写。Service 生命周期中常用的回调方法如表 5-1 所示。

表 5-1 Service 生命周期回调方法

方 法	描 述
IBinder onBind(Intent intent)	Service 中定义的抽象方法，Service 子类必须实现该方法。该方法返回一个 IBinder 对象，应用程序通过该方法的返回对象与服务通信
void onCreate()	当 Service 被创建后回调该方法
void onDestroy()	当 Service 被销毁前回调该方法
void onStartCommand(Intent intent, int flags, int startId)	当客户端调用 startService(Intent intent)方法启动服务时回调该方法
Boolean onUnbind(Intent intent)	当该 Service 上绑定的所有客户端都断开链接时将会回调该方法

其中，onBind()方法是 Service 类中的抽象方法，必须在创建的 Service 类中给出 onBind()方法的实现。

2. 注册 Service

创建好的 Service 类需要在 AndroidManifest. xml 文件中进行注册才能被使用。在 AndroidManifest. xml 文件中注册 Service，需要在< application >标签中使用< service >标签。< service >标签的语法格式如下：

```
<service android:name="String"
        [android:enabled=["true"|"false"]]
        [android:exported=["true"|"false"]]
        [ android:icon="drawable resource"]
        [ android:label="string resource"]
        [android:permission="string"]
        [ android:process="string"]
        ……
>
</service>
```

根据需要可以在< service >标签中设置相关属性，常用属性说明如表 5-2 所示。

表 5-2 < service >标签常用属性

属 性	描 述
android:name	定义的 Service 类的类全名，该属性的取值为 packageName. className
android:enabled	是否能被系统实例化，true 表示可以，false 表示不可以。如果 enabled 属性被设置为 false，该服务就会被禁用，而不能被实例化。默认值为 true
android:exported	服务是否可以被其他应用程序组件调用，true 表示可以，false 表示不可以
android:icon	服务的图标，该属性必须设置为包含图片定义的可绘制资源的引用
android:label	服务的名称
android:permission	启动服务的权限，其他组件必须包含该权限，调用该服务

续表

属　性	描　述
android:process	是否需要在单独的进程中运行，当设置为 android:process=":remote"时，代表 Service 在单独的进程中运行。注意":"很重要，它的意思是指要在当前进程名称前面附加上当前的包名，所以"remote"和":remote"不是同一个意思，前者的进程名称为：remote，而后者的进程名称为：App-packageName:remote

在配置< service >时，可以为< service/>标签配置< intent-filter >用于说明该 Service 可以被哪些 Intent 启动。在 Android 中，启动的服务的方式有两种，分别是调用 Context. startService()方法启动服务和调用 Context. bindService()方法绑定服务。

5.1.2 startService 启动服务

当使用 Context. startService(Intent intent)方法启动服务时，访问者与服务之间没有关联，无法进行数据通信。在服务启动后，即使启动它的组件销毁了，服务依然会在后台运行。使用 startService()启动的服务通常用于完成一些单一的任务或者不要求有返回结果的任务，例如从网络下载东西或在后台播放音乐。

使用 Context. startService(Intent intent)方法启动服务时，如果 Service 尚未被创建，系统会首先创建服务对象，然后调用 Service 的 onCreate()方法，接着调用 onStartCommand()方法。如果调用 Context. startService()方法前服务已经被创建，系统将会直接调用服务的 onStartCommand()方法。因此，当有多个组件同时调用 Context. startService(Inent intent)方法启动服务时，不会多次创建服务，onCreate()方法只会执行一次，但是会多次调用 onStartCommand()方法。如果一个 Service 是采用 Context. startService()方法启动，这个 Service 将一直运行直到服务内部调用 stopSelf()或者其他应用组件调用 Context. stopService()方法停止服务为止，当服务结束前，Android 系统会调用服务的 onDestroy()方法用于释放服务占用的资源。

5.1.3 案例(一)

在 Chapter05Application 项目中创建一个简单的 Service 类——MyFirstService，并重写其中的 onBind()、onCreate()、onStartCommand()、onDestroy()和 onUnbind()方法，在这些方法中实现输出日志信息的简单功能。MyFirstService 代码如文件清单 5-2 所示。

文件清单 5-2　MyFirstService. java

```
import android.app.Service;
import android.content.Intent;
import android.os.IBinder;
import android.util.Log;

public class MyFirstService extends Service {
```

```
    private static String LOG_SERVICE="MyFirstService";

    @Override
    public IBinder onBind(Intent intent) {
        // TODO Auto-generated method stub
        Log.i(LOG_SERVICE, "Service is bind");
        return null;
    }

    @Override
    public void onCreate() {
        // TODO Auto-generated method stub
        super.onCreate();
        Log.i(LOG_SERVICE, "Service is created");
    }

    @Override
    public int onStartCommand(Intent intent, int flags, int startId) {
        // TODO Auto-generated method stub
        Log.i(LOG_SERVICE, "Service is started");
        return super.onStartCommand(intent, flags, startId);

    }

    @Override
    public void onDestroy() {
        // TODO Auto-generated method stub
        super.onDestroy();
        Log.i(LOG_SERVICE, "Service will destory");
    }

    @Override
    public boolean onUnbind(Intent intent) {
        // TODO Auto-generated method stub
        Log.i(LOG_SERVICE, "Service is unbind");
        return super.onUnbind(intent);
    }
}
```

上述代码创建一个简单的没有什么具体功能的 Service 服务，如果需要 Service 实现某些功能，只需要在这些回调方法中写入具体业务逻辑代码即可。对于创建的 MyFirstService 服务，还需要在 AndroidManifest. xml 中进行注册后才能被其他组件调用。因此我们在 AndroidManifest. xml 中添加如下注册的代码：

```
<service
        android:name=".MyFirstService"
        android:enabled="true"
        android:exported="true" />
```

完成注册后，MyFirstService 服务就可以被其他应用程序组件启动了。我们在 Chapter05Application 项目中创建一个 Activity——StartServiceActivity，该 Activity 对应的布局文件中包含两个按钮，一个按钮用于启动 Service，另一个按钮用于关闭 Service。StartServiceActivity 的布局文件 activity_start_service.xml 相关代码如文件清单 5-3 所示。

文件清单 5-3　activity_start_service.xml

```
<?xml version="1.0" encoding="utf-8"?>
<RelativeLayout xmlns:android="http://schemas.android.com/apk/res/android"
    xmlns:tools="http://schemas.android.com/tools"
    android:layout_width="match_parent"
    android:layout_height="match_parent"
    android:paddingBottom="@dimen/activity_vertical_margin"
    android:paddingLeft="@dimen/activity_horizontal_margin"
    android:paddingRight="@dimen/activity_horizontal_margin"
    android:paddingTop="@dimen/activity_vertical_margin"
    tools:context="com.nsu.zyl.chapter05application.StartServiceActivity">

    <Button
        android:layout_width="match_parent"
        android:layout_height="wrap_content"
        android:text="startService"
        android:id="@+id/btnStart"
        android:layout_alignParentTop="true"
        android:layout_centerHorizontal="true" />

    <Button
        android:layout_width="match_parent"
        android:layout_height="wrap_content"
        android:text="stopService"
        android:id="@+id/btnStop"
        android:layout_below="@+id/btnStart"
        android:layout_centerHorizontal="true" />
</RelativeLayout>
```

在 StartServiceActivity 中为两个按钮分别设置监听器，当用户点击 startService 按钮时，使用 startService(Intent intent)方法启动服务；当用户点击 stopService 按钮时，使用 stopService(Intent intent)方法停止服务。StartServiceActivity 源代码如文件清单 5-4 所示。

文件清单 5-4　StartServiceActivity.java

```
import android.app.Activity;
import android.content.Intent;
import android.os.Bundle;
import android.view.View;
import android.view.View.OnClickListener;
```

```
import android.widget.Button;

public class StartServiceActivity extends Activity {
    private Button btnStartService, btnStopService;
    private Intent intent;

    @Override
    protected void onCreate(Bundle savedInstanceState) {
        super.onCreate(savedInstanceState);
        setContentView(R.layout.activity_start_service);
        btnStartService = (Button) findViewById(R.id.btnStartService);
        btnStopService = (Button) findViewById(R.id.btnStopService);
        intent =new Intent(StartServiceActivity.this, MyFirstService.class);
        btnStartService.setOnClickListener(new OnClickListener() {

            @Override
            public void onClick(View v) {
                // TODO Auto-generated method stub
                startService(intent);
            }
        });
        btnStopService.setOnClickListener(new OnClickListener() {

            @Override
            public void onClick(View v) {
                // TODO Auto-generated method stub
                stopService(intent);
            }
        });
    }

}
```

编辑 Chapter05Application 项目中的 MainActivity 文件，重写其中的 onListItemClick(ListView l, View v, int position, long id)方法，当用户点击 MainActivity 页面上第一个选项时，应用跳转至 StartServiceActivity 页面。重写后的 onListItemClick(ListView l, View v, int position, long id)方法如下：

```
protected void onListItemClick(ListView l, View v, int position, long id) {
        super.onListItemClick(l, v, position, id);
        Intent intent=new Intent();
        switch (position){
            case 0:
              intent.setClass(getApplicationContext(),StartServiceActivity.class);
              break;
        }
        startActivity(intent);

    }
```

运行该程序，在 MainActivity 页面点击第一个列表项，进入 StartServiceActivity 页面，点击该页面中的 startService 按钮启动 Service，然后再点击 stopService 按钮关闭 Service。在 Logcat 面板中对应的日志输出如图 5-3 所示。

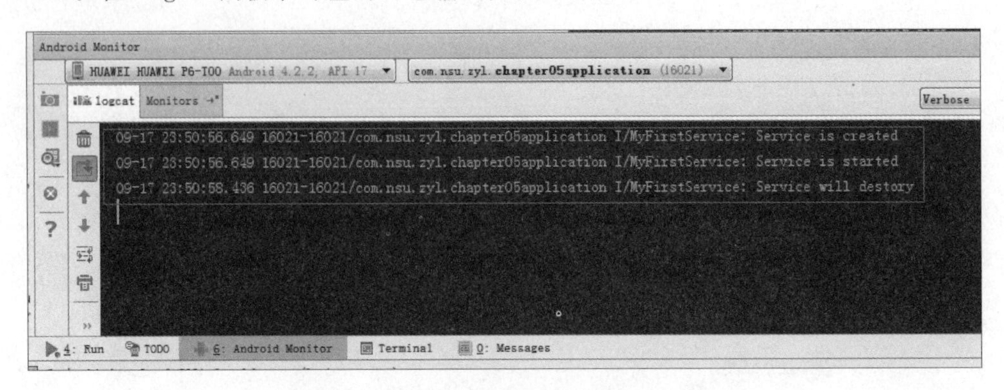

图 5-3　启动和关闭 Service Logcat 的输出

在不关闭 Service 的情形下，连续 4 次点击 startService 按钮，程序将连续 4 次启动 Service，在 Logcat 面板中可以看到的输出如图 5-4 所示。

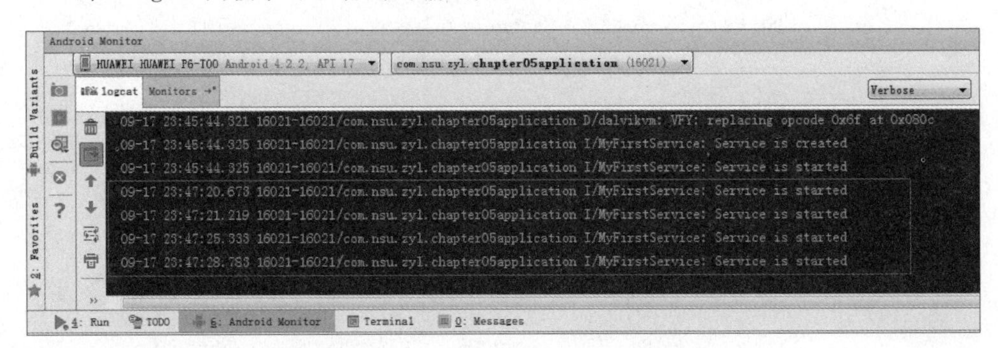

图 5-4　连续启动 Service Logcat 的输出

从图 5-4 的 Logcat 输出可以看到，第一次调用 startService 方法时，onCreate 方法、onStartCommand 方法将依次被调用；而多次调用 startService 时，只有 onStartCommand 方法被调用。最后调用 stopService 方法停止服务时 onDestroy 方法被回调，如图 5-5 所示。

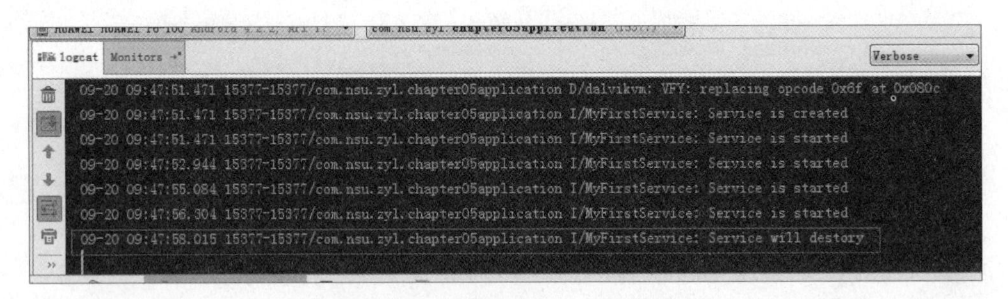

图 5-5　关闭服务 Logcat 的输出

5.1.4 bindService 启动服务

当程序通过 startService()和 stopService()方法启动和关闭 Service 时,Service 和访问者之间无法进行通信和数据交换。如果 Service 和访问者之间需要进行方法调用或数据交换,应该使用 bindService()和 unbindService()方法绑定和解除绑定 Service。

通过 Context.bindService(Intent intent,ServiceConnection conn,int flags)绑定服务时,需要传入 3 个参数:

(1) Intent intent: 指定要启动的 Service。

(2) ServiceConnection conn: ServiceConnection 接口对象,用于监听访问者与 Service 之间的连接情况。当访问者与 Service 连接成功时将回调该 ServiceConnection 对象的 onServiceConnected(ComponentName name,IBinder service)方法。该方法中的 IBinder 对象是服务中 onBind()方法的返回值;当 Service 与访问者之间由于异常终止而断开连接时,将回调 ServiceConnection 对象的 onServiceDisconnected(ComponentName name)方法。需要注意的是,如果调用者通过调用 unbindServcie()方法断开与 Service 连接时,ServiceConnection 对象的 onServiceDisconnected()方法不会被调用。

(3) int flags: 指定绑定服务时是否自动创建 Service(如果 Service 未创建)。该参数可以指定为 0(不自动创建)或 BIND_AUTO_CREATE(自动创建)。

由此可知,当客户端通过 bindService()方法绑定服务时,需要做到两点: ①在服务端必须实现 onBind()方法,该方法返回的 IBinder 对象定义了客户端用来与服务进行交互的编程接口; ②在客户端提供 ServiceConnection 接口的实现类,用于监视客户端与服务之间的连接情况。

当多个客户端同时绑定到一个服务上,只有第一个客户端绑定时,系统才会调用服务的 onBind()方法来获取 IBinder 对象,系统会向后续请求绑定的客户端传送这同一个 IBinder 对象,但不再调用 onBind()方法。当最后一个客户端解除绑定后,系统会销毁服务(除非服务同时是通过 startService()启动的)。

使用 bindService()绑定服务时,服务器需要定义 onBind()方法返回的接口,有 3 种方式可以定义这个接口: 继承 Binder 类,使用 Messenger 和使用 AIDL。

1. 继承 Binder 类

在实际开发过程中,如果服务与客户端运行在同一个进程中且服务是应用程序私有的,应该通过扩展 Binder 类来创建接口,并使用 onBind()方法返回它的实例。当客户端利用 bindService 的方法对 Service 进行绑定时,客户端获取 Binder 对象,并用 Binder 对象直接访问 Binder 甚至 Service 中可用的公共方法。

实现步骤:

(1) 在服务中,创建一个 Binder 类实例对象,它满足如下某一个条件。

① 包含客户端能调用的公共方法;

② 返回当前 Service 实例;

③ 返回服务管理的其他类的实例,其中包含客户端能调用的公共方法。

(2) 从 onBind()回调方法中返回 Binder 类的实例。

(3) 在客户端，从 onServiceConnected()回调方法接收 Binder 类实例，并且使用提供的方法与服务通信。

2. 使用 Messenger

在实际开发过程中，如果服务需要与远程进程进行通信，可以考虑使用 Messenger 来为服务创建接口。在这种方式下，服务需定义一个响应不同类型的 Message 对象的 Handler 对象。此 Handler 是 Messenger 与客户端能够共用同一个 IBinder 的基础，它使得客户端可以用消息对象 Message 向服务发送指令。此外，客户端还可以定义自己的 Messenger，以便服务能够回送消息。这是进行进程间通信(IPC)最为简便的方式，因为 Messenger 会把所有的请求放入一个独立线程中的队列，这样就不必对服务进行线程安全设计了。

实现步骤：

(1) 在服务类中创建一个 Handler 对象，负责对客户端发送过来的 Message 进行响应。

(2) 根据创建的 Handler 对象创建一个 Messenger 对象。

(3) 使用 Messenger 的 getBinder()方法得到一个 IBinder 对象，在服务的 onBind()方法中将其返回客户端。

(4) 客户端在 onServiceConnected()方法中，根据 IBinder 参数创建一个 Messenger 对象(引用服务的 Handler)。

(5) 客户端中利用得到的 Messenger 对象将 Message 对象发送给服务。

(6) 服务在其 Handler 的 handleMessage()方法中处理接收每个 Message。

经过上述步骤，实现客户端向 Service 单向通信，客户端使用 Messenger 对象给 Service 发送 Message 后，Service 中的 Handler 将对消息进行响应。如果要实现双向通信，需要完成如下步骤：

(7) 在客户端中创建一个 Handler 对象，用于处理 Service 发过来的消息。

(8) 根据客户端中的 Handler 对象创建一个客户端自己的 Messenger 对象。

(9) 在第(5)步中获得的 Service 的 Messenger 对象，并通过它来给 Service 发送消息，这时将客户端的 Messenger 对象赋给待发送的 Message 对象的 replayTo 字段。

(10) 在 Service 的 Handler 处理 Message 时将客户端的 Messenger 解析出来，并使用客户端的 Messenger 对象给客户端发送消息。

以上就实现了客户端和 Service 的双向通信，客户端和 Service 都有自己的 Handler 和 Messenger 对象，使得对方可以给自己发送消息。

3. 使用 AIDL

AIDL(Android Interface Definition Language，Android 接口定义语言)是一种 IDL 语言，用于生成在 Android 设备上两个进程之间进行通信的代码。AIDL 可以实现进程间通信(IPC)，并且在实现 IPC 的基础上允许多线程访问使用 AIDL 进行进程间通信时，进程之间的通信信息首先会被转换成 AIDL 协议消息，然后发送给对方，对方接收到 AIDL 消息后再转换成相应的对象。Android 官方建议只有当你允许来自不同的客户端访问你的服务，并且需要处理多线程问题时，你才必须使用 AIDL。

AIDL 使用简单的语法声明接口，描述其方法及方法的参数和返回值。这些参数和返回值可以是任何类型。AIDL 支持的数据类型有 5 种：Java 基本数据类型，String，CharSequence，List 和 Map。其他数据类型必须使用 import 语句导入，即使它们是在同一个 package 中。

5.1.5 案例(二)

首先在 Chapter05Application 项目中创建 BindServiceActivity，在该 Activity 对应的布局页面上放置 3 个按钮，当用户分别点击这 3 个按钮时，将分别启动 3 种服务，并分别以继承 Binder、Messenger 和 AIDL 这 3 种方式实现 Activity 与 Service 的通信。

BindServiceActivity 对应的布局文件 activity_bind_service.xml 如文件清单 5-5 所示。

文件清单 5-5 activity_bind_service.xml

```
<?xml version="1.0" encoding="utf-8"?>
<RelativeLayout xmlns:android="http://schemas.android.com/apk/res/android"
    xmlns:tools="http://schemas.android.com/tools"
    android:layout_width="match_parent"
    android:layout_height="match_parent"
    android:paddingBottom="@dimen/activity_vertical_margin"
    android:paddingLeft="@dimen/activity_horizontal_margin"
    android:paddingRight="@dimen/activity_horizontal_margin"
    android:paddingTop="@dimen/activity_vertical_margin"
    tools:context=".BindServiceActivity">

    <Button
        android:layout_width="match_parent"
        android:layout_height="wrap_content"
        android:text="继承 Binder 类"
        android:id="@+id/btnBinder"
        android:layout_alignParentTop="true"
        android:layout_centerHorizontal="true"
        android:onClick="onBinderButtonClick"/>

    <Button
        android:layout_width="match_parent"
        android:layout_height="wrap_content"
        android:text="使用 Messenger"
```

```
        android:id="@+id/btnMessenger"
        android:layout_below="@+id/btnBinder"
        android:layout_centerHorizontal="true"
        android:onClick="onMessengerButtonClick"/>

    <Button
        android:layout_width="match_parent"
        android:layout_height="wrap_content"
        android:text="使用 AIDL"
        android:id="@+id/btnAIDL"
        android:layout_below="@+id/btnMessenger"
        android:layout_centerHorizontal="true"
        android:onClick="onAIDLButtonClick"
        />

    <TextView
        android:layout_width="match_parent"
        android:layout_height="wrap_content"
        android:textAppearance="?android:attr/textAppearanceLarge"
        android:text="当前服务进度："
        android:id="@+id/textView"
        android:layout_centerVertical="true"
        android:layout_centerHorizontal="true"
        />
</RelativeLayout>
```

通过为 Button 按钮指定 android:onClick 属性，指定该按钮被点击时将会执行的方法。由此可知，第一个按钮被点击时将执行 void onBinderButtonClick(View view)方法，第二个按钮被点击时将执行 void onMessengerButtonClick(View view)方法，第三个按钮被点击时将执行 void onAIDLButtonClick（View view）方法。因此，在 BindServiceActivity 中需要定义上述 3 个方法。BindServiceActivity 最初的代码如文件清单 5-6 所示。

文件清单 5-6　BindServiceActivity.java

```
package com.nsu.zyl.chapter05application;

import android.content.ComponentName;
import android.content.Context;
import android.content.Intent;
import android.content.ServiceConnection;
import android.os.Handler;
import android.os.IBinder;
```

```
import android.os.Message;
import android.os.Messenger;
import android.os.RemoteException;
import android.support.v7.app.AppCompatActivity;
import android.os.Bundle;
import android.view.View;
import android.widget.TextView;

public class BindServiceActivity extends AppCompatActivity {
    private int progress=0;
    private TextView txtView;
    @Override
    protected void onCreate(Bundle savedInstanceState) {
        super.onCreate(savedInstanceState);
        setContentView(R.layout.activity_my_binder_service);
        txtView=(TextView) findViewById(R.id.textView);
    }
    public void onBinderButtonClick(View view){

    }
    public void onMessengerButtonClick(View view){

    }
    public void onAIDLButtonClick(View view){

    }

}
```

为了让 Chapter05Application 运行时能够跳转至 BindServiceActivity，需要继续重写 MainActivity 中的 onListItemClick(ListView l, View v, int position, long id)方法，当点击 MainActivity 页面上第二个选项时，应用跳转至 BindServiceActivity 页面。重写后的 onListItemClick(ListView l, View v, int position, long id)方法如下：

```
protected void onListItemClick(ListView l, View v, int position, long id) {
        super.onListItemClick(l, v, position, id);
        Intent intent=new Intent();
        switch (position){
            case 0:
              intent.setClass(getApplicationContext(),StartServiceActivity.class);
              break;
            case 1:
              intent.setClass(getApplicationConext(),BindServiceActivity.class);
              break;
        }
        startActivity(intent);

    }
```

同理，当需要在 MainActivity 页面跳转至其他示例时，也需要重写 MainActivity 类中的 onListItemClick()方法，在 onListItemClick 方法中根据当前被点击的列表项跳转至不同的示例页面，后续示例将不再给出对 onListItemClick()方法的重写。

在 Chapter05Application 项目中定义服务类 ProgressBinderService，并且在 ProgressBinderService 类中定义 ProgressBinder 类继承自 Binder 类，在 ProgressBinder 类中定义 getService()方法返回当前服务对象，定义 showProgress()方法返回当前服务进度值 progress。ProgressBinderService 代码如文件清单 5-7 所示。

文件清单 5-7　ProgressBinderService. java

```
package com.nsu.zyl.chapter05application;
import android.app.Service;
import android.content.Intent;
import android.os.Binder;
import android.os.IBinder;
public class ProgressBinderService extends Service {
    private int progress=0;
    private ProgressBinder progressBinder;
    public ProgressBinderService() {
    }
    @Override
    public void onCreate() {
        super.onCreate();
        progressBinder=new ProgressBinder();
    }
    @Override
    public IBinder onBind(Intent intent) {
        // TODO: Return the communication channel to the service.
        return progressBinder;
    }
    class ProgressBinder extends Binder {
        public Service getService(){
            return ProgressBinderService.this;
        }
    }
    public int showProgress(){
        progress+=1;
        return progress;
    }
}
```

当点击 BindServiceActivity 页面中第一个按钮时，将会执行 BindServiceActivity 中的 public void onBinderButtonClick(View view)方法，我们将在这个方法中启动 ProgressBinderService 服务，并在 BindServiceActivity 页面的 TextView 中显示启动服务当前的进度值。为了实现客户端与服务的通信，我们在 onBinderButtonClick()方法中使用 bindService 启动服务，当服务连接成功时，在 onServiceConnected()方法中，通过传过来的 ProgressBinderService. ProgressBinder 对象得到启动的服务对象，然后调用服务对象的 showProgress()方法得到当前服务的进度值。重写的 onBinderButtonClick()方

法代码如下：

```
public void onBinderButtonClick(View view){
            Intent  intent  =  new  Intent  ( BindServiceActivity.  this,
ProgressBinderService.class);
        bindService(intent, new ServiceConnection() {
            @Override
        public void onServiceConnected (ComponentName componentName, IBinder
iBinder) {
                ProgressBinderService.ProgressBinder progressBinder=
                    (ProgressBinderService.ProgressBinder)iBinder;
                ProgressBinderService progressBinderService=
                    (ProgressBinderService) progressBinder.getService();
                progress=progressBinderService.showProgress();
                txtView.setText ("ProgressBinderService 服务当前进度: "+progress+
"% ");
            }

            @Override
            public void onServiceDisconnected(ComponentName componentName) {

            }
        },Context.BIND_AUTO_CREATE);

    }
```

点击 BindServiceActivity 页面中的第一个按钮，启动 ProgressBinderService 服务获取服务当前 progress 值功能，每次点击该按钮均能获取服务的当前 progress 值，并在 TextView 中显示出来，如图 5-6 所示。

图 5-6 获取 ProgressBinderService 的进度值

定义服务类 ProgressMessengerService，在该类中定义 ProgressHandler 用于处理访问端发送过来的消息。定义好的 ProgressMessengerService 类的源代码如文件清单 5-8 所示。

文件清单 5-8 ProgressMessengerService. java

```
package com.nsu.zyl.chapter05application;

import android.app.Service;
import android.content.Intent;
import android.os.Handler;
import android.os.IBinder;
import android.os.Message;
import android.os.Messenger;
import android.os.RemoteException;

public class ProgressMessengerService extends Service {
    private int progress=0;
    private Messenger messenger=new Messenger(new ProgressHandler());
    final private int MESSAGE_FROM_SERVICE_TO_ACTIVITY=0x100;
    final private int MESSAGE_FROM_ACTIVITY_TO_SERVICE=0x200;
    public ProgressMessengerService() {
    }

    @Override
    public void onCreate() {
        super.onCreate();
    }

    @Override
    public IBinder onBind(Intent intent) {
        return messenger.getBinder();
    }
    class ProgressHandler extends Handler{
        @Override
        public void handleMessage(Message msg) {
            super.handleMessage(msg);
            switch (msg.what){
                case MESSAGE_FROM_ACTIVITY_TO_SERVICE:
                    progress+=1;
                    Message message=Message.obtain();
                    message.arg1=progress;
                    message.what=MESSAGE_FROM_SERVICE_TO_ACTIVITY;
                    try {
                        msg.replyTo.send(message);
                    } catch (RemoteException e) {
                        e.printStackTrace();
                    }
```

```
                    break;
                }
            }
        }
    }
```

当用户点击 BindServiceActivity 页面第二个按钮时，实现启动 Progress-MessengerService 服务，将服务当前进度值返回到 BindServiceActivity，并在 TextView 中显示出来。重写的 public void onMessengerButtonClick(View view)方法如下：

```
public void onMessengerButtonClick(View view){
        Intent intent=new Intent(BindServiceActivity.this,
            ProgressMessengerService.class);
        bindService(intent, new ServiceConnection() {
            @Override
            public void onServiceConnected(ComponentName componentName, IBinder
iBinder) {
                Messenger messenger=new Messenger(iBinder);
                Messenger messenger2=new Messenger(new MyHandler());
                Message message=Message.obtain();
                message.what=MESSAGE_FROM_ACTIVITY_TO_SERVICE;
                message.replyTo=messenger2;
                try {
                    messenger.send(message);
                } catch (RemoteException e) {
                    e.printStackTrace();
                }
            }

            @Override
            public void onServiceDisconnected(ComponentName componentName) {

            }
        },Context.BIND_AUTO_CREATE);
    }
```

其中，MyHandler 类是在 BindServiceActivity 中定义的内部类，主要用于处理服务器端发送到客户端的信息。MyHandler 的源代码如下：

```
class MyHandler extends Handler{
        @Override
        public void handleMessage(Message msg) {
            super.handleMessage(msg);
            switch (msg.what){
                case MESSAGE_FROM_SERVICE_TO_ACTIVITY:
                    progress=msg.arg1;
```

```
                    txtView.setText("ProgressMessengerService 服务当前进度: "+
progress+"% ");
                    break;
                }
            }
        }
```

点击 BindServiceActivity 页面中的第二个按钮，将启动 ProgressMessengerService 服务，并且通过 Messenger 实现访问端与服务的通信，每次点击该按钮均能获取服务的当前 progress 值，并在 TextView 中显示出来，如图 5-7 所示。

图 5-7　获取 ProgressMessengerService 进度值

在 Chapter05Application 项目中 main 目录下新建 aidl 文件夹，在创建好的 aidl 文件夹下新建以项目包名命名的文件夹，然后在该文件夹下新建 IProgressAidlInterface.aidl 文件。创建好的 IProgressAidlInterface.aidl 所在的目录结构如图 5-8 所示。

IProgressAidlInterface.aidl 文件的内容如文件清单 5-9 所示。

文件清单 5-9　IProgressAidlInterface.aidl

```
package com.nsu.zyl.chapter05application;

interface IProgressAidlInterface {
    int showProgress();
}
```

在 Android Studio 中重新编译工程，在 app→build→generated→source→aidl→debug→包名下即可找到一个和 AIDL 文件名相同的文件，这个就是 Android Studio 自动生成的文件，该文件不允许修改，如文件清单 5-10 所示。

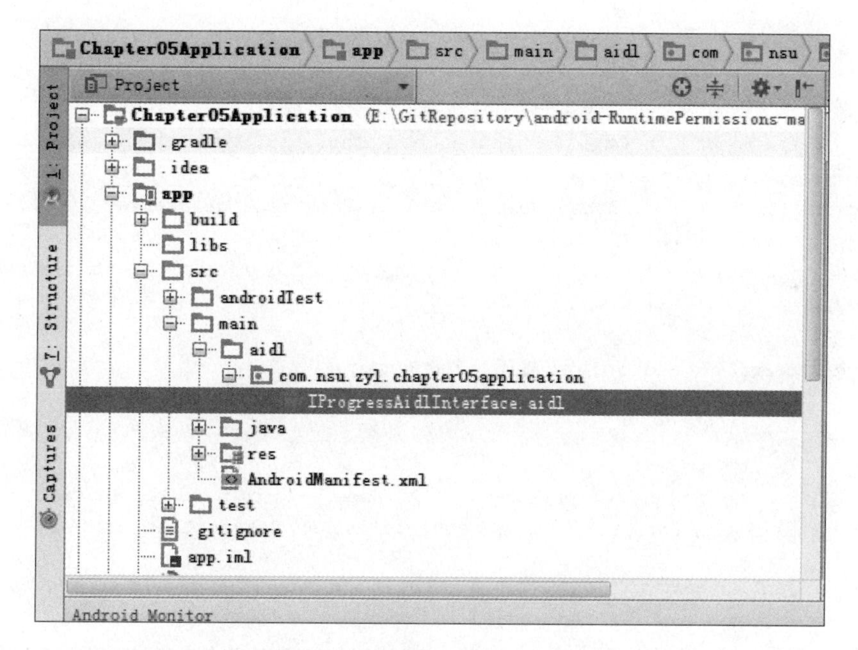

图 5-8 IProgressAidlInterface

文件清单 5-10 IMyAidlInterface.java

```
package com.nsu.zyl.chapter05application;
// Declare any non-default types here with import statements

public interface IProgressAidlInterface extends android.os.IInterface
{
/** Local-side IPC implementation stub class. */
public static abstract class Stub extends android.os.Binder implements com.nsu.
zyl.chapter05application.IProgressAidlInterface
{
private static final java.lang.String DESCRIPTOR = "com.nsu.zyl.
chapter05application.IProgressAidlInterface";
/** Construct the stub at attach it to the interface. */
public Stub()
{
this.attachInterface(this, DESCRIPTOR);
}
/**
 * Cast an IBinder object into an com.nsu.zyl.chapter05application.
IProgressAidlInterface interface,
 * generating a proxy if needed.
 */
public static com.nsu.zyl.chapter05application.IProgressAidlInterface
asInterface(android.os.IBinder obj)
{
if ((obj==null)) {
```

```
return null;
}
android.os.IInterface iin =obj.queryLocalInterface(DESCRIPTOR);
if (((iin! = null) && (iin instanceof com. nsu. zyl. chapter05application.
IProgressAidlInterface))) {
return ((com.nsu.zyl.chapter05application.IProgressAidlInterface)iin);
}
return new com.nsu.zyl.chapter05application.IProgressAidlInterface.Stub.Proxy
(obj);
}
@Override public android.os.IBinder asBinder()
{
return this;
}
@Override public boolean onTransact(int code, android.os.Parcel data, android.
os.Parcel reply, int flags) throws android.os.RemoteException
{
switch (code)
{
case INTERFACE_TRANSACTION:
{
reply.writeString(DESCRIPTOR);
return true;
}
case TRANSACTION_showProgress:
{
data.enforceInterface(DESCRIPTOR);
int _result =this.showProgress();
reply.writeNoException();
reply.writeInt(_result);
return true;
}
}
return super.onTransact(code, data, reply, flags);
}
private static class Proxy implements com. nsu. zyl. chapter05application.
IProgressAidlInterface
{
private android.os.IBinder mRemote;
Proxy(android.os.IBinder remote)
{
mRemote =remote;
}
@Override public android.os.IBinder asBinder()
{
return mRemote;
}
public java.lang.String getInterfaceDescriptor()
```

```
{
return DESCRIPTOR;
}
@Override public int showProgress() throws android.os.RemoteException
{
android.os.Parcel _data = android.os.Parcel.obtain();
android.os.Parcel _reply = android.os.Parcel.obtain();
int _result;
try {
_data.writeInterfaceToken(DESCRIPTOR);
mRemote.transact(Stub.TRANSACTION_showProgress, _data, _reply, 0);
_reply.readException();
_result = _reply.readInt();
}
finally {
_reply.recycle();
_data.recycle();
}
return _result;
}
}
static final int TRANSACTION_showProgress = (android.os.IBinder.FIRST_CALL_
TRANSACTION + 0);
}
public int showProgress() throws android.os.RemoteException;
}
```

这个 Java 文件中有一个抽象内部类 Stub(继承了 Binder 类)实现了定义的接口,并提供了一个 asInterface 方法将 IBinder 对象转化为定义的接口类型。实现定义的接口就转化为实现其抽象内部类 Stub(Service 端的业务函数全部在这里实现)。这个 Stub 将会作为远程 Service 的回调类。

上一步定义好 AIDL 接口后,接下来定义一个 Service 的实现类 ProgressAIDLService,该 Service 的 onBind()方法所返回的 IBinder 对象是 IProgressAidlInterface.Stub 的子类的实例,如文件清单 5-11 所示。

文件清单 5-11　ProgressAIDLService.java

```
package com.nsu.zyl.chapter05application;

import android.app.Service;
import android.content.Intent;
import android.os.IBinder;
import android.os.RemoteException;

public class ProgressAIDLService extends Service {
    int progress=0;
    public ProgressAIDLService() {
```

```
    }

    @Override
    public IBinder onBind(Intent intent) {
        return new MyAIDLBinder();
    }
    class MyAIDLBinder extends IProgressAidlInterface.Stub{
        @Override
        public int showProgress() throws RemoteException {
            progress+=1;
            return progress;
        }
    }
}
```

上面的程序定义了 MyAIDLBinder 类，该类继承了 IProgressAidlInterface. Stub 类，也就是实现了 IBinder 接口和 IProgressAidlInterface 接口，因此重写 onBind()方法时可以返回该类的实例。

AIDL 接口定义了两个进程间的通信接口，不仅服务器端需要 AIDL 接口，客户端同样需要定义 AIDL 接口。因此当使用 AIDL 实现客户端与服务器端通信时，需要把服务器端定义的. aidl 文件复制到客户端应用对应的目录中。客户端绑定远程服务与绑定本地服务类似，需要创建一个 ServiceConnection 接口对象，然后以 ServiceConnection 对象作为参数，调用 Context 的 bindService()方法绑定远程服务。

在当前项目中点击 BindServiceActivity 页面第三个按钮将执行 onAIDLButtonClick()方法，在该方法实现启动 ProgressAIDLService 服务，获取服务的当前 progress 值，并且在 TextView 中显示出来(图 5-9)。onAIDLButtonClick()方法的源代码如下：

```
public void onAIDLButtonClick(View view){
        Intent intent=new Intent(BindServiceActivity.this,ProgressAIDLService.
class);
        bindService(intent, new ServiceConnection() {
            @Override
            public void onServiceConnected(ComponentName componentName, IBinder
iBinder) {
                IProgressAidlInterface iProgressAidlInterface=
                    IProgressAidlInterface.Stub.asInterface(iBinder);
                try {
                    progress=iProgressAidlInterface.showProgress();
                    txtView. setText ( " ProgressAIDLService 服务当前进度: " +
progress+"% ");
                } catch (RemoteException e) {
                    e.printStackTrace();
                }
```

```
            }

            @Override
            public void onServiceDisconnected(ComponentName componentName) {

            }
        },Context.BIND_AUTO_CREATE);
    }
```

需要注意,绑定远程 Service 的 ServiceConnection 不能直接获取 Service 的 onBind()方法返回的对象,它只能返回 onBind()方法所返回的对象的代理,因此在代码中做如下处理:

```
IProgressAidlInterface iProgressAidlInterface = IProgressAidlInterface.Stub.
asInterface(iBinder);
```

图 5-9　通过 AIDL 获取 ProgressAIDLService 服务进度值

完成 public void onBinderButtonClick(View view)、public void onMessengerButtonClick(View view)和 public void onAIDLButtonClick(View view)3 个方法重写后的 BindServiceActivity 的原代码如文件清单 5-12 所示。

文件清单 5-12　BindServiceActivity. java

```
package com.nsu.zyl.chapter05application;

import android.content.ComponentName;
```

```
import android.content.Context;
import android.content.Intent;
import android.content.ServiceConnection;
import android.os.Handler;
import android.os.IBinder;
import android.os.Message;
import android.os.Messenger;
import android.os.RemoteException;
import android.support.v7.app.AppCompatActivity;
import android.os.Bundle;
import android.view.View;
import android.widget.TextView;

public class BindServiceActivity extends AppCompatActivity {
    private int progress=0;
    final private int MESSAGE_FROM_SERVICE_TO_ACTIVITY=0x100;
    final private int MESSAGE_FROM_ACTIVITY_TO_SERVICE=0x200;
    private TextView txtView;
    @Override
    protected void onCreate(Bundle savedInstanceState) {
        super.onCreate(savedInstanceState);
        setContentView(R.layout.activity_my_binder_service);
        txtView=(TextView) findViewById(R.id.textView);
    }
    public void onBinderButtonClick(View view){
        Intent intent = new Intent (BindServiceActivity.this, ProgressBinderService.
class);
        bindService(intent, new ServiceConnection() {
            @Override
            public void onServiceConnected(ComponentName componentName, IBinder
iBinder) {
                ProgressBinderService.ProgressBinder progressBinder=
                        (ProgressBinderService.ProgressBinder)iBinder;
                ProgressBinderService progressBinderService=
                        (ProgressBinderService) progressBinder.getService();
                progress=progressBinderService.showProgress();
                txtView.setText ("ProgressBinderService 服务当前进度："+progress+
"% ");
            }

            @Override
            public void onServiceDisconnected(ComponentName componentName) {

            }
        },Context.BIND_AUTO_CREATE);

    }
    public void onMessengerButtonClick(View view){
```

```
        Intent intent=new Intent(BindServiceActivity.this,
            ProgressMessengerService.class);
        bindService(intent, new ServiceConnection() {
            @Override
            public void onServiceConnected(ComponentName componentName, IBinder
 iBinder) {
                Messenger messenger=new Messenger(iBinder);
                Messenger messenger2=new Messenger(new MyHandler());
                Message message=Message.obtain();
                message.what=MESSAGE_FROM_ACTIVITY_TO_SERVICE;
                message.replyTo=messenger2;
                try {
                    messenger.send(message);
                } catch (RemoteException e) {
                    e.printStackTrace();
                }
            }

            @Override
            public void onServiceDisconnected(ComponentName componentName) {

              }
        },Context.BIND_AUTO_CREATE);
    }
    public void onAIDLButtonClick(View view){
        Intent intent=new Intent(BindServiceActivity.this,ProgressAIDLService.
 class);
        bindService(intent, new ServiceConnection() {
            @Override
            public void onServiceConnected(ComponentName componentName, IBinder
 iBinder) {
                IProgressAidlInterface
 iProgressAidlInterface= IProgressAidlInterface.Stub.asInterface(iBinder);
                try {
                    progress=iProgressAidlInterface.showProgress();
                    txtView. setText ( " ProgressAIDLService 服务当前进度: " +
progress+"% ");
                } catch (RemoteException e) {
                    e.printStackTrace();
                }
            }

            @Override
            public void onServiceDisconnected(ComponentName componentName) {

            }
        },Context.BIND_AUTO_CREATE);
```

```
    }

    class MyHandler extends Handler{
        @Override
        public void handleMessage(Message msg) {
            super.handleMessage(msg);
            switch (msg.what){
                case MESSAGE_FROM_SERVICE_TO_ACTIVITY:
                    progress=msg.arg1;
                     txtView.setText("ProgressMessengerService 服务当前进度: "+
progress+"% ");
                    break;
            }
        }
    }
}
```

AndroidManifest.xml 的内容如文件清单 5-13 所示。

文件清单 5-13　AndroidManifest.xml

```
<?xml version="1.0" encoding="utf-8"?>
<manifest xmlns:android="http://schemas.android.com/apk/res/android"
    package="com.nsu.zyl.chapter05application">
    <application
        android:allowBackup="true"
        android:icon="@mipmap/ic_launcher"
        android:label="@string/app_name"
        android:supportsRtl="true"
        android:theme="@style/AppTheme">
        <activity android:name=".MainActivity" >
        <intent-filter>
            <action android:name="android.intent.action.MAIN" />
            <category android:name="android.intent.category.LAUNCHER" />
        </intent-filter>
        </activity>
        <service
            android:name=".MyFirstService"
            android:enabled="true"
            android:exported="true" />
        <activity android:name=".StartServiceActivity" />
        <activity android:name=".BindServiceActivity"/>
        <service
            android:name=".ProgressBinderService"
            android:enabled="true"
            android:exported="true" />
        <service
```

```
            android:name=".ProgressMessengerService"
            android:enabled="true"
            android:exported="true" />
        <service
            android:name=".ProgressAIDLService"
            android:enabled="true"
            android:exported="true" />
    </application>
</manifest>
```

5.1.6 Service的生命周期

根据Service启动方式的不同，Service生命周期有两种路径，使用startService()方法启动服务时，Service的生命周期如图5-10(a)所示；使用bindService()方法来启动服务时，Service的生命周期如图5-10(b)所示。

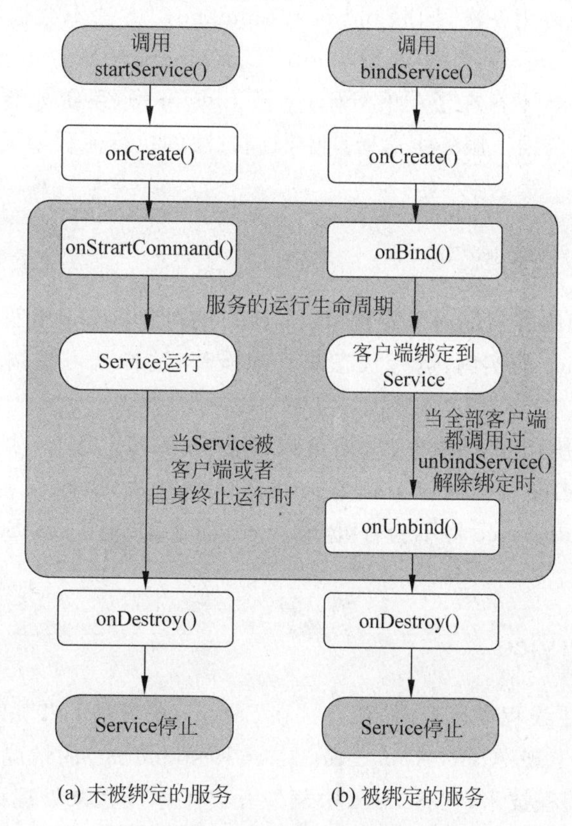

图5-10 Service生命周期

当使用context.startService()方法启动服务时，如果Service还没有创建，则Android先调用onCreate()方法创建服务，然后调用onStartCommand()方法启动服务。如果Service已经创建，则只调用onStartCommand()方法，当有多个服务启动同一个服务时，Service 的 onCreate()方法只会执行一次，Service 的 onStartCommand()方法可能会被调用多次(对应调用 startService 的次数)，并且系统只会创建 Service 的一个实例。客户端通过调用 stopService()方法结束服务时会回调 Service 的 onDestroy()方法，在该方法中主要完成资源的释放等操作。如果是调用者自己直接退出而没有调用 stopService()方法，Service 会一直在后台运行，该 Service 的调用者再启动后可以通过调用 stopService 方法关闭 Service。当然如果系统资源不足，Android 系统也可能结束服务。

使用 context. startService()方法启动服务，服务生命周期中回调方法的执行流程为：

```
    context.startService () → onCreate () → onStartCommand () → Service running →
context.stopService () → onDestroy () → Service Stop
```

当使用 context. bindService()方法启动服务时，不管 bindService 调用几次，onCreate 方法都只会调用一次，同时 onStartCommand()方法不会被调用。当连接建立之后，Service 将会一直运行，除非调用 Context. unbindService 断开连接或者之前调用 bindService 的 Context 不存在了(如 Activity 被 finish 时)，系统才会自动停止 Service，对应 onDestroy 将被调用。服务生命周期中回调方法的执行流程如下：

```
    context.bindService () → onCreate () → onBind () → Service running → onUnbind () →
onDestroy () → Service.stop
```

onBind()将返回给客户端一个实现 IBind 接口的对象，IBind 允许客户端回调服务的方法，例如得到 Service 的实例、运行状态或其他操作，这样就可把调用者与服务绑定在一起。

此外，可以混合使用这两种方式，即可以既启动又绑定服务，并且不管如何调用，onCreate 始终只会调用一次，对应 startService 调用多少次，Service 的 onStart 便会调用多少次。调用 unbindService 将不会停止 Service，而必须调用 stopService 或 Service 的 stopSelf 来停止服务。

5.1.7 IntentService

Service 默认在主线程中运行，如果直接在 Service 中处理一些耗时逻辑，会影响设备的运行流畅度，容易出现 ANR(Application Not Responding)的情况。对于这种情况可以考虑使用多线程编程技术，在 Service 中额外开启工作线程来处理那些耗时逻辑，或者使用 IntentService。

IntentService 是继承于 Service 并处理异步请求的一个类，在 IntentService 内有一个工作线程来处理耗时操作，启动 IntentService 的方式和启动传统 Service 一样。同时，当任务执行完后，IntentService 会自动停止，而不需要手动控制。

IntentService 使用队列来管理请求 Intent，客户端可以通过 startService(Intent)方

法传递请求给 IntentService，IntentService 将该 Intent 放入队列中，然后单独开启一个线程来处理所有 Intent 请求对象（通过 startService 的方式发送过来的）所对应的任务，这样以免事务处理阻塞主线程。执行完所有 Intent 请求对象所对应的工作之后，如果没有新的 Intent 请求达到，则自动停止 Service；否则执行下一个 Intent 请求所对应的任务。

使用 IntentService 需要实现它的一个抽象方法：onHandleIntent(Intent intent)，在这个方法里，可以获得访问者传来的 Intent 对象，并在其中实现异步操作。对于 IntentService 而言，因为其继承自 Service 类，所以其他 Service 的生命周期方法在此类中也适用。

在 Chapter05Application 项目中创建 MyServiceActivity，MyServiceActivity 对应的布局中包含两个按钮，我们在 MyServiceActivity 中实现当用户点击这两个按钮时分别启动普通的 Service 和 IntentService 功能。程序布局代码很简单，这里不再给出，MyServiceActivity 代码如文件清单 5-14 所示。

文件清单 5-14 MyServiceActivity. java

```
package com.nsu.zyl.chapter05application;

import android.content.Intent;
import android.support.v7.app.AppCompatActivity;
import android.os.Bundle;
import android.view.View;
import android.widget.Button;

public class MyServiceActivity extends AppCompatActivity {
    private Button btnStartService,btnStartIntentService;

    @Override
    protected void onCreate(Bundle savedInstanceState) {
        super.onCreate(savedInstanceState);
        setContentView(R.layout.activity_my_service);
        btnStartIntentService= (Button)findViewById(R.id.btnStartIntentService);
        btnStartService= (Button)findViewById(R.id.btnStartService);

        btnStartIntentService.setOnClickListener(new View.OnClickListener() {
            @Override
            public void onClick(View view) {
                Intent intent = new Intent (MyServiceActivity. this, MyIntentService.
class);
                startService(intent);
            }
        });
        btnStartService.setOnClickListener(new View.OnClickListener(){
            @Override
            public void onClick(View view) {
                Intent intent = new Intent (MyServiceActivity. this, MyService.
class);
```

```
                startService(intent);

            }
        });
    }
}
```

在上面 Activity 的两个事件处理方法中分别启动 MyService 和 MyIntentService，其中 MyService 继承自 Service，MyIntentService 继承自 IntentService。MyService 代码如文件清单 5-15 所示。

文件清单 5-15　MyService.java

```
public class MyService extends Service {
    private static String TAG="MyService";
    public MyService() {
    }

    @Override
    public IBinder onBind(Intent intent) {
        // TODO: Return the communication channel to the service.
        throw new UnsupportedOperationException("Not yet implemented");
    }

    @Override
    public int onStartCommand(Intent intent, int flags, int startId) {
        Log.i(TAG,"begin onStartCommand in "+this);
        try {
            Log.i(TAG,"MyService 当前线程 ID: "+Thread.currentThread().getId());
            Thread.sleep(10* 1000);
        } catch (InterruptedException e) {
            e.printStackTrace();
        }
        Log.i(TAG,"end onStartCommand in "+this);

        return super.onStartCommand(intent, flags, startId);
    }
}
```

MyIntentService 继承 IntentService 不需要实现 onBind( ) 方法，只需要实现 onHandleIntent( ) 方法即可，在该方法中可以定位该 Service 需要完成的任务。MyIntentService 代码如文件清单 5-16 所示。

文件清单 5-16　MyIntentService.java

```
public class MyIntentService extends IntentService {
    private static String TAG="MyIntentService";
```

```
    public MyIntentService() {
        super("MyIntentService");
    }


    @Override
    protected void onHandleIntent(Intent intent) {
        Log.i(TAG,"begin onHandleIntent() in "+this);
        try {
            Log.i(TAG,"MyIntentService 当前线程 ID: "+Thread.currentThread().
getId());
            Thread.sleep(10*1000);
        } catch (InterruptedException e) {
            e.printStackTrace();
        }
        Log.i(TAG,"end onHandleIntent() in "+this);
    }
}
```

运行该程序,点击“启动 Service”按钮,将会激发 startService()方法,该方法启动 MyService,对应的 Logcat 截图如图 5-11 所示,从 Logcat 输出信息可知 MyService 与主线程属于同一个线程。

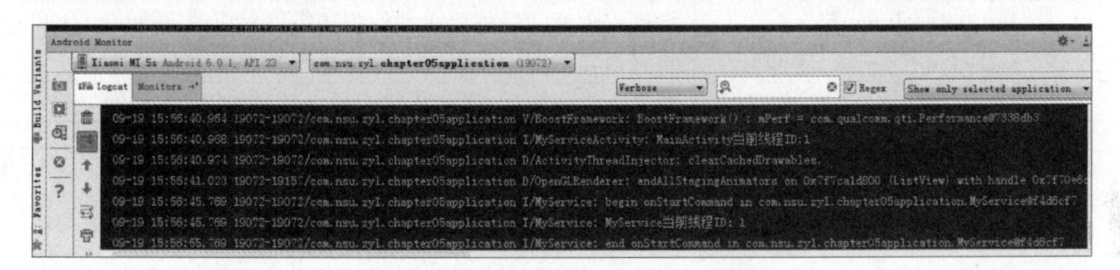

图 5-11 启动 MyService Logcat 输出

点击“启动 IntentService”按钮,将会启动 MyIntentService,对应的 Logcat 输出信息截图如图 5-12 所示,从 Logcat 输出信息可知 MyIntentService 与主线程属于不同的线程。

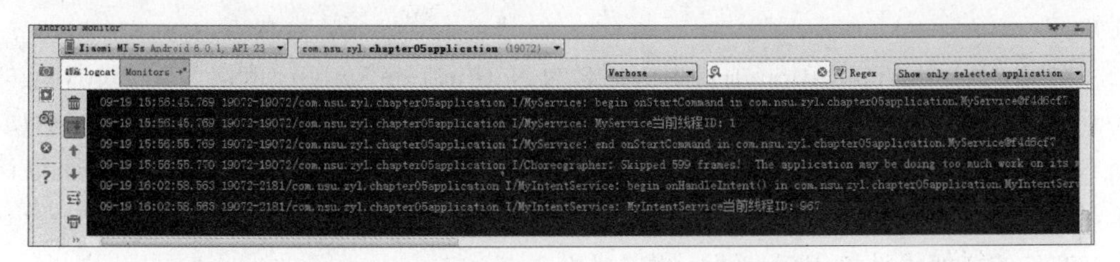

图 5-12 启动 MyIntentService Logcat 输出

5.2 系统Service的用法

Android系统本身提供很多系统服务,开发者可通过调用Content. getSystemService(String name)方法获取系统服务。

5.2.1 NotificationManager

NotificationManager是一个非常重要的系统服务,可以通过调用getSystemService(NOTIFICATION_SERVICE)方法来获得NotificationManager服务,程序一般通过NotificationManager向系统发送全局通知。

Android 3.0版本以后可以方便地使用Notification. Builder类来创建Notification对象。Notification. Builder提供的常用方法如表5-3所示。

表5-3 Notification. Builder常用方法

方　法	描　述
setContentTitle()	设置通知标题
setContentText()	设置通知内容
setSmallIcon()	设置通知小图标
setLargeIcon()	设置通知大图标
setTick()	设置通知在状态栏上的提示文本
setContentIntent()	设置点击通知后将要启动的程序组件对应的PendingIntent
setDefault()	设置通知的提示效果(振动、音乐等)
setAutoCancle()	设置点击通知后,通知从状态栏自动删除

Notification的基本使用流程是:

(1) 调用getSystemService(NOTIFICATION_SERVICE)方法获得Notification-Manager对象。

(2) 创建一个通知栏的Builder构造器。

(3) 对Builder构造器设置各种属性,例如标题、内容、图标和动作等。

(4) 调用Builder的build()方法创建Notification对象。

(5) 调用NotificationManager的notify()方法发送通知或者调用cancle()方法取消通知。

在Chapter05Application项目中创建名为NotificationActivity和名为ToActivity的Activity类,在NotificationActivity中将演示如何使用NotificationManager来发送和取消Notification,ToActivity用于显示通知的详细内容。NotificationActivity页面很简单,只包含两个按钮,分别对应发送通知和取消通知,在此就不给出NotificationActivity布局文件的详细代码和ToActivity相关代码。

NotificationActivity运行效果如图5-13所示。

NotificationActivity对应的Java代码如文件清单5-17所示。

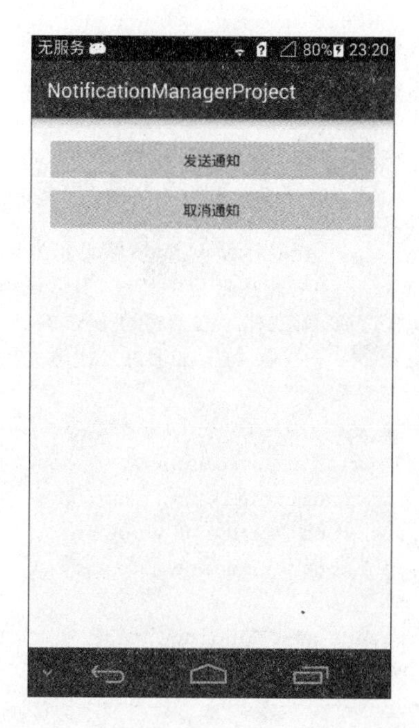

图 5-13　NotificationActivity 运行界面

文件清单 5-17　NotificationActivity. java

```
package com.nsu.zyl.chapter05application;

import android.app.Notification;
import android.app.NotificationManager;
import android.app.PendingIntent;
import android.content.Intent;
import android.support.v7.app.AppCompatActivity;
import android.os.Bundle;
import android.view.View;
import android.widget.Button;

public class NotificationActivity extends AppCompatActivity {

    private Button btnSend,btnCancel;
    private static int NOTIFICATION_ID = 0x100;
    private int id=0;
    private NotificationManager nm;

    @Override
    protected void onCreate(Bundle savedInstanceState) {
        super.onCreate(savedInstanceState);
        setContentView(R.layout.activity_notification);
        btnSend=(Button)findViewById(R.id.btnSend);
        btnCancel=(Button)findViewById(R.id.btnCancel);
        nm = (NotificationManager) getSystemService(NOTIFICATION_SERVICE);
```

```
        btnSend.setOnClickListener(new View.OnClickListener() {
            @Override
            public void onClick(View view) {
                // TODO Auto-generated method stub
                //创建一个启动其他Activity的Intent
                String title="今日新闻";
                String content="据外媒报道,谷歌去年发布的Android样板机Pixel异常
火爆,虽说该机打破了以往Nexus手机高配低价的传统,但凭借很多谷歌黑科技以及Android优
先权魅力不减。现在谷歌高管透露,第二代Pixel手机已经在路上了。";
                String ticker="谷歌Pixel手机二代曝光";

                Intent intent =new Intent(NotificationActivity.this,
                        ToActivity.class);
                intent.putExtra("title",title);
                intent.putExtra("content",content);
                intent.putExtra("ticker",ticker);

                PendingIntent pi =PendingIntent.getActivity(
                        NotificationActivity.this, 0, intent, 0);

                //创建Notification.Builder对象
                Notification.Builder builder =new Notification.Builder(
                        NotificationActivity.this);
                //为Notification.Builder对象设置各种属性
                builder.setAutoCancel(true).setTicker(ticker)
                        .setSmallIcon(R.drawable.notification_template_icon_bg)
                        .setContentTitle(title).setContentText(content)
                        .setWhen(System.currentTimeMillis())
                        .setContentIntent(pi);
                //调用Builder构造器创建Notification对象
                Notification notification =builder.build();
                //发送通知
                id=id+1;
                nm.notify(id, notification);
            }
        });
        btnCancel.setOnClickListener(new View.OnClickListener() {
            @Override
            public void onClick(View view) {
                nm.cancel(id);
            }
        });
    }
}
```

当用户在NotificationActivity中点击“发送通知”按钮(图5-14),系统将发送通知到通知栏,用户在通知栏点击通知信息,可以执行启动其他组件的相关操作。

图 5-14　发送通知演示效果

5.2.2　系统短信服务

SmsManager 是 Android 提供的短信管理服务，用于管理短信操作，SmsManager 提供了一系列 sendXxxMessage()方法用于发送短信。SmsManager 常用方法如表 5-4 所示。

表 5-4　SmsManager 常用方法

方　法	描　述
ArrayList<String> divideMessage (String text)	当短信超过 SMS 消息的最大长度时，将短信分割为几块
static SmsManager getDefault ()	获取 SmsManager 的默认实例
voidsendDataMessage (String desAddress, String scAddress, short destPort, byte[] data, PendingIntent sentIntent, PendingIntent deliveryIntent)	发送一个基于 SMS 的数据到指定的应用程序端口
voidsendMultipartTextMessage (String destAddress, String scAddress, ArrayList<String> parts, ArrayList <PendingIntent> sentIntents, ArrayList<PendingIntent> deliveryIntents)	发送一个基于 SMS 的多部分文本，调用者应用已经通过调用 divideMessage (String text)将消息分割成正确的大小
voidsendTextMessage (String desAddress, String scAddress, String text, PendingIntent sentIntent, PendingIntent deliveryIntent)	发送一个基于 SMS 的文本

在 Chapter05Application 项目中新建 SendSMSActivity，该 SendSMSActivity 页面中包含一个文本框让用户输入收件人号码，一个文本框让用户输入短信文本，点击“发

送”按钮将短信发送出去。SendSMSActivity 源代码如文件清单 5-18 所示。

文件清单 5-18　SendSMSActivity. java

```
package com.nsu.zyl.chapter05application;
import android.Manifest;
import android.app.PendingIntent;
import android.content.Intent;
import android.content.pm.PackageManager;
import android.os.Build;
import android.support.annotation.NonNull;
import android.support.v4.content.ContextCompat;
import android.support.v7.app.AppCompatActivity;
import android.os.Bundle;
import android.telephony.SmsManager;
import android.view.View;
import android.widget.Button;
import android.widget.EditText;
import android.widget.Toast;

public class SendSMSActivity extends AppCompatActivity {

    private EditText edtNumber,edtContext;
    private Button btnSend;
    private SmsManager smsManager;
    private int MY_PERMISSION_REQUEST_CODE=0,requestCode=0;

    @Override
    protected void onCreate(Bundle savedInstanceState) {
        super.onCreate(savedInstanceState);
        setContentView(R.layout.activity_send_sms);
        smsManager=SmsManager.getDefault();
        edtNumber=(EditText)findViewById(R.id.edtPhoneNumber);
        edtContext=(EditText)findViewById(R.id.edtMessage);
        btnSend=(Button)findViewById(R.id.btnSend);
        btnSend.setOnClickListener(new View.OnClickListener(){

            @Override
            public void onClick(View v) {
                // TODO Auto-generated method stub
                PendingIntent pi = PendingIntent. getActivity (SendSMSActivity.
this, requestCode, new Intent(), 0);
                if (Build.VERSION.SDK_INT >=Build.VERSION_CODES.M) {
                    if (ContextCompat.checkSelfPermission(SendSMSActivity.this,
Manifest.permission.SEND_SMS) !=PackageManager.PERMISSION_GRANTED) {
                        requestPermissions (new String[]{Manifest. permission.
SEND_SMS},requestCode);
                    }else{
```

```
                        smsManager.sendTextMessage(edtNumber.getText().toString(),
null, edtContext.getText().toString(), pi, null);
                        Toast.makeText(SendSMSActivity.this, "短信发送成功",
Toast.LENGTH_LONG).show();
                    }
                }

            }});
    }
    @Override
    public void onRequestPermissionsResult(int requestCode, @NonNull String[]
permissions, @NonNull int[] grantResults) {
                super. onRequestPermissionsResult (requestCode, permissions,
grantResults);
        if (requestCode==MY_PERMISSION_REQUEST_CODE) {
            boolean isAllGranted=true;

            //判断是否所有的权限都已经授予了
            for (int grant : grantResults) {
                if (grant !=PackageManager.PERMISSION_GRANTED) {
                    isAllGranted=false;
                    break;
                }
            }

            if (isAllGranted) {
                smsManager.sendTextMessage(edtNumber.getText().toString(),
null, edtContext.getText().toString(), pi, null);
                Toast.makeText(SendSMSActivity.this, "短信发送成功", Toast.
LENGTH_LONG).show();

            } else {
                Toast.makeText(SendSMSActivity.this, "本应用需要发送短信权限,请确
认是否打开", Toast.LENGTH_LONG).show();
            }
        }

    }

}
```

上述代码中使用 sendTextMessage(String destinationAddress, String scAddress, String text, PendingIntent sentIntent, PendingIntent deliverIntent)方法发送短信，其中参数含义依次是：

(1) String destinationAddress：收信人的电话号码。

(2) String scAddress：短信中心的号码，null 的话使用当前默认的短信服务中心。

(3) String text：短信内容。

(4) PengdingIntentsentIntent：发送是否成功的回执，用于监听短信是否发送成功。

(5) PendingIntentdeliverIntent：接收是否成功的回执，用于监听短信对方是否接收成功。

使用 SMSManager 发送短信，需获得发送短信的权限，因此需在 AndroidManifest.xml 进行如下权限设置：

```
<uses-permission android:name="android.permission.SEND_SMS">
```

对于 Android 6.0 版本以上系统，还需要在源代码中判断当前应用是否已经被授权使用 android.permission.SEND_SMS 权限，如果没有授予该权限需要首先申请该权限，如果已经被授予了该权限则直接执行发送短信操作。

5.3 BroadcastReceiver

广播是 Android 提供的一种运用在应用程序之间传递消息的机制。BroadcastReceiver 是一种用于接收广播消息的组件，利用 BroadcastReceiver 可以方便地实现组件之间的通信。本节主要从广播发送和广播接收对广播机制进行讲解，其中广播发送是通过调用 Context.sendBroadcast(Intent intent)或 Context.sendOrderBroadcast(Intent intent)方法实现，广播接收则通过 BroadcastReceiver 实现。一个广播可以被多个订阅该广播的 BroadcastReceiver 接收。当程序发出一个 Broadcast Intent 后，所有匹配该 Intent 的 BroadcastReceiver 都可能被启动。

5.3.1 BroadcastReceiver 的创建

BroadcastReceiver 用于监听程序(包括系统程序和用户开发的应用程序)发送的 Broadcast Intent。创建 BroadcastReceiver 类只需要继承 BroadcastReceiver 类，重写 BroadcastReceiver 中的抽象方法 onReceive(Context context, Intent intent)即可。默认情况下，BroadcastReceiver 运行在 UI 线程中，因此在 onReceive()方法中不能执行超过 10s 的耗时操作，否则将造成 ANR。如果需要完成耗时操作，可以通过 Intent 启动一个 Service 来完成该操作。

5.3.2 BroadcastReceiver 的注册

实现 BroadcastReceiver 后，需要指定该 BroadcastReceiver 能匹配的 Intent，有两个方法可以实现 BroadcastReceiver 的注册。

(1) 代码注册。直接在代码中通过调用 Context 的 registerReceiver (Broadcast-Receiver receiver, IntentFilter filter)方法来动态注册 BroadcastReceiver。采用代码注册时，注意在退出程序前需要调用 unregisterReceiver(BroadcastReceiver receiver)解除注册。

代码注册又称动态注册，如果在 onCreate()方法中注册，需要在 onDestroy()方法中注销，如果在 onResume()方法中注册，就需要在 onPause()方法中注销。由于动态注册依赖于注册广播的组件的生命周期，因此又称为非常驻型广播接收器。

(2) 在 AndroidManifest. xml 中注册。在 AndroidManifest. xml 注册又称为静态注册，在 AndroidManifest. xml 文件中 BroadcastReceiver 对应< receiver >元素。采用 AndroidManifest. xml 注册的广播接收器在应用程序关闭状态下，依然能够接收到其他应用程序发送的广播。当接收到其他应用程序发出的广播，该程序会自动重新启动。采用 AndroidManifest. xml 注册的广播又称为常驻型广播接收器。

BroadcastReceiver 的配置规则如下：

```
<receiver android:name=""
          android:enabled=""
          android:permission=""
          android:process=""
          android:exported=""
          ....
          >
      <intent-filter  android:priority="">
         <action android:name=""></action>
         <category></category>
         <data></data>
      </intent-filter>
      <intent-filter>
         <action android:name=""></action>
         <category></category>
           <data></data>
           </intent-filter>
           …
</receiver>
```

上述配置中涉及的相关属性说明如下：

1. android:exported

用于标识 BroadcastReceiver 是否能够接收其他 APP 发出的广播。默认值与有无< intent-filter >相关，如没有配置< intent-filter >则默认为 false，如配置< intent-filter >则默认为 true。

2. android:name

BroadcastReceiver 的类全名。

3. android:permission

如果设置，则具有相应权限的广播发送方发送的广播才能够被当前 BroadcastReceiver 所接收。

4. android:process

BroadcastReceiver运行所处的进程。默认为APP的进程,可以指定独立的进程。

与定义隐式意图一样,广播接收器也需要注册一个< intent-filter >,在过滤器中指定要接收的广播事件。在< intent-filter >中可以设置android:propriety=""指定广播的优先级,优先级范围在-1000～1000,这个值越大代表接收的优先级越高。

5.3.3 广播的类型

在Android系统中,根据广播的传递方式不同,可以分为普通广播和有序广播两类。

(1) 普通广播(Normal Broadcast),调用Context.sendBroadcast()方法发送,在广播发送后,所有监听该广播的广播接收器异步接收到广播消息,它们没有先后顺序,因此普通广播无法被拦截。

(2) 有序广播(Ordered Broadcast),调用Context.sendOrderedBroadcast()方法发送,所有监听该广播的广播接收器将按照顺序接收广播,接收次序与广播接收器的优先级有关,优先级高的接收器首先接收到广播,执行onReceive()方法后,可以继续传播到一下个广播接收器,也可以终止广播的传递。如果优先级高的广播接收器终止广播传递,优先级低的广播接收器将无法接收到广播。广播接收器的优先级是在清单文件中注册广播接收器时定义的android:priority来控制的。如果两个广播接收器的优先级相同,则先注册的广播接收器接收广播。

对于有序广播,优先接收广播的接收者可以通过调用setResultExtra(Bundle)方法将处理结果存入Broadcast中,然后传给下一个接收者,下一个接收者通过调用代码Bundle bundle=getResultExtra(true)可以获取上一个接收者存入的数据。

5.3.4 案例

在Chapter05Application项目中创建广播接收器MyBroadcastReceiver,在MyBroadcastReceiver类中重写onReceive()方法,在该方法中利用Toast弹出消息提示框并把消息内容显示出来。MyBroadcastReceiver的代码如文件清单5-19所示。

文件清单5-19 MyBroadcastReceiver.java

```
package com.nsu.zyl.chapter05application;
import android.content.BroadcastReceiver;
import android.content.Context;
import android.content.Intent;
import android.widget.Toast;

public class MyBroadcastReceiver extends BroadcastReceiver {
    public MyBroadcastReceiver() {
    }
```

```
    @Override
    public void onReceive(Context context, Intent intent) {
        Toast.makeText(context, "接收到发送过来的广播,广播内容:" + intent.
getStringExtra("content"), Toast.LENGTH_LONG).show();
    }
}
```

广播接收器 MyBroadcastReceiver 对应在 AndroidManifest.xml 中的配置如下：

```
<receiver
            android:name=".MyBroadcastReceiver"
            android:enabled="true"
            android:exported="true" >
</receiver>
```

在 Chapter05Application 项目中创建 MyBroadcastReceiverActivity，该 Activity 对应的布局文件包含两个按钮，分别用于发送普通广播和有序广播。该 Activity 的布局文件比较简单，在此就不给出详细代码了。

在 onCreate()方法中利用代码注册的方式指定 MyBroadcastReceiver 广播接收器接收 action="com.broadcast.action"的广播。在 onDestroy()方法中调用 unregisterReceiver (BroadcastReceiver receiver)方法注销注册。当 Activity 结束后，通过代码动态注册的广播接收器将不再接收广播。

为 MyBroadcastReceiverActivity 页面中第一个按钮设置点击事件监听器，在监听器的事件处理方法中调用 sendBraodcast(Intent intent)方法发送一个 action="com.broadcast.action"广播，这个广播被 MyBroadcastReceiver 接收器接收后执行 onReceive()方法。MyBroadcastReceiverActivity 的代码如文件清单 5-20 所示。

文件清单 5-20　MyBroadcastReceiverActivity.java

```
package com.nsu.zyl.chapter05application;

import android.content.BroadcastReceiver;
import android.content.Intent;
import android.content.IntentFilter;
import android.support.v7.app.AppCompatActivity;
import android.os.Bundle;
import android.view.View;
import android.widget.Button;
public class MyBroadcastReceiverActivity extends AppCompatActivity {

    private Button btnSendBroadcast,btnSendOrderBroadcast;
    public static final String BROADCAST_ACTION="com.broadcast.action";
public static final String BROADCAST_ORDER_ACTION="com.broadcast.order.action";
    private BroadcastReceiver mybr;
```

```
    @Override
    protected void onCreate(Bundle savedInstanceState) {
        super.onCreate(savedInstanceState);
        setContentView(R.layout.activity_my_broadcast_receiver);
        btnSendBroadcast=(Button)findViewById(R.id.btnSendBroadcast);
        btnSendOrderBroadcast=(Button)findViewById(R.id.btnSendOrderBroadcast);
        mybr=new MyBroadcastReceiver();
        IntentFilter intentFilter=new IntentFilter();
        intentFilter.addAction(BROADCAST_ACTION);
        registerReceiver(mybr, intentFilter);
        btnSendBroadcast.setOnClickListener(new View.OnClickListener() {
            @Override
            public void onClick(View view) {

                Intent intent=new Intent();
                intent.setAction(BROADCAST_ACTION);
                intent.putExtra("content","最新广播,Android 8.0 发布,命名为奥利
奥!");
                sendBroadcast(intent);
            }
        });
        btnSendOrderBroadcast.setOnClickListener(new View.OnClickListener() {
            @Override
            public void onClick(View view) {

            }
        });

    }

    @Override
    protected void onDestroy() {
        super.onDestroy();
        unregisterReceiver(mybr);
    }

}
```

运行 MyBroadcastReceiverActivity，点击“发送广播”按钮发送的广播被 MyBroadcastReceiver 广播接收器接收并弹出消息提示框。运行显示结果如图 5-15 所示。

若采用静态注册的方式监听 MyBroadcastReceiverActivity 中发送的广播，就不需要在 onCreate()方法中添加动态注册代码。对 MyBroadcastReceiverActivity 做如下修改，内容如文件清单 5-21 所示。

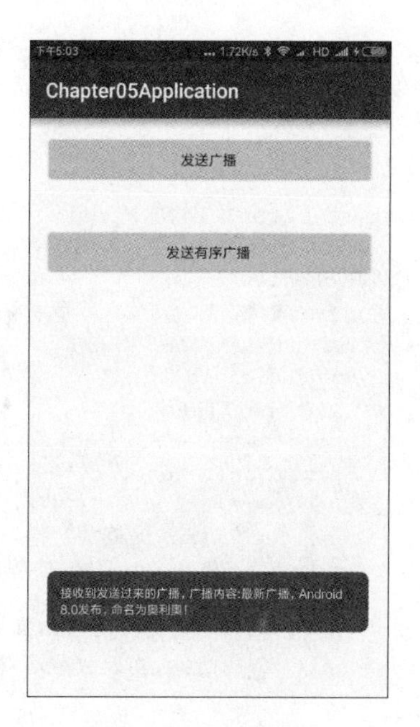

图 5-15 运行结果

文件清单 5-21 MyBroadcastReceiverActivity. java

```
package com.nsu.zyl.chapter05application;
import android.content.BroadcastReceiver;
import android.content.Intent;
import android.content.IntentFilter;
import android.support.v7.app.AppCompatActivity;
import android.os.Bundle;
import android.view.View;
import android.widget.Button;
public class MyBroadcastReceiverActivity extends AppCompatActivity {
    private Button btnSendBroadcast,btnSendOrderBroadcast;
    public static final String BROADCAST_ACTION="com.broadcast.action";
    public static final String BROADCAST_ORDER_ACTION= "com.broadcast.order.
action";
//    private BroadcastReceiver mybr;
    @Override
    protected void onCreate(Bundle savedInstanceState) {
        super.onCreate(savedInstanceState);
        setContentView(R.layout.activity_my_broadcast_receiver);
        btnSendBroadcast=(Button)findViewById(R.id.btnSendBroadcast);
        btnSendOrderBroadcast=(Button)findViewById(R.id.btnSendOrderBroadcast);
//        mybr=new MyBroadcastReceiver();
//        IntentFilter intentFilter=new IntentFilter();
//        intentFilter.addAction(BROADCAST_ACTION);
//        registerReceiver(mybr, intentFilter);
        btnSendBroadcast.setOnClickListener(new View.OnClickListener() {
            @Override
            public void onClick(View view) {
                Intent intent=new Intent();
```

```
                intent.setAction(BROADCAST_ACTION);
                intent.putExtra("content","最新广播,Android 8.0发布,命名为奥利奥!");
                sendBroadcast(intent);
            }
        });
        btnSendOrderBroadcast.setOnClickListener(new View.OnClickListener() {
            @Override
            public void onClick(View view) {
                Intent intent=new Intent();
                intent.setAction(BROADCAST_ORDER_ACTION);
                Bundle extras=new Bundle();
                extras.putInt("count",0);
                extras.putString("data","2017-8");
                sendOrderedBroadcast(intent,//意图动作,指定action动作
                    null, //receiverPermission,接收这条广播具备什么权限
                    null,//resultReceiver,最终的广播接受者,广播一定会传给他
                    null, //scheduler,handler对象处理广播的分发
                    0,//initialCode,初始代码
                    "据传Android 8.0发布,命名为奥利奥!", //initialData,初始数据
                    extras//initialExtras,初始数据不够,通过bundle携带更多数据
                );
            }
        });
    }
    @Override
    protected void onDestroy() {
        super.onDestroy();
    }
}
```

从上述修改的发送广播的程序中指定发送广播时所用的Intent的Action为com.broadcast.action,因此需要在AndroidManifest.xml中为MyBroadcastReceiver增加<intent-filter>配置:

```
<receiver
    android:name=".MyBroadcastReceiver"
    android:enabled="true"
    android:exported="true">
    <intent-filter>
        <action android:name="com.broadcast.action" />
    </intent-filter>
</receiver>
```

当用户点击发送有序广播按钮时使用sendOrderedBroadcast()方法发送一条有序广播,为了接收发送的有序广播,我们在Chapter05Application项目中定义两个广播接收器,第一个广播接收器代码如文件清单5-22所示。

文件清单5-22　MyOrderReceiver1.java

```
package com.nsu.zyl.chapter05application;
import android.content.BroadcastReceiver;
```

```
import android.content.Context;
import android.content.Intent;
import android.os.Bundle;
import android.widget.Toast;
public class MyOrderReceiver1 extends BroadcastReceiver {
    public MyOrderReceiver1() {
    }
    @Override
    public void onReceive(Context context, Intent intent) {
        Bundle bundle=getResultExtras(true);
        int count=bundle.getInt("count",0);
        Toast.makeText(context,"已经有"+(count+1)+"个广播接收器接收到有序广播,广
播内容为:"+ getResultData(),Toast.LENGTH_LONG).show();
        setResultData("经过确认,Android 8.0已经发布,命名为奥利奥!");
        bundle.putInt("count",(count+1));
        bundle.putString("data","2017-8-25");
        setResultExtras(bundle);
    }
}
```

在第一个广播接收器中,首先以 Toast 消息框的形式显示已经有多少广播接收器接收到广播消息,然后将数量+1 调用 setResultExtra()方法向广播中存入这个数据,这个信息将会被后面的广播接收器接收到。

第二个广播接收器的代码如文件清单 5-23 所示。

文件清单 5-23 MyOrderReceiver2. java

```
package com.nsu.zyl.chapter05application;
import android.content.BroadcastReceiver;
import android.content.Context;
import android.content.Intent;
import android.os.Bundle;
import android.widget.Toast;
public class MyOrderReceiver2 extends BroadcastReceiver {
    @Override
    public void onReceive(Context context, Intent intent) {
        Bundle bundle=getResultExtras(true);
        String data=bundle.getString("data");
        int count=bundle.getInt("count",0);
        Toast.makeText(context,"已经有"+(count+1)+"个广播接收器接收到有序广播,广
播内容为:"+getResultData()+"发布时间为:"+data,Toast.LENGTH_LONG).show();
    }
}
```

在第二个广播接收器中以 Toast 消息提示框的形式显示广播的内容和第一个广播接收器存入的信息。

为了保证第一个广播接收器先于第二个广播接收器接收到广播,分别为两个广播接收器设置优先级,其中第一个广播接收器的优先级要大于第二个广播接收器的优先级。AndroidManifest. xml 的内容如文件清单 5-24 所示。

文件清单 5-24 AndroidManifest.xml

```
<?xml version="1.0" encoding="utf-8"?>
<manifest xmlns:android="http://schemas.android.com/apk/res/android"
    package="com.nsu.zyl.chapter05application">

    <uses-permission android:name="android.permission.SEND_SMS" />


    <application
        android:allowBackup="true"
        android:icon="@mipmap/ic_launcher"
        android:label="@string/app_name"
        android:theme="@style/AppTheme"

        >
        <activity android:name=".MainActivity">
            <intent-filter>
                <action android:name="android.intent.action.MAIN" />

                <category android:name="android.intent.category.LAUNCHER" />
            </intent-filter>
        </activity>
        <service
            android:name=".MyFirstService"
            android:enabled="true"
            android:exported="true" />

        <activity android:name=".StartServiceActivity" />
        <activity android:name=".BindServiceActivity" />

        <service
            android:name=".ProgressBinderService"
            android:enabled="true"
            android:exported="true" />
        <service
            android:name=".ProgressMessengerService"
            android:enabled="true"
            android:exported="true" />
        <service
            android:name=".ProgressAIDLService"
            android:enabled="true"
            android:exported="true" />
        <service
            android:name=".MyIntentService"
            android:exported="false" />
        <service
            android:name=".MyService"
            android:enabled="true"
            android:exported="true" />

        <activity android:name=".MyServiceActivity" />

        <activity android:name=".NotificationActivity" />
        <activity android:name=".ToActivity" />
        <activity android:name=".SendSMSActivity" />
```

```
        <receiver
            android:name=".MyBroadcastReceiver"
            android:enabled="true"
            android:exported="true">
            <intent-filter >
                <action android:name="com.broadcast.action"/>
            </intent-filter>
        </receiver>

        <activity android:name=".MyBroadcastReceiverActivity" />

        <receiver
            android:name=".MyOrderReceiver1"
            android:enabled="true"
            android:exported="true">
            <intent-filter android:priority="100">
                <action android:name="com.broadcast.order.action"></action>
            </intent-filter>

        </receiver>
        <receiver
            android:name=".MyOrderReceiver2"
            android:enabled="true"
            android:exported="true">
            <intent-filter android:priority="10">
                <action android:name="com.broadcast.order.action"></action>
            </intent-filter>

        </receiver>
    </application>

</manifest><
```

运行程序首先看到图 5-16 显示的内容,然后看到图 5-17 显示的内容。

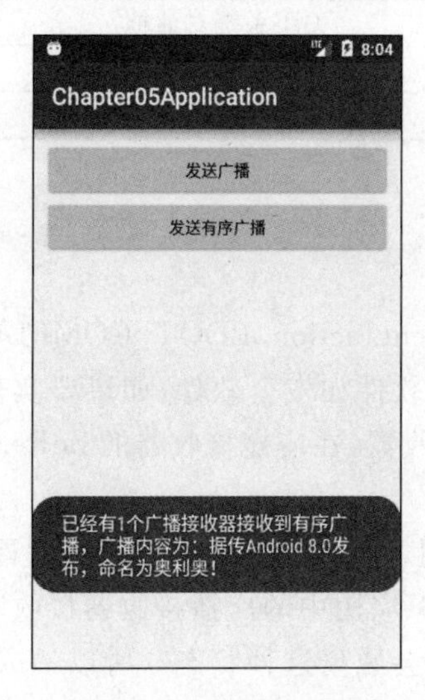

图 5-16 发送普通广播

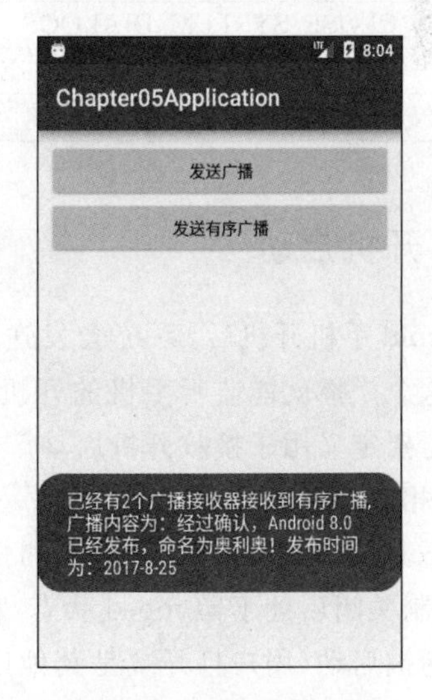

图 5-17 发送有序广播

如果在第一个广播接收器的 onReceive()方法中调用 abortBroadcast()方法，则第二个广播接收器将无法接收到该广播。

5.4 监听系统广播

Android 系统中内置多个系统广播，只要涉及手机的基本操作，基本都会发出相应的系统广播，如开机启动、网络状态改变、拍照、电量变化等。系统广播在系统内部有特定事件发生时由系统自动发出。如果需要在系统特定时刻执行某些操作，就可以通过 BroadCastReceiver 监听特定的系统广播，然后让应用随系统执行这些操作。

常见系统广播如表 5-5 所示。

表 5-5 常见的系统广播

广播常量	描述
ACTION_TIME_CHANGED	系统时间被改变
ACTION_DATE_CHANGED	系统日期被改变
ACTION_TIMEZONE_CHANGED	系统时区被改变
ACTION_BOOT_COMPLETED	系统启动完成
ACTION_PACKAGE_ADDED	系统添加包
ACTION_PACKAGE_CHANGED	系统的包改变
ACTION_PACKAGE_REMOVED	系统的包被删除
ACTION_PACKAGE_RESTARTED	系统的包被重启
ACTION_PACKAGE_DATA_CLEARED	系统的包数据被清空
ACTION_BATTERY_CHANGED	电池电量改变
ACTION_BATTERY_LOW	电池电量低
ACTION_POWER_CONNECTED	系统连接电源
ACTION_POWER_DISCONNECTED	系统与电源断开
ACTION_CLOSE_SYSTEM_DIALOGS	关闭系统对话框
ACTION_CONFIGURATION_CHANGED	当前设备配置被修改
ACTION_SHUTDOWN	系统被关闭

5.4.1 开机启动

Android 手机开机后，系统会发送 android. intent. action. BOOT_COMPLETED 广播，监听这个广播就能监听手机的开机。对于某些软件如安全软件，如果要实现开机启动服务，需要定义用于接收开机启动广播的广播接收器，在广播接收器的 onReceive()方法中启动相应服务。

Android 3.1 版本以后，系统为加强安全性控制，应用程序安装后或是(设置)应用管理中被强制关闭后处于 stopped 状态，在这种状态下应用中的广播接收器接收不到任何广播，直到被启动(用户打开或是其他应用调用)才会脱离这种状态。所以 Android 3.1 版本之后应用程序无法在安装后自己启动，没有 UI 的程序必须通过其他应用激活才能启动，如它的 Activity、Service、Content Provider 被其他应用调用。

开机启动的步骤如下：

(1) 定义广播接收器，在 onReceive()方法中启动服务。

```
public class BootCompletedReceiver extends BroadcastReceiver {
    @Override
    public void onReceive(Context context, Intent intent) {
        // TODO Auto-generated method stub
        Log.d("LibraryTestActivity", "recevie boot completed ... ");
        context.startService(new Intent(context, StartService.class));
    }
}
```

(2) 在 AndroidManifest. xml 文件中注册广播接收器。

```
<receiver android:name=".BootCompletedReceiver">
    <intent-filter>
        <action android:name="android.intent.action.BOOT_COMPLETED" />
    </intent-filter>
</receiver>
```

(3) 为了让程序访问系统的开机事件，需要在 AndroidManifest. xml 文件中声明权限。

```
<uses-permission android:name="android.permission.RECEIVE_BOOT_COMPLETED" />
```

(4) 安装该应用到手机，启动一次该应用，下次开机就能收到 BOOT_COMPLETED 广播实现开机启动了。

5.4.2 系统短信拦截

当短信到达时，Android 系统将发送一个有序广播，该广播的 Intent 对应的 Action 值为 android. provider. Telephony. SMS_RECEIVED，监听这个广播就可以获取短信信息。由于该广播是有序广播，所以可在广播接收器中拦截短信。

拦截短信步骤：

(1) 定义短信广播接收器，用于拦截包含“贷款”“旺铺”“中奖”“发票”和“开盘”等内容的短信。MessageReceiver 代码如文件清单 5-25 所示。

文件清单 5-25 MessageReceiver. java

```
public class MessageReceiver extends BroadcastReceiver{
        private String[] words={"贷款","旺铺","中奖","发票","开盘"};
        public void onReceive(Context context, Intent intent){
            if(intent. getAction (). equals ( " android. provider. Telephony. SMS _
RECEIVED")){
```

```
            Object[] objs=(Object[])Intent.getExtras().get("puus");
            for(Object obj:objs){
                SmsMessage smsMessage=SmsMessage.createFromPdu((byte[])obj);
                String body=smsMessage.getMessageBody();
                String sender=smsMessage.getOriginatingAddress();
                for(int i=0;i<words.lenght;i++){
                if(body.contain(words[i])){
                    abortBroadcast();
                }
            }
        }
    }
  }
}
```

上述代码通过 Intent 获取所有的短信数据，然后遍历短信，将 Pdu 对象转换成 SmsMessage 对象，并通过 SmsMessage 对象获取每条短信的内容，判断内容是否包含广告关键字信息，如果包含则终止当前的广播。

(2) 在 AndroidManifest. xml 文件中注册该广播接收器。

```
<receiver android:name="SmsReceiver">
    <intent-filter android:priority="1000">
        <action android:name="android.provider.Telephony.SMS_RECEIVED"/>
</intent-filter>
</receiver>
```

在< intent-filter >中为 Action 设置的 android. provider. Telephony. SMS_RECEIVED 属性就是接收短信的广播。为了让本程序在系统的短信接收程序之前被启动，所以将该 BroadcastReceiver 的优先级设置得高一些，这样就可以在系统的短信接收程序之前被触发。

(3) 在 AndroidManifest. xml 文件中添加权限。由于短信操作需要用户的权限许可，所以需要在 AndroidManifest. xml 文件中配置接收短信的用户权限。

```
<uses-permission android:name="android.permission.RECEIVE_SMS">
```

5.4.3 手机电量提醒

当手机电量改变时，Android 系统会发送 Intent 的 Action 为 ACTION_BATTERY_CHANGED 常量的广播，当手机电量过低时，Android 系统又会发送 Intent 的 Action 为 ACTION_BATTERY_LOW 常量的广播。因此可以通过监听上述广播实现对手机电量的提醒。

拦截短信步骤：

(1) 定义短信广播接收器，获取当前手机电量信息。

```
public class BatteryReceiver extends BroadcastReceiver{
        public void onReceive(Context context, Intent intent){
            Bundle bundle=bundle.getIntent("level");
            int current=bundle.getInt("level");
            int total=bundle.getInt("scale");
            Toast.makeText(context,"剩余电量"+(current*1.0/total)*100 +"% ").
show();
        }
}
```

(2) 在 AndroidManifest.xml 文件中注册该广播接收器。

```
<receiver android:name=".BroadcastReceiver">
    <intent-filter>
        <action android:name=" android.intent.action.BATTERY_LOW"/>
</intent-filter>
</receiver>
```

本章小结

本章主要讲述服务和广播机制。关于服务，首先讲解服务的创建和配置方式，接着讲解服务的两种启动方式和服务的生命周期，最后通过示例展示在程序中使用服务。关于广播机制，首先讲解广播的创建和注册，然后演示广播接收器的使用。

习　　题

1. 在创建服务时必须实现 Service 中的________方法。
2. Service 的启动方式有两种，分别为________和________。
3. 代码注册广播需要使用________方法，解除广播需要使用________方法。
4. 用于发送有序广播的方法是________。
5. 简述 Service 两种启动方式下的生命周期过程。
6. 简述常驻型广播接收器与非常驻型广播接收器的区别。
7. 编写程序，提示用户手机电量变化，当电量小于10%时开始进行提示。
8. 编写程序，实现音乐播放功能。

第6章 chapter 6

Android 多线程

主要内容：Android 多线程概述，Handler 线程通信机制，AsyncTask 异步任务

建议课时：4 课时

知识目标：(1) 了解 Android 多线程机制；

(2) 掌握 Handler 线程通信和 AsyncTask 操作。

能力目标：(1) 初步具备提高程序用户体验的能力；

(2) 具备 Android 多线程开发的能力。

在 AndroidStudio 集成开发环境中创建项目 Chapter06MultiThread，在该项目中将通过案例的形式展示 ANR 错误、非 UI 线程使用 Post 方法委托 UI 线程更改界面、非 UI 线程使用 Message 方法通知 UI 线程更改界面、模拟下载任务、AsyncTask 异步下载等。Chapter06MultiThread 项目的主页如图 6-1 所示。

图 6-1　Chapter06MultiThread 项目的主页

6.1 Android 多线程概述

1. 进程与线程

一个进程就是一个执行中的程序，而每一个进程都有自己独立的一块内存空间、一组系统资源。在进程的概念中，每一个进程的内部数据和状态都是完全独立的。

多线程指的是在单个程序中可以同时运行多个不同的线程，执行不同的任务。线程与进程相似，是一段完成某个特定功能的代码，是程序中单个顺序的流控制。但与进程不同的是，多个线程共享一块内存空间和一组系统资源，所以系统在各个线程之间切换时，资源占用要比进程小得多，正因如此，线程也称为轻量级进程。一个进程中可以包含多个线程，多线程意味着一个程序的多行语句可以看上去几乎在同一时间内运行。

2. Java 多线程机制

Java 的线程类是 java. lang. Thread 类。当生成一个 Thread 类的对象之后，一个新的线程就产生了。Java 中每个线程都是通过 Thread 对象的 run()方法来完成其操作，run( )方法称为线程体。

在 Java 中实现多线程一般遵循以下步骤。

(1) 创建线程类，继承 Thread 类或实现 Runnable 接口。

(2) 通过 Thread 类构造器来创建线程对象：

```
Thread( )
Thread(Runnable target)
```

(3) 通过 start()方法激活线程对象。

在 Android 应用开发中可以使用 Java 里的多线程完成一些操作。除此之外，Android 也有自身的多线程机制。

在 Android 的设计思想中，为了给用户提供流畅的操作体验，一些耗时的任务不能直接在 UI 线程中运行，必须启动一个后台线程去执行这些任务，通常这些任务最终又会直接或者间接地需要访问和控制 UI 控件。由于 Android 规定除了 UI 线程外，其他线程都不可以直接对 UI 控件访问和操控，所以就产生了如何在其他线程中通知 UI 线程访问和控制 UI 控件的问题。为了解决这一问题，我们需要先对 Android 多线程机制有深入的了解。

6.1.1 UI 线程及 Android 的单线程模型原则

在 Android 中，当应用启动时系统会创建一个主线程(Main Thread)。这个主线程负责向 UI 组件分发事件(包括绘制事件)，以及应用程序和 Android 的 UI 组件的交互。

所以 Main Thread 也称为 UI Thread，即 UI 线程。

Android 系统不会为每个组件单独创建线程，在同一个进程里的 UI 组件都会在 UI 线程里进行实例化，系统对每一个组件的调用都从 UI 线程分发出去。也就是说，响应系统回调的方法（比如响应用户的动作和各种生命周期回调）永远都在 UI 线程里运行。

6.1.2 ANR 问题

既然 Android 系统已经有了 UI 线程，为什么还要使用多线程呢？

在 Android 中使用多线程主要是为了提高用户体验或者避免 ANR（Application is not Responding）问题。大家在使用手机时，应该都会遇到程序无响应的情况。在事件处理代码中，如果事件响应时间超过 5s，即会出现 ANR 对话框，提示用户程序无响应，选择等待还是强制关闭。程序因为响应较慢会导致用户体验很差。

在 Android 系统中，为了防止应用程序反应较慢导致系统无法正常运行做如下限定：

（1）当用户输入事件（如 Activity 的事件处理）在 5s 内无法得到响应，那么系统会认为程序没有响应，从而弹出 ANR 对话框；

（2）BroadcastReciever 超过 10s 没执行完也会弹出 ANR 对话框。

我们来看下面这个案例，案例源代码如文件清单 6-1 所示。

文件清单 6-1 ANRActivity.java

```
public class ANRActivity extends Activity{

    private Button btnCount;
    private TextView tvCount;
    private int count =0;

    @Override
    protected void onCreate(Bundle savedInstanceState) {
        super.onCreate(savedInstanceState);
        setContentView(R.layout.anr_view);
        tvCount = (TextView) findViewById(R.id.txt_count);
        btnCount = (Button) findViewById(R.id.btn_count);
        btnCount.setOnClickListener(new OnClickListener() {
            @Override
            public void onClick(View v) {
                //1、主线程耗时太久
                while (count <100) {
                    count +=10;
                    try {
                        Thread.sleep(1000);
                    } catch (Exception e) {
                        e.printStackTrace();
                    }
```

```
                    tvCount.setText("当前 count 的值为："+count);
                }
}
        });
    }
```

由于ANRActivity页面非常简单，只有一个按钮控件和一个文本控件，所以在这里就不给出anr_view.xml文件的内容。在ANRActivity运行时，我们期望点击“开始计数”按钮后，能够每秒动态显示当前的count值。实际情况却是点击了“开始”计数按钮之后，程序失去了响应，弹出了ANR对话框，如图6-2所示。

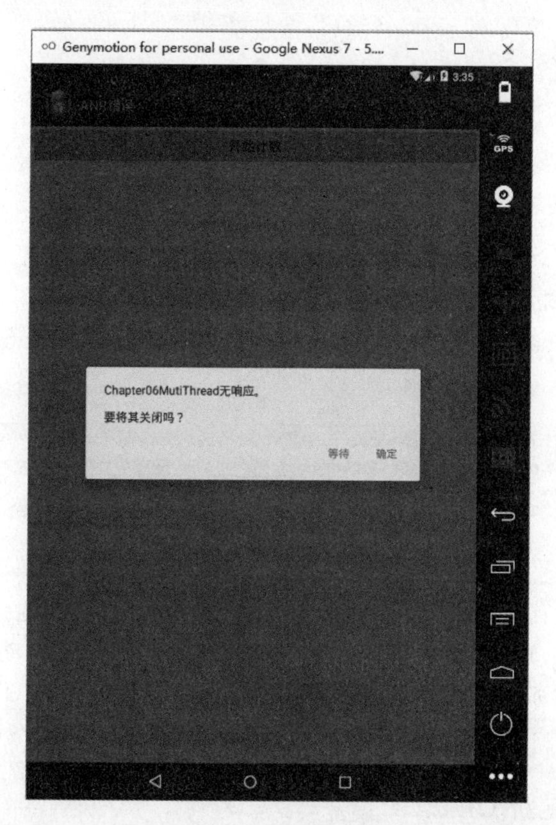

图6-2　ANR对话框

为什么会出现这个问题呢？分析程序逻辑，按钮的点击事件处理方法里有一段耗时10s的任务代码，导致Activity响应时间超过5s，从而系统弹出ANR对话框。如果一款应用程序在使用过程中频繁地出现ANR对话框，那一定非常糟糕，这个应用程序也将面临被抛弃的命运。

鉴于此，我们在进行事件处理时需要牢记一个原则：所有可能耗时的操作都放到其他线程去处理。我们用下面这段代码来替换上述案例中加黑部分的代码，直接另起一个线程来更新TextView控件的文本内容。结果又会如何呢？

```
//2、非 UI 线程中更新 UI
                new Thread(){
                    @Override
                    public void run() {
                        while (count <100) {
                            count +=10;
                            //在新线程中直接更新 UI
                            tvCount.setText("当前 count 的值为："+count);
                            try {
                                Thread.sleep(1000);
                            } catch (Exception e) {
                                e.printStackTrace();
                            }
                        }
                    }
                }.start();
```

很遗憾，当我们运行程序后会发现，程序直接崩掉了。为什么呢？我们已经另起线程来处理耗时任务了，为什么还不能得到想要的结果呢？

为了搞清错误原因，我们仔细检查 Logcat 日志信息，会发现图 6-3 所示的错误日志信息：Only the original thread that created a view hierarchy can touch its views(只有 UI 线程才能更新 UI)。

```
10-28 11:25:29.169: E/AndroidRuntime(350): android.view.ViewRoot
$CalledFromWrongThreadException: Only the original thread that created a view
hierarchy can touch its views.
```

图 6-3 UI 线程问题

Android 规定只有 UI 线程才能更新 UI，除了 UI 线程外，其他线程都不可以对那些 UI 控件访问和操控。UI 的职责是显示 UI 控件、处理 UI 事件、启动子线程、停止子线程和更新 UI，子线程的职责是计算逝去的时间和向主线程发出更新 UI 消息。

那么现在问题来了，我们该如何跨线程更新 UI 呢？

6.1.3 跨线程更新 UI

在 Android 中，有如下 3 种方式可以解决跨线程更新 UI 的问题。

方式 1：其他线程委托 UI 线程更新 UI。

方式 2：通过 Handler 发送 Message 给 UI 线程，令 UI 线程根据 Message 消息更新 UI。

方式 3：使用 Android 提供的 AsyncTask。

其中方式 1 比较简单，可采用如下几种方法来实现：

(1) Activity. runOnUiThread(Runnable)

(2) View. post(Runnable)

(3) View. postDelay(Runnable ，long)

```
//3.其他线程委托 UI 线程去更新 UI
new Thread(new Runnable() {
    @Override
    public void run() {
        while (count <100) {
            count +=10;
            tvCount.post(new Runnable() {
                @Override
                public void run() {
                    tvCount.setText("当前 count 的值为:"+count);

                }
            });
            try {
                Thread.sleep(1000);
            } catch (Exception e) {
                e.printStackTrace();
            }
        }
    }
}).start();
```

我们用上述代码来替换文件清单 6-1 中加黑部分的代码,在子线程中采用 View.post()方法委托 UI 线程去更新 TextView 控件的文本内容,会得到图 6-4 所示的运行效果。

图 6-4 跨线程更新 UI

其他线程委托 UI 线程更新 UI 的方式比较简单，通常只能适用一些简单的场景，因此我们的重点将是 Handler 和 AsyncTask 这两种方式。

6.2 Handler 线程通信机制

Android 中只有在 UI 线程才能操作 UI 线程中的对象，为了将非 UI 线程中的数据传送到 UI 线程，可以使用一个 Handler 运行在 UI 线程中。

Handler 是 Android Framework 中管理线程的一部分，一个 Handler 对象负责接收消息然后处理消息。

我们可以为一个新的线程创建一个 Handler，也可以创建一个 Handler 然后将它和已有线程连接。如果将一个 Handler 和 UI 线程连接，处理消息的代码就将会在 UI 线程中执行。

在 Handler 中，要重写 handleMessage()方法。Android 系统会在 Handler 管理的相应线程收到新消息时调用 handleMessage()方法来处理这个消息。一个特定线程的所有 Handler 对象都会收到同样的方法。

6.2.1 Handler 线程通信模型

Handler 机制主要包括 4 个关键对象，即 Message、Handler、MessageQueue、Looper。其通信模型如图 6-5 所示。

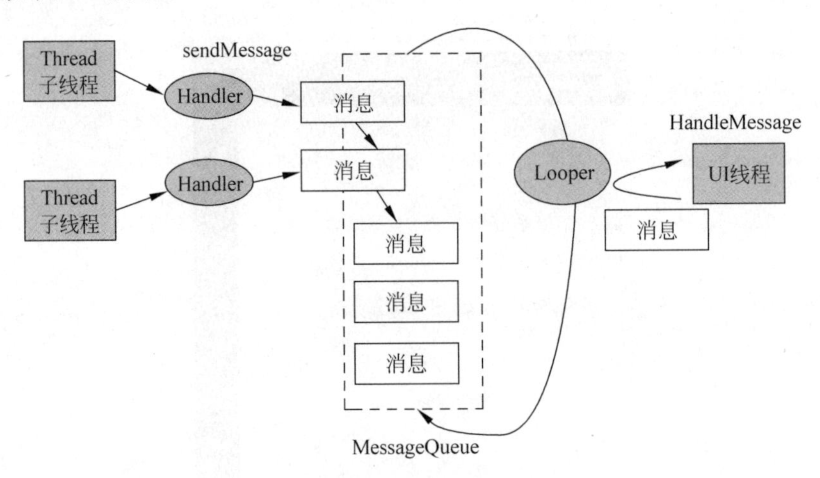

图 6-5 Handler 线程通信模型

1. Message

Message 是线程间通信的消息载体，在线程之间传递消息时，可以在内部携带少量的信息，用于在不同的线程之间交换数据。就像两个码头之间运输货物，Message 充当集装箱的功能，里面可以存放任何想要传递的消息。

Message常用属性如表6-1所示。

表6-1 Message属性

属性类型	属性名	属性说明
public int	arg1	若只需要在Message中放入少量的整型变量,使用arg1和arg2比setData()节约资源
public int	arg2	同上
public Object	obj	可以存放任意的数据,适用于存放简单的数据,效率同样比setData()高
public int	what	用来匹配Message的标识符,很重要

如果线程间通信的数据比较复杂,比如需要使用键值对来存放大量数据,此时需要使用setData()和getData()方法,用Bundle对象来封装数据。

```
Message msg =Message.obtain();
msg.what =101;
Bundle bundle =new Bundle();
bundle.putInt("number",12);
bundle.putString("Name","Rice");
bundle.putString("Hobby","Swimming");
msg.setData(bundle);
```

封装好后,使用Handler对象将此Message发送出去。

2. Handler

Handler在Android里有两个主要用途：将Message或Runnable对象发送给其他线程(发送Message)；处理来自其他线程的Message(处理Message),通过它实现其他线程与Main线程之间的消息通信。主要方法如下：

(1) 发送Message：sendEmptyMessage(int)、sendMessage(Message)、sendMessageAtTime(Message,long)、sendMessageDelayed(Message,long)。

(2) 处理Message：handleMessage(Message)

3. MessageQueue

MessageQueue(消息队列)用来存放Message对象的数据结构,按照“先进先出”的原则存放消息。存放并非实际意义的保存,而是将Message对象以链表的方式串联起来。MessageQueue对象不需要我们自己创建,而是由Looper对象对其进行管理,一个线程最多只可以拥有一个MessageQueue。我们可以通过Looper.myQueue()获取当前线程中的MessageQueue。

4. Looper

Looper负责管理线程的消息队列和消息循环,是MessageQueue的管理者。在一个线程中，如果存在 Looper 对象，则必定存在 MessageQueue 对象，并且只存在一个 Looper 对象和一个 MessageQueue 对象。倘若线程中存在 Looper 对象，则可以通过 Looper. myLooper()获取，此外还可以通过 Looper. getMainLooper()获取当前应用系统中主线程的 Looper 对象。假如，Looper 对象位于应用程序主线程中，则 Looper. myLooper()和 Looper. getMainLooper()获取的是同一个对象。

Android 中除了 UI 线程外，创建的工作线程默认是没有消息循环和消息队列的。如果想让该线程具有消息队列和消息循环，并具有消息处理机制，就需要在线程中首先调用 Looper. prepare()来创建消息队列，然后调用 Looper. loop()进入消息循环。

6.2.2 Post 方式

在 Handler 中，关于 Post 方式的方法有：

(1) boolean post(Runnable r)：把一个 Runnable 对象入队到消息队列中，UI 线程从消息队列中取出这个对象后，立即执行。

(2) boolean postAtTime(Runnable r, long uptimeMillis)：把一个 Runnable 入队到消息队列中，UI 线程从消息队列中取出这个对象后，在特定的时间执行。

(3) boolean postDelayed(Runnable r, long delayMillis)：把一个 Runnable 入队到消息队列中，UI 线程从消息队列中取出这个对象后，延迟 delayMills 秒执行。

(4) void removeCallbacks(Runnable r)：从消息队列中移除一个 Runnable 对象。

接下来，用 Post 方式来实现一个霓虹灯效果，如文件清单 6-2 所示。

文件清单 6-2 PostActivity. java

```
public class PostActivity extends Activity{

    private Button btnStop;
    private Button btnPostChange;
    //定义颜色数组
    private int[] colors =new int[]{
            R.color.color1, R.color.color2, R.color.color3,
            R.color.color4, R.color.color5
    };
    //定义控件数组
    private int[] names =new int[]{
            R.id.textView1, R.id.textView2, R.id.textView3,
            R.id.textView4, R.id.textView5
    };
    private TextView[] views =new TextView[names.length];
    private Handler handler =new Handler();
    private int currentColor =0;
    private Timer timer;

    @Override
    protected void onCreate(Bundle savedInstanceState) {
```

```
        super.onCreate(savedInstanceState);
        setContentView(R.layout.post_layout);
        initView();
    }

    private void initView() {
        btnStop = (Button) findViewById(R.id.btn_post_msg);
        btnPostChange = (Button) findViewById(R.id.btn_post_change);
        for (int i = 0; i < names.length; i++) {
            views[i] = (TextView) findViewById(names[i]);
        }

        btnStop.setOnClickListener(new OnClickListener() {

            @Override
            public void onClick(View v) {
                // TODO Auto-generated method stub
                timer.cancel();
            }
        });

        btnPostChange.setOnClickListener(new OnClickListener() {

            @Override
            public void onClick(View v) {
                timer = new Timer();
                timer.schedule(new TimerTask() {

                    @Override
                    public void run() {
                        handler.post(new Runnable() {

                            @Override
                            public void run() {
                              for (int i = 0; i < names.length; i++) {
                                views[i]. setBackgroundResource (colors [(i +
currentColor)
                                        % names.length]);
                              }
                                currentColor++;
                            }
                        });
                    }
                }, 0, 200);
            }
        });
    }
}
```

运行效果如图6-6所示，页面采用帧布局，点击“霓虹灯效果”按钮，页面闪烁；点击“停止”按钮，页面停止。

图6-6 Handler Post 霓虹灯

补充：Java中定时器Timer的用法

Timer类用来实现某一个时间或某一段时间后安排某一个任务执行一次或定期重复执行，该功能和TimerTask配合使用。

void cancel()：终止此计时器，丢弃所有当前已安排的任务。

void schedule(TimerTask task, long delay, long period)：安排指定的任务从指定的延迟后开始进行重复的固定延迟执行。

6.2.3 Message方式

在Handler中，与Message发送消息相关的方法有：

(1) Message obtainMessage()：获取一个Message对象。

(2) boolean sendMessage()：发送一个Message对象到消息队列中，并在UI线程取到消息后，立即执行。

(3) boolean sendMessageDelayed()：发送一个Message对象到消息队列中，在UI线程取到消息后，延迟执行。

(4) boolean sendEmptyMessage(int what)：发送一个空的Message对象到队列中，并在UI线程取到消息后，立即执行。

(5) void removeMessage()：从消息队列中移除一个未响应的消息。

(6) boolean sendEmptyMessageDelayed(int what,long delayMillis)：发送一个空Message到消息队列中，延迟执行。

接下来，用Message方式来模拟一个任务下载。理解Message方式之后，大家自行使用Message方式来实现前面的案例及霓虹灯效果，如文件清单6-3所示。

文件清单6-3　HandlerActivity.java

```
public class HandlerActivity extends Activity{

    private TextView tvContent;
    private ProgressBar progressBar;
    private Button btnDownload;
    private Handler handler;

    @Override
    protected void onCreate(Bundle savedInstanceState) {
        super.onCreate(savedInstanceState);
        setContentView(R.layout.download_view);
        initView();

        handler =new Handler(){
            @Override
            public void handleMessage(Message msg) {
                switch (msg.what) {
                case 0x100:
                    tvContent.setText("已接收数据 "+msg.arg1+"% ");
                    progressBar.setProgress(msg.arg1);
                    break;
                case 0x101:
                    tvContent.setText("接收数据完成！");
                    break;
                }
            }
        };
    }

    private void initView() {
        tvContent = (TextView) findViewById(R.id.tv_down_content);
        progressBar = (ProgressBar) findViewById(R.id.pb_down);
        btnDownload = (Button) findViewById(R.id.btn_down);
        btnDownload.setOnClickListener(new OnClickListener() {

            @Override
            public void onClick(View v) {
                //Handler应用程序主线程与子线程之间的通信
                new Thread(new Runnable() {
                    @Override
```

```
                    public void run() {
                        for (int i =1; i <=10; i++) {
                            try {
                                Thread.sleep(1000);
                            } catch (InterruptedException e) {
                                e.printStackTrace();
                            }
                            Message msg =Message.obtain();
                            msg.what =0x100;
                            msg.arg1 =i *10;
                            handler.sendMessage(msg);
                        }
                        //更新完成,发送空消息,只包含状态码
                        handler.sendEmptyMessage(0x101);
                    }
                }).start();
            }
        });
    }
}
```

运行效果如图 6-7 和图 6-8 所示。点击 ProgressBar 开始加载,TextView 显示进度。

图 6-7 模拟下载页面原型

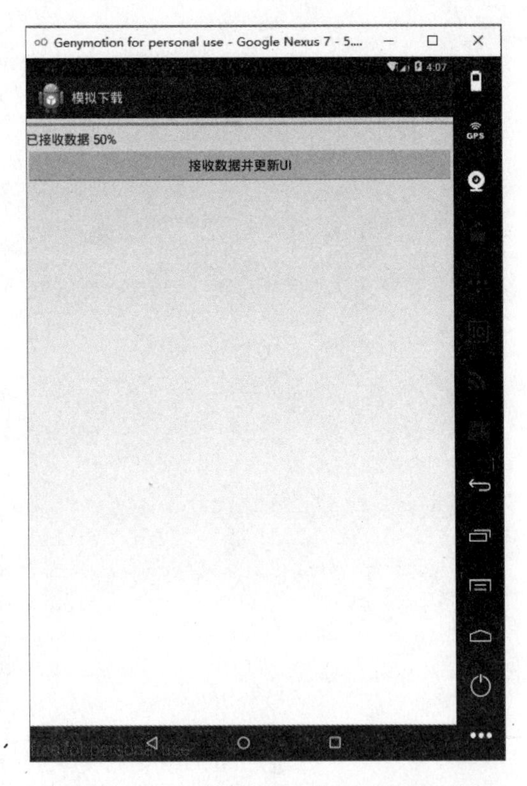

图 6-8 Handler Message 模拟下载

Handler 使用要点：Handler 可以在主线程中给子线程发送消息，也可以在子线程中给主线程发送消息。

在子线程中给主线程发送消息：在主线程中创建一个 Handler 对象，然后重写 handleMessage()方法，在子线程中调用 handler.sendEmptyMessage()方法，然后在 handleMessage()方法中就可以执行我们想要执行的操作了，例如更新 UI。

在主线程中给子线程发送消息：在子线程中 new 出来一个 Handler 对象，然后重写 handleMessage 方法，在主线程调用 sendEmptyMessage()方法；因为 Handler 要处理消息，但是在子线程中默认没有 Looper 对象，我们要通过 Looper.prepare()来创建一个 Looper 对象，在这里 Looper 对象被创建时会创建一个 MessageQueue，这样就有了 Looper，在执行完操作之后要执行以下 Looper.loop()方法去轮询这些消息，让 Handler 来处理这些消息。在 Android 源代码中，Looper.loop()方法是一个永远为 true 的 while 循环，所以要在子线程中执行的动作，都要在 Looper.loop()方法之前执行。

6.3 AsyncTask

线程的开销较大，如果每个任务都要创建一个线程，那么程序的效率要低很多。线程无法管理，匿名线程创建并启动后就不受程序的控制了。如果有很多个请求发送，那么就会启动非常多的线程，系统将不堪重负。另外，在新线程中更新 UI 还必须引入 handler，这让代码看上去非常臃肿。

AsyncTask(异步任务)的特点是任务在主 UI 线程之外运行，而回调方法是在主 UI 线程中，从而有效地避免使用 Handler 带来的麻烦。

6.3.1 AsyncTask 简化多线程开发

AsyncTask 专门用于完成非 UI 线程更新 UI 的任务。本质上也是开启新线程执行耗时操作，然后将结果发送给 UI 线程。

优点：简化代码，减少编写线程间通信代码这一烦琐且易出错的过程。

1. 构造参数详解

AsyncTask 类定义了 3 种泛型类型即 Params、Progress 和 Result：AsyncTask < Params, Progress, Result >。

Params 是启动任务执行的输入参数。

Progress 是后台任务执行的进度。

Result 是后台执行任务返回的结果。

AsyncTask 为抽象类，在使用时必须先子类化。实现 AsyncTask < Params, Progress, Result >子类，需要实现相关的 4 个方法：onPreExecute()方法，doInBackground(Params…)方法，onProgressUpdate(Progress...)方法，onPostExecute(Result)方法。其中，doInBackground(Params…)和 onProgressUpdate(Progress….)的参数为数组。

onPreExecute()：开始执行前的准备工作。

doInBackground(Params...)：开始执行后台处理，处理耗时事务，并把结果返回。在后台处理中，调用 publishProgress(Progress)方法来更新实时的任务进度。

onProgressUpdate(Progress...)：在 publishProgress(Progress)方法被调用后，UI 线程将调用这个方法从而在界面上展示任务的进展情况。

onPostExecute(Result)：后台处理执行完成后的操作，后台处理返回的值将在 onPostExecute 方法作为参数，传送结果给 UI 线程。

我们来看下面这个自定义的 AsyncTask，理解各泛型参数的意义：

```
private class task extends AsyncTask<String, Integer, Bitmap>
```

AsyncTask <>的参数类型由用户设定，这里设为 String，Integer，Bitmap。

String 代表输入到任务的参数类型，即 doInBackground()的参数类型，调用 execute()方法时传入的参数类型，可能表示要下载的图片的 URL。

Integer 代表处理过程中的参数类型，也就是 doInBackground()执行过程中产出的参数类型，通过 publishProgress()发消息，传递给 onProgressUpdate()，一般用来更新界面。

Bitmap 代表任务结束的产出类型，也就是 doInBackground()的返回值类型和 onPostExecute()的参数类型，这里代表下载后的图片。

2. 要遵守的准则

(1) Task 的实例必须在 UI 线程中创建。

(2) Execute 方法必须在 UI 线程中调用。

(3) 不要手动调用 onPreExecute()、onPostExecute(Result)、doInBackground(Params...)、onProgressUpdate(Progress...)这几个方法，需要在 UI 线程中实例化这个 Task 来调用。

(4) 该 Task 只能被执行一次，否则多次调用时会出现异常。

(5) doInBackground 方法和 onPostExecute 的参数必须对应，这两个参数在 AsyncTask 声明的泛型参数列表中指定，第一个为 doInBackground 接受的参数，第二个为显示进度的参数，第三个为 doInBackground 返回和 onPostExecute 传入的参数。

6.3.2 AsyncTask 的使用

接下来，使用 AsyncTask 来模拟任务下载。在理解 AsyncTask 用法的基础上，大家自行使用 AsyncTask 实现前述案例霓虹灯效果，如文件清单 6-4 所示。

文件清单 6-4　AsyncTaskActivity. java

```
public class AsyncTaskActivity extends Activity{
    private Button btnDownload;
    private Button btnCancel;
```

```
    private ProgressBar progressBar;
    private TextView tvContent;
    private DownloadTask downloadTask;

    @Override
    protected void onCreate(Bundle savedInstanceState) {
        super.onCreate(savedInstanceState);
        setContentView(R.layout.async_task_view);
        initView();
    }
    private void initView() {
        btnDownload = (Button) findViewById(R.id.btn_download_async);
        btnCancel = (Button) findViewById(R.id.btn_cancle_async);
        progressBar = (ProgressBar) findViewById(R.id.pb_async);
        tvContent = (TextView) findViewById(R.id.tv_async);

        btnDownload.setOnClickListener(new OnClickListener() {

            @Override
            public void onClick(View v) {
                if (downloadTask ==null) {
                    downloadTask =new DownloadTask();
                }
                downloadTask.execute(100);
            }
        });

        btnCancel.setOnClickListener(new OnClickListener() {

            @Override
            public void onClick(View v) {
                if (downloadTask !=null &&
downloadTask.getStatus() ==AsyncTask.Status.RUNNING) {
                    downloadTask.cancel(true);
                    downloadTask =null;
                }

            }
        });
    }

    private class DownloadTask extends AsyncTask<Integer, Integer, String>{
        //后面尖括号内分别是
        //参数(例子里是线程休息时间)
        //进度(publishProgress 用到)
        //返回值 类型
        //第一个执行方法
        @Override
```

```
protected void onPreExecute() {
    // TODO Auto-generated method stub
    super.onPreExecute();
    //已经开始下载,则下载 Button 处于不可点击的状态
    btnDownload.setEnabled(false);
    btnCancel.setEnabled(true);
    tvContent.setText("点击按钮开始下载");
}

//第二个执行方法,onPreExecute()执行完后执行
@Override
protected String doInBackground(Integer... params) {
    // TODO Auto-generated method stub
    for (int i =0; i <=100; i++) {
        publishProgress(i);
        try {
            Thread.sleep(params[0]);
        } catch (InterruptedException e) {
            e.printStackTrace();
        }
        if (isCancelled()) {
            return null;
        }
    }
    return "下载完毕";
}
//这个函数在调用 publishProgress 时触发,虽然调用时只有一个参数
//但是这里取到的是一个数组,所以要用 progress[0]来取值
//第 n+1 个参数就用 progress[n]来取值
@Override
protected void onProgressUpdate(Integer... progress) {
    // TODO Auto-generated method stub
    super.onProgressUpdate(progress);
    if (isCancelled()) {
        return;
    }
    progressBar.setProgress(progress[0]);
    tvContent.setText("当前已下载 "+progress[0] +"% ");
}

// doInBackground 返回时触发,即 doInBackground 执行完后触发
//这里的 result 就是上面 doInBackground 执行后的返回值,所以这里是"执行完毕"
@Override
protected void onPostExecute(String result) {
    // TODO Auto-generated method stub
    super.onPostExecute(result);
    tvContent.setText(result);
    btnDownload.setEnabled(true);
```

```
            btnCancel.setEnabled(false);
            downloadTask =null;
        }

        /**
         * AsyncTask.cancel()并不意味着 AsyncTask 对象立即停止
         * 会在 doInBackground()返回后触发 onCancelled,而不是 onPostExecute
         * 可以在 doInBackground 定期调用 AsyncTask.isCancelled()来检查,以便及早返回
         */
        @Override
        protected void onCancelled() {
            // TODO Auto-generated method stub
            super.onCancelled();
            btnDownload.setEnabled(true);
            btnCancel.setEnabled(false);
            progressBar.setProgress(0);
            tvContent.setText("已停止");
        }
    }
}
```

运行效果如图 6-9 和图 6-10 所示，点击“下载”按钮，ProgressBar 开始加载，TextView 显示进度；点击“取消下载”按钮，页面停止。

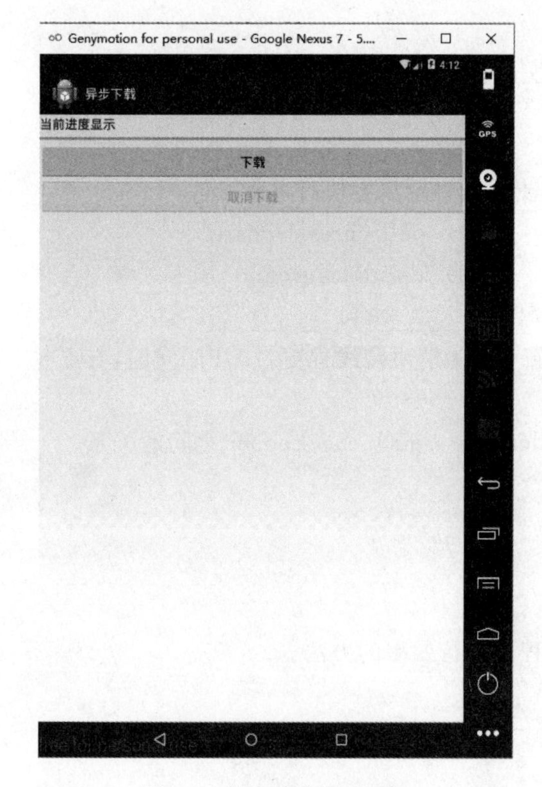

图 6-9　模拟下载页面原型

图 6-10　AsyncTask 模拟下载

本章小结

当用户与我们的应用交互时，事件处理方法的执行快慢决定了应用的响应性是否良好，常见的策略如下：

(1) 同步，需要等待返回结果。例如，用户点击了“注册”按钮，需要等待服务端返回结果，那么需要有一个进度条来提示用户其程序正在运行没有死掉。一般与服务端交互的都要有进度条，例如系统自带的浏览器，URL 跳转时会有进度条。

(2) 异步，不需要等待返回结果。例如微博中的收藏功能，点击“收藏”按钮后是否成功执行完成告诉我就行了，我不想等它，这里最好实现为异步的。

(3) ExecutorServic 线程池。用线程池来管理的好处是，可以保证系统稳定运行，适于大量线程、高工作量的情景下使用。假如要展示 1000 张图片，如果创建 1000 个线程去加载，系统肯定会死掉。用线程池就可以避免这个问题，可以用 5 个线程轮流执行，5 个一组，执行完的线程不直接回收而是等待下次执行，这样对系统的开销就可以大大减小。

习　题

1. 在 AsyncTask 中，下列哪个方法负责执行那些比较耗时的后台计算工作？(　　)

   A. execute　　　　B. onPostExecute

   C. doInBackground　　　　D. run

2. 在 AsyncTask 中，下列的哪个方法负责执行展示后台执行进度？(　　)

   A. publishProgress　　　　B. onProgressUpdate

   C. onPostExecute　　　　D. doInBackground

3. 程序运行过程中，导致 ANR 的两个原因是________和________。

4. Android 系统中只有 UI 线程才能更新 UI，为解决跨线程更新 UI 的问题，有哪些方法？

5. 解释在单线程模型中 Message、Handler、MessageQueue、Looper 之间的关系。

6. 以 Post 方式实现案例 1 的功能。

7. 以 Message 方式实现案例 1 的功能。

8. 以 Message 方式实现霓虹灯效果。

9. 以 AsyncTask 实现霓虹灯效果。

10. 写出几种你认为可以提高 Android 程序运行效率的方法。

第7章 chapter 7

Android 网络编程

主要内容：Android Http 通信，Android Socket 通信，Android 网络数据解析，WebView 与 WebService

建议课时：8 课时

知识目标：(1) 掌握 Android HTTP 的几种方法；

(2) 掌握 Android TCP Socket 通信的基本原理；

(3) 掌握 XML 文件和 JSON 数据解析的常规方法；

(4) 了解 WebView 和 WebService 的实现原理。

能力目标：(1) 初步具备 Android 浏览器应用开发的能力；

(2) 具备从网络获取数据并解析的能力；

(3) 具备开发 Android 网络应用的能力。

在 Android Studio 集成开发环境中创建项目 Chapter07Network，在该项目中我们将通过案例的形式展示如何在 Android 程序中采用 Http、Socket 等方式进行网络通信，如何对通信数据进行解析。为了验证 Android 的 Socket 通信，我们事先开发 Java 后台程序用于接收 Socket。

7.1 通信概述

Android 应用层采用 Java 语言编写，因此 Java 支持的网络编程方式，Android 都可以支持，同时 Google 还引入了 Apache 的 HTTP 扩展包。另外，针对 Wi-Fi，NFC 分别提供单独开发的 API。

Android 的网络编程分为两种：Http 网络通信和 Socket 网络通信。

本章重点介绍以下几类 Android 网络编程：

(1) 针对直接 URL 的 HttpURLConnection 编程。

(2) 早期 Google 集成了 Apache HTTP 客户端，可使用 HTTP 进行网络编程。

(3) 针对 TCP/IP 的 Socket、ServerSocket 的编程。

(4) 针对 UDP 的 DatagramSocket、DatagramPackage 编程。

(5) 使用 WebService。Android 可以通过开源包如 Jackson 去支持 Xmlrpc 和

Jsonrpc，另外也可以用 Ksoap2 去实现 WebService。

(6) 直接使用 WebView 视图组件显示网页。基于 WebView 进行开发，Google 已经提供了一个基于 chrome-lite 的 Web 浏览器，直接就可以进行上网浏览网页。

7.2 Android Http 通信

Android 对于 Http 网络通信，提供了标准的 Java 接口——HttpURLConnection 接口和 Apache 的接口——HttpClient 接口。

7.2.1 URL 加载网络资源

URL(Uniform Resource Locator)表示统一资源定位器，是指向互联网资源的指针。URL 可以由协议名、主机、端口和资源组成，资源可以是简单的文件或目录，也可以是更复杂对象的引用。

典型的 URL 地址的语法格式为：

```
protocol://host:port/resourceName
```

例如：

```
http://10.0.2.2:8080/abc/login.jsp
```

URL 类提供了多个构造器用于创建 URL 对象。

获得对象后，可以使用表 7-1 所示的方法来访问该 URL 的资源。

表 7-1　URL 常用方法

方　法	说　明
String getFile()	获取此 URL 的资源名
String getHost()	获取此 URL 的主机名
String getPath()	获取此 URL 的路径部分
int getPort()	获取此 URL 的端口号
String getProtocol()	获取此 URL 的协议名称
String getQuery()	获取此 URL 的查询字符串部分
URLConnection openConnection()	表示到 URL 所引用的远程对象的连接
InputStream openStream()	打开与此 URL 的连接，返回用于读取该资源的 InputStream

使用 URL 加载网络资源的一般步骤如下：

(1) 获取 URL 对象。

(2) 调用 openStream()方法打开 URL 的连接，获取 URL 资源输入流。

(3) 通过输入流 InputStream 进行文件读写。

(4) 关闭输入流。

现在以使用URL加载网络资源的方式实现一个下载网络图片的应用。在图7-1所示的界面上，点击“下载图片”按钮，这时会在界面显示一张网络图片，并把图片下载到本地。

图7-1 URL方式下载网络图片

由于连接网络，下载图片是一个耗时的操作，所以这里结合多线程，采用Handler的方式来实现。

核心代码如文件清单7-1所示。

文件清单7-1 URLActivity.java

```
public class URLActivity extends Activity {

    private Button btn_getPic;
    private ImageView img;
    private Bitmap bmp;
    //网络图片地址,可根据需要修改,也可通过EditText控件获取
    private static final String IMG_PATH =
"http://files.jb51.net/file_images/game/201605/2016051009271447.jpg";

    private Handler handler =new Handler(){

        @Override
        public void handleMessage(Message msg) {
```

```
            super.handleMessage(msg);
            if(msg.what ==0x123){
                //使用 ImageView 显示该图片
                img.setImageBitmap(bmp);
            }
        }
    };

    @Override
    protected void onCreate(Bundle savedInstanceState) {
        super.onCreate(savedInstanceState);
        setContentView(R.layout.down_pic_layout);
        btn_getPic = (Button)findViewById(R.id.btn_get_pic);
        img = (ImageView)findViewById(R.id.image_pic);

        btn_getPic.setOnClickListener(new View.OnClickListener() {
            @Override
            public void onClick(View v) {
                new Thread(new Runnable() {
                    @Override
                    public void run() {
                        try {
                            //定义一个 URL 对象
                            URL url =new URL(IMG_PATH);
                            //打开该 URL 对应的资源的输入流
                            InputStream is =url.openStream();
                            //从 InputStream 中解析出图片
                            bmp =BitmapFactory.decodeStream(is);
                            //发送消息,通知 UI 组件显示该图片
                            handler.sendEmptyMessage(0x123);
                            is.close();
                            //再次打开 URL 对应的资源的输入流
                            is =url.openStream();
                            //打开手机文件对应的输出流
                            OutputStream os =openFileOutput("wanhua.jpg", 0);
                            byte[] buff =new byte[1024];
                            int hasRead =0;
                            //将 URL 对应的资源下载到本地
                            while((hasRead =is.read(buff)) >0){
                                os.write(buff, 0 , hasRead);
                            }
                            is.close();
                            os.close();
                        } catch (MalformedURLException e) {
                            e.printStackTrace();
                        } catch (FileNotFoundException e) {
                            e.printStackTrace();
                        } catch (IOException e) {
```

```
                            e.printStackTrace();
                        }
                    }
                }).start();
            }
        });
    }
}
```

7.2.2 HttpURLConnection 加载网络资源

程序通过 URLConnection 实例向该 URL 发送请求，读取 URL 引用的资源。通常建立一个 URL 连接，并发送请求、读取此 URL 引用的资源需要以下几个步骤：

(1) 通过调用 URL 对象的 openConnection()方法来创建 URLConnection 对象。

(2) 设置 URLConnection 的参数和普通请求属性。

(3) 如果只是发送 GET 方式请求，使用 connect 方法建立连接即可；如果发送 POST 方式的请求，则需要获取 URLConnection 实例对应的输出流来发送请求参数。关于 GET 方式与 POST 方式的区别，读者可自行查阅相关资料。

(4) 远程资源变为可用，程序可以访问远程资源的头字段，或通过输入流读取远程资源的数据。

URLConnection 设置请求头字段：

setAllowUserInteraction 设置该 URLConnection 的 allowUserInteraction 请求头字段的值。

setDoInput 设置该 URLConnection 的 doInput 字段的值。

setDoOutput 设置该 URLConnection 的 doOutput 字段的值。

setIfModifiedSince 设置该 URLConnection 的 IfModifiedSince 字段的值。

setUseCaches 设置该 URLConnection 的 useCaches 字段的值。

HttpURLConnection 继承了 URLConnection，并做了进一步改进，操作更加便捷。

常用方法：

(1) int getResponseCode()：获取服务器的响应代码。

(2) String getResponseMessage()：获取服务器的响应消息。

(3) String getRequestMethod()：获取发送请求的方法。

(4) void setRequestMethod(String method)：设置发送请求的方法。

以下使用 HttpURLConnection 与 AsyncTask 相结合的方式再实现一个下载网络图片的应用。如图 7-2 所示，开始下载前，"开始下载"按钮可以点击，"取消下载"按钮不可点击。点击"开始下载"按钮，ImageView 控件显示加载中的图片，进度条显示当前图片下载的进度，此时"开始下载"按钮不可点击，"取消下载"按钮则可以点击。点击"取消下载"按钮，可终止下载任务。可显示一张网络图片，并把图片下载到本地。HttpURLActivity 的代码如文件清单 7-2 所示。

文件清单 7-2　HttpURLActivity. java

```
public class HttpURLActivity extends Activity {

    private static final String IMG_PATH =
"http://files.jb51.net/file_images/game/201605/2016051009271447.jpg";
    private Button btn_getPic;
    private Button btn_abort;
    private ImageView img;
    private ProgressBar progressBar;
    private TextView tv_progress;
    private ImageLoader loader;

    @Override
    protected void onCreate(Bundle savedInstanceState) {
        super.onCreate(savedInstanceState);
        setContentView(R.layout.http_url_download_pic_layout);

        btn_getPic = (Button)findViewById(R.id.btn_load_pic);
        btn_abort = (Button)findViewById(R.id.btn_abort);
        img = (ImageView)findViewById(R.id.imageView1);
        progressBar = (ProgressBar)findViewById(R.id.progressBar1);
        tv_progress = (TextView)findViewById(R.id.tv_progress);


        btn_getPic.setOnClickListener(new View.OnClickListener() {

            @Override
            public void onClick(View v) {
                loader =new ImageLoader();
                //execute 方法必须在 UI thread 中调用
                loader.execute(IMG_PATH);
            }
        });
        btn_abort.setOnClickListener(new View.OnClickListener() {

            @Override
            public void onClick(View v) {
                loader.cancel(true);
                btn_getPic.setEnabled(true);
                btn_abort.setEnabled(false);
            }
        });
        btn_abort.setEnabled(false);
    }

    @Override
```

```
        protected void onPreExecute() {
            super.onPreExecute();
            btn_getPic.setEnabled(false);
            btn_abort.setEnabled(true);
            progressBar.setVisibility(View.VISIBLE);
            progressBar.setProgress(0);
            img.setImageResource(R.mipmap.jiazai);
        }

        @Override
        protected Bitmap doInBackground(String... url) {
            if (isCancelled())
                return null;

            try {
                URL u;
                HttpURLConnection conn = null;
                InputStream in = null;
                OutputStream out = null;
                final String filename = "local_temp_image";

                try {
                    u = new URL(url[0]);
                    conn = (HttpURLConnection) u.openConnection();
                    conn.setDoInput(true);
                    conn.setDoOutput(false);
                    conn.setConnectTimeout(10*1000);

                    in = conn.getInputStream();
                    out = openFileOutput(filename, MODE_PRIVATE);
                    byte[] buff = new byte[8192];
                    int seg = 0;
                    final long total = conn.getContentLength();
                    long current = 0;

                    //每次读入 buff 大小
                    while (!isCancelled() && (seg = in.read(buff)) != -1) {
                        out.write(buff, 0, seg);
                        //当前已读出的字节
                        current += seg;
                        int progress = (int) ((float)current/(float)total * 100f);
                        //调用 onProgressUpdate(Integer... progress)更新进度
                        publishProgress(progress);

                        SystemClock.sleep(1000);
                    }
```

```
                }finally{
                    if (conn !=null) {
                        conn.disconnect();              //断开连接
                    }
                    if (in !=null) {
                        in.close();                     //关闭输入流
                    }
                    if (out !=null) {
                        out.close();                    //关闭输出流
                    }
                }
                return BitmapFactory.decodeFile(getFileStreamPath(filename)
.getAbsolutePath());

            }catch (Exception e) {
                e.printStackTrace();
            }
            return null;
        }

        @Override
        protected void onProgressUpdate(Integer... progress) {
            super.onProgressUpdate(progress);
            if (isCancelled())
                return;;
            progressBar.setProgress(progress[0]);
            tv_progress.setText("当前已加载了"+progress[0]+"% ");
        }

        @Override
        protected void onPostExecute(Bitmap image) {
            super.onPostExecute(image);
            if (isCancelled())
                return;
            if (image !=null) {
                img.setImageBitmap(image);
            }
            progressBar.setProgress(100);
            progressBar.setVisibility(View.GONE);
            tv_progress.setText("图片加载成功！");
            btn_getPic.setEnabled(true);
            btn_abort.setEnabled(false);
        }
    }
}
```

程序运行效果如图 7-2 所示。

图 7-2 HttpURLConnection 下载网络图片

7.2.3 HttpClient 加载网络资源

为了更好地处理向 Web 站点的请求，Apache 开源组织提供了一个 HttpClient 项目。HttpClient 是一个简单的 HTTP 客户端，可用于发送 HTTP 请求，接收 HTTP 响应。它对 Java 提供的方法做了一些封装，将 HttpURLConnection 中的输入/输出流操作，封装成 HttpPost(HttpGet)和 HttpResponse，从而简化了操作。

HttpClient 的使用方法：

(1) 创建 HttpClient 对象。

(2) 如果需要发送 GET 请求，则创建 HttpGet 对象；如果需要发送 POST 请求，则创建 HttpPost 对象。当使用 POST 方式时，需要对字符进行编码。

(3) 要发送参数，使用 setParams()方法来添加参数，对于 HttpPost 对象，也可调用 setEntity()设置参数。

(4) 调用 HttpClient 的 execute()方法发送请求，返回一个 HttpResponse。

(5) 调用 HttpResponse 的 getAllHeaders()、getHeaders(String name)等方法可以获取服务器的响应头；调用 HttpResponse 的 getEntity()方法可以获取 HttpEntity 对象，该对象包装了服务器的响应内容。

接下来，使用 HttpClient 与多线程相结合的方式来实现图片的下载。如图 7-3 所示，点击“下载图片”按钮，另起线程加载图片，同时弹出进度条对话框，图片加载完成后，显示对应网络地址的图片。相关代码如文件清单 7-3 所示。

文件清单 7-3 HttpClientActivity.java

```
public class HttpClientActivity extends Activity{
```

```
    private Button btn_getPic;
    private ImageView img;
    private ProgressDialog dialog;
    private static final String IMG_PATH =
"http://img5.duitang.com/uploads/item/201406/12/20140612151239_vtVe4.jpeg";
    private Handler handler = new Handler();

    @Override
    protected void onCreate(Bundle savedInstanceState) {
        super.onCreate(savedInstanceState);
        setContentView(R.layout.down_pic_layout);

        btn_getPic = (Button)findViewById(R.id.btn_get_pic);
        img = (ImageView)findViewById(R.id.image_pic);

        dialog = new ProgressDialog(this);
        dialog.setTitle("提示");
        dialog.setMessage("正在拼命下载,请稍后...");
        dialog.setCancelable(false);

        btn_getPic.setOnClickListener(new View.OnClickListener() {
            @Override
            public void onClick(View v) {
                //另起线程下载
                new Thread(new MyThread()).start();
                dialog.show();
            }
        });

    }

    private class MyThread implements Runnable {
        @Override
        public void run() {
            HttpClient httpClient = new DefaultHttpClient();
            HttpGet httpGet = new HttpGet(IMG_PATH);
            HttpResponse httpResponse = null;

            try {
                httpResponse = httpClient.execute(httpGet);
                if (httpResponse.getStatusLine().getStatusCode() == 200) {
                    byte[] data = EntityUtils.toByteArray(httpResponse.getEntity());
                    final Bitmap bmp = BitmapFactory
.decodeByteArray(data, 0, data.length);

                    handler.post(new Runnable() {
```

```
                    @Override
                    public void run() {
                        // TODO Auto-generated method stub
                        img.setImageBitmap(bmp);
                    }
                });
                dialog.dismiss();
            }else {
                Toast.makeText(HttpClientActivity.this, "连接错误",
Toast.LENGTH_LONG).show();
            }
        } catch (IOException e) {
            e.printStackTrace();
        }
    }
  }
}
```

图 7-3 HttpClient 下载图片

需要注意的是，在 Android 6.0 版本（API Level 23）中，Google 已经移除了 Apache HttpClient 相关类，推荐使用 HttpURLConnection，如果要继续使用，可在 Android studio 对应的 module 下的 build.gradle 文件中加入：

```
android {
useLibrary 'org.apache.http.legacy'
}
```

7.3 Android Socket 通信

Socket又称套接字，在程序内部提供了与外界通信的端口，也就是端口通信。通过建立Socket连接，可为通信双方的数据传输提供通道。

根据不同的底层协议，Socket可以实现基于TCP/IP协议的通信，也可以实现基于UDP协议的通信。基于TCP/IP协议的Socket类型为流套接字(streamsocket)，将TCP作为其端对端协议，提供了一个可信赖的字节流服务。基于UDP协议的Socket类型为数据报套接字(datagramsocket)，提供数据打包发送服务。图7-4所示为Socket基本通信模型。

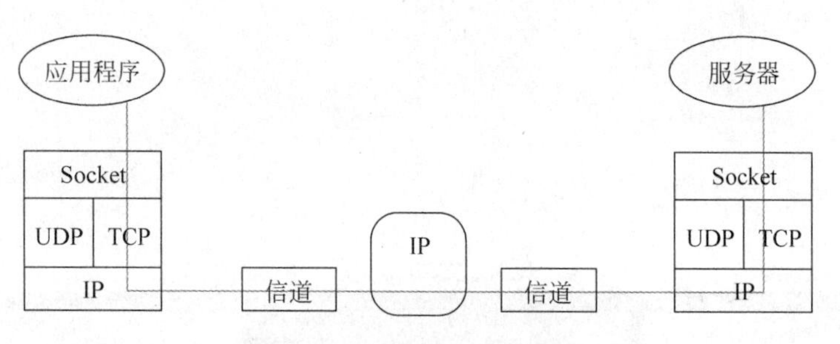

图7-4 Socket基本通信模型

7.3.1 TCP Socket 通信

服务器端首先声明一个ServerSocket对象并且指定端口号，然后调用ServerSocket的accept()方法接收客户端的数据。accept()方法在没有数据接收时处于堵塞状态(Socket socket=serversocket.accept())，一旦接收到数据，通过inputstream读取接收的数据。

客户端创建一个Socket对象，指定服务器端的IP地址和端口号(Socket socket=newSocket("172.168.10.108",8080);)，通过InputStream读取数据，获取服务器发出的数据(OutputStream outputstream=socket.getOutputStream())，最后将要发送的数据写入OutputStream即可进行TCP协议的Socket数据传输。

1. Socket 服务器编程

服务器端编程步骤：

(1) 创建服务器端套接字并绑定到一个端口。

(2) 套接字设置监听模式等待连接请求。

(3) 接收连接请求后进行通信。

(4) 返回，等待下一个连接请求。

Java中使用ServerSocket类来接收其他通信实体的连接请求，该类专门用来创建TCP套接字服务器，常用的构造函数如下：

ServerSocket(int port)：用指定端口号(0～65535 范围的一个整数值)创建 TCP 套接字服务器。

ServerSocket(int port, int backlog)：增加一个用来改变连接队列长度的参数，当队列中的连接请求达到了队列的最大容量时，服务器进程所在的主机会拒绝新的连接请求。只有当服务器进程通过 ServerSocket 的 accept()方法从队列中取出连接请求，使队列腾出空位时，队列才能继续加入新的连接请求。

ServerSocket(int port, int backlog, InetAddress localAddr)：当机器存在多个 IP 时，通过 localAddr 来为 ServerSocket 绑定指定 IP。

建立了 ServerSocket 之后，需要调用 accept()方法来接收客户端的连接。

Socket accept()：如果接收到一个客户端 Socket 的连接请求，该方法返回一个与连接客户端 Socket 对应的 Socket，否则该方法将一直处于等待状态。通常情况下，采用循环不断接收客户端的请求。当 Socket 使用完后，需调用 close()方法关闭该 Socket。如文件清单 7-4 所示。程序运行效果如图 7-5 所示。

文件清单 7-4 ServerListener.java

```java
public class ServerListener extends Thread{

    @Override
    public void run() {
        try {
            ServerSocket serverSocket =new ServerSocket(44444);
            while(true){
                //block
                Socket socket =serverSocket.accept();
                //建立联系
                JOptionPane.showMessageDialog(null, "有客户连接到本机的服务器");
                //将 Socket 传递给新的线程
                ChatSocket cs =new ChatSocket(socket);
                cs.start();
                ChatManager.getChatManager().add(cs);
            }
        } catch (IOException e) {
            e.printStackTrace();
        }
    }
}

public class ChatSocket extends Thread {
    Socket socket;
    public ChatSocket(Socket s) {
        this.socket =s;
    }

    public void out(String out) {
```

```
        try {
            socket.getOutputStream().write((out+"\n").getBytes("UTF-8"));
        } catch (UnsupportedEncodingException e) {
            // TODO Auto-generated catch block
            e.printStackTrace();
        } catch (IOException e) {
            System.out.println("断开一个客户");
            ChatManager.getChatManager().remove(this);
            e.printStackTrace();
        }
    }
//读取客户端信息,将读取的内容发送到集合中
    @Override
    public void run() {
        try {
            BufferedReader br =new BufferedReader(new
InputStreamReader(socket.getInputStream(),"UTF-8"));
            String line =null;
            while ((line =br.readLine()) !=null) {
                System.out.println(line);
                ChatManager.getChatManager().sent(this, line);
            }
            //关闭当前流
            br.close();
            System.out.println("断开了一个客户");
            ChatManager.getChatManager().remove(this);
        } catch (UnsupportedEncodingException e) {
            // TODO Auto-generated catch block
            e.printStackTrace();
        } catch (IOException e) {
            System.out.println("断开了一个客户");
            ChatManager.getChatManager().remove(this);
            e.printStackTrace();
        }
    }
}

public class ChatManager {

    private ChatManager() {
    };

    private static final ChatManager cm =new ChatManager();
    Vector<ChatSocket>vector =new Vector<>();

    public static ChatManager getChatManager() {
        return cm;
    }
```

```
    public void add(ChatSocket cs) {
        vector.add(cs);
    }

    public void sent(ChatSocket cs, String out) {
        for (int i =0; i <vector.size(); i++) {
            ChatSocket csChatSocket =vector.get(i);
            if (!cs.equals(csChatSocket)) {
                csChatSocket.out(out);
            }
        }
    }

    public void remove(ChatSocket cs) {
        vector.add(cs);

    }
}

public class MySocket {

    public static void main(String[] args) {
      new ServerListener().start();
    }

}
```

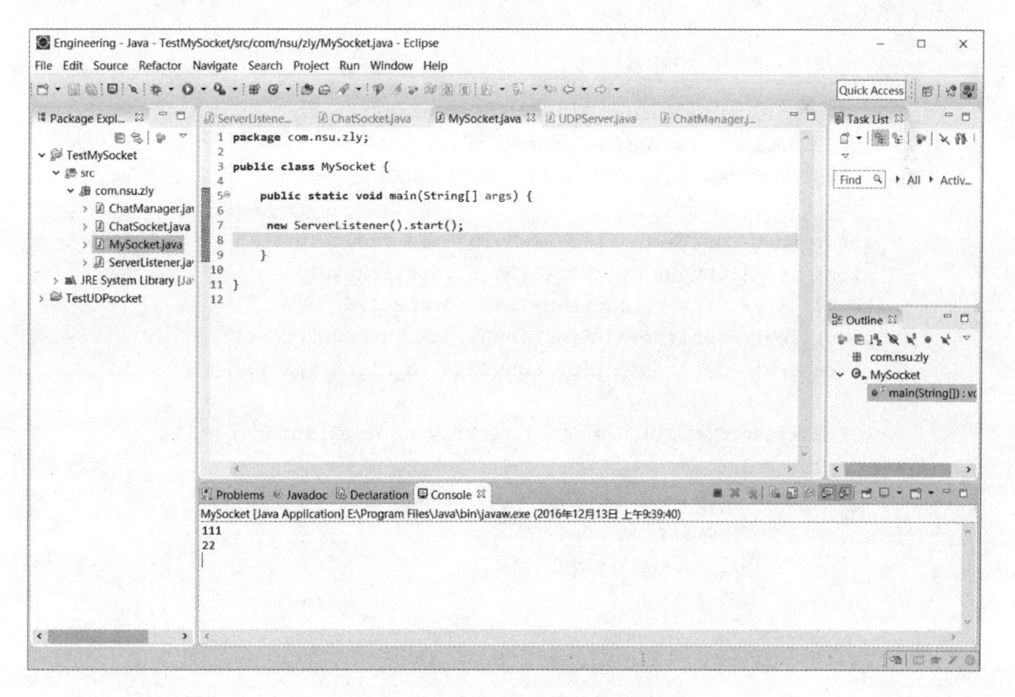

图 7-5 Socket 基本通信模型演示图

2. Socket 客户端编程

客户端编程步骤：

(1) 创建客户端套接字(指定服务器端 IP 与端口)。

(2) 连接(Android 创建 Socket 时会自动连接)。

(3) 与服务器端进行通信。

(4) 关闭套接字。

客户端使用 Socket 的构造器连接指定服务器，常用的有以下两个构造器。

(1) Socket(InetAddress remoteAddress,int port)：创建连接到指定远程主机、远程端口的 Socket。

(2) Socket(InetAddress remoteAddress, int port, InetAddress localAddr , int localPort)：创建连接到指定远程主机、远程端口的 Socket，并指定本地 IP 地址和本地端口号，适用于本地存在多个 IP 的情形。如文件清单 7-5 所示。

文件清单 7-5 Sendout. java

```
public class Sendout extends Activity {
    private EditText input_ip;
    private Button click_ip;
    private TextView text1;
    private EditText input_text;
    private Button click_text;

    public Sendout() {
    }

    @Override
    protected void onCreate(Bundle savedInstanceState) {
        super.onCreate(savedInstanceState);
        setContentView(R.layout.activity_sendout);

        input_ip = (EditText) findViewById(R.id.input_ip);
        click_ip = (Button) findViewById(R.id.click_ip);
        text1 = (TextView) findViewById(R.id.text1);
        input_text = (EditText) findViewById(R.id.input_text);
        click_text = (Button) findViewById(R.id.click_text);

        click_ip.setOnClickListener(new View.OnClickListener() {
            @Override
            public void onClick(View view) {
                connect();
            }
        });

        click_text.setOnClickListener(new View.OnClickListener() {
```

```
            @Override
            public void onClick(View view) {
                send();
            }
        });
    }

    Socket socket =null;
    BufferedReader reader =null;
    BufferedWriter writer =null;

    public void connect() {

        AsyncTask<Void, String, Void> read =new AsyncTask<Void, String, Void>() {
            @Override
            protected Void doInBackground(Void... voids) {
                try {
                    socket =new Socket(input_ip.getText().toString(), 44444);
                    writer =new BufferedWriter(new
OutputStreamWriter(socket.getOutputStream()));
                    reader =new BufferedReader(new
InputStreamReader(socket.getInputStream()));
                    publishProgress("@刘萌萌");
                } catch (UnknownHostException e1) {
                    Toast.makeText(Sendout.this, "无法建立连接",
Toast.LENGTH_SHORT).show();
                } catch (IOException e1) {
                    Toast.makeText(Sendout.this, "无法建立连接",
Toast.LENGTH_SHORT).show();
                    e1.printStackTrace();
                }
                try {
                    String line;
                    while ((line =reader.readLine()) !=null) {
                        publishProgress(line);
                    }
                } catch (IOException e) {
                    e.printStackTrace();
                }
                return null;
            }

            protected void onProgressUpdate(String... values) {
                if(values[0].equals("@刘萌萌")){
                    Toast.makeText(Sendout.this, "连接成功",
Toast.LENGTH_SHORT).show();
                }
                text1.append("对方说" +values[0] +"\n");
```

```
                super.onProgressUpdate(values);
            }
        };
        read.execute();
    }

    public void send() {
        try {
            text1.append("我说" +input_text.getText().toString() +"\n");
            writer.write(input_text.getText().toString() +"\n");
            writer.flush();
            input_text.setText("");
        } catch (IOException e) {
            e.printStackTrace();
        }
    }
}
```

图 7-6 和图 7-7 所示为 TCP Socket 通信示意图。

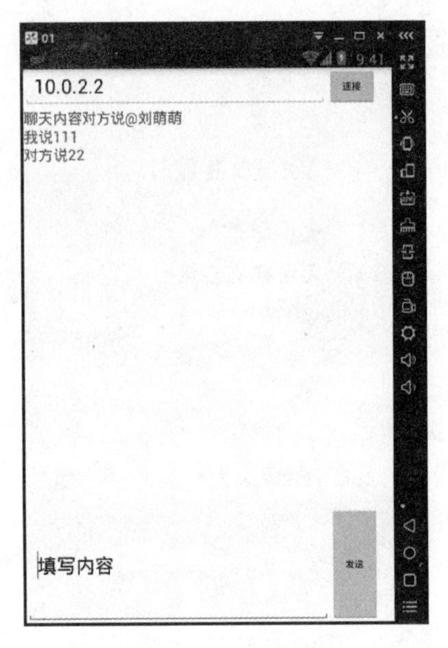

图 7-6　TCP Socket 通信(一)

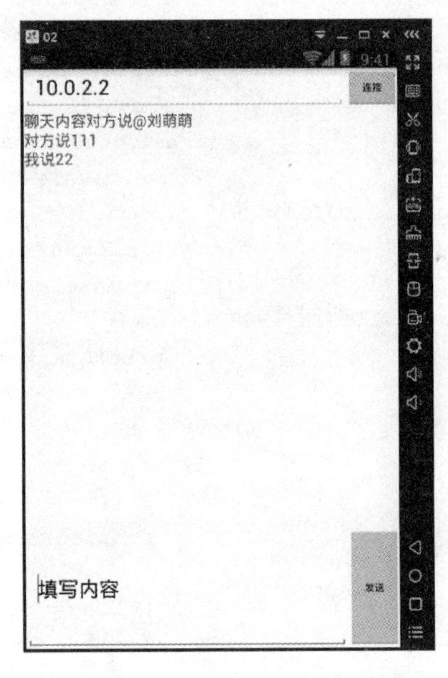

图 7-7　TCP Socket 通信(二)

7.3.2　UDP Socket 通信

UDP Socket 服务器编程要点：

(1) 服务器端首先创建一个 DatagramSocket 对象，并且指点监听的端口：

```
DatagramSocket socket =new DatagramSocket(ip)
```

（2）创建一个空的 DatagramSocket 对象用于接收数据：

```
byte data[]=new byte[1024];
DatagramPacket inPacket =new DatagramPacket(data,data.length))
```

（3）使用 DatagramSocket 的 receive()方法接收客户端发送的数据，receive()与 ServerSocket 的 accept()类似，在没有数据接收时处于堵塞状态。

```
socket.receive(inPacket);
```

UDP Socket 客户端编程要点：

（1）客户端创建 DatagramSocket 对象，并且指点监听的端口：

```
DatagramSocket socket =new DatagramSocket(port)
```

（2）创建一个 InetAddress 对象，这个对象类似于一个网络的发送地址：

```
InetAddress serveraddress= InetAddress.getByName("172.168.1.120"))
```

（3）定义要发送的一个字符串，创建一个 DatagramPacket 对象，并指定将这个数据包发送到网络的某个地址以及端口号：

```
String str= "hello";
byte data[]=str.getByte();
DatagramPacket packet=new DatagramPacket(data,data.length,serveraddress,4567);
```

（4）最后使用 DatagramSocket 的 send()方法发送数据：

```
socket.send(packet);
```

示例核心源代码如文件清单 7-6 所示。

文件清单 7-6　UDP_1.java

```
public class UPD_1 extends Activity {
    private EditText edtSendInfo, edtReceiveInfo, edtSendIP, edtSendPort,
edtReceivePort;
    private CheckBox chkSendHex, chkReceiveHex;
    private String sendInfo, receiveInfo;
    private byte[] buf;
    private Button btnListen;
    private Boolean listenStatus =false;
    private DatagramSocket socket;
    public Handler receiveHandler;

    @Override
```

```
    protected void onCreate(Bundle savedInstanceState) {
        super.onCreate(savedInstanceState);
        setContentView(R.layout.activity_main);
        edtSendInfo = (EditText) findViewById(R.id.edtSendInfo);
        edtReceiveInfo = (EditText) findViewById(R.id.edtReceiveInfo);
        edtSendIP = (EditText) findViewById(R.id.edtSendIP);
        edtSendPort = (EditText) findViewById(R.id.edtSendPort);
        edtReceivePort = (EditText) findViewById(R.id.edtReceivePort);
        chkSendHex = (CheckBox) findViewById(R.id.chkSendHex);
        chkReceiveHex = (CheckBox) findViewById(R.id.chkReceiveHex);
        btnListen = (Button) findViewById(R.id.btnListen);

        receiveHandler =new Handler() {
            public void handleMessage(Message msg) {
                edtReceiveInfo.setText(receiveInfo);
            }
        };
    }

    //UDP 数据发送线程
    public class udpSendThread extends Thread {
        @Override
        public void run() {
            try {
                if (chkSendHex.isChecked()) {
                    buf =hexStringToBytes(edtSendInfo.getText().toString());
                } else {
                    buf =edtSendInfo.getText().toString().getBytes();
                }
                if (listenStatus ==false) {
                    socket =new DatagramSocket(Integer.parseInt(
edtSendPort.getText().toString()));
                }
                InetAddress serverAddr = InetAddress.getByName(
edtSendIP.getText().toString());
                DatagramPacket outPacket = new DatagramPacket (buf, buf. length,
serverAddr, Integer.parseInt(edtSendPort.getText().toString()));
                socket.send(outPacket);
                socket.close();

            } catch (Exception e) {
                // TODO Auto-generated catch block
            }
        }
    }

    //UDP 数据接收线程,服务器编程
    public class udpReceiveThread extends Thread {
        @Override
        public void run() {
            try {
```

```
                socket = new DatagramSocket(Integer.parseInt(
                                    edtReceivePort.getText().toString()));
                listenStatus = true;
                while (listenStatus) {
                    byte[] inBuf = new byte[1024];
                    DatagramPacket inPacket = new DatagramPacket(inBuf, inBuf.length);
                    socket.receive(inPacket);
                    if (chkReceiveHex.isChecked()) {
                        receiveInfo = bytes2HexString(inBuf, inPacket.getLength());
                    } else {
                        receiveInfo = new String(inPacket.getData());
                    }
                    Message msg = new Message();
                    receiveHandler
                            .sendMessage(msg);
                }
            } catch (Exception e) {
                // TODO Auto-generated catch block
            }
        }
    }

    //发送按钮点击事件
    public void SendButtonClick(View source) {
        new udpSendThread().start();
    }

    //监听按钮点击事件
    public void ListenButtonClick(View source) {
        if (listenStatus == false) {
            btnListen.setText("停止监听");
            new udpReceiveThread().start();
        } else {
            btnListen.setText("开始监听");
            socket.close();
            listenStatus = false;
            new udpReceiveThread().interrupt();
        }
    }

    //十六进制字符串转byte[]
    public static byte[] hexStringToBytes(String str) {
        if (str == null || str.equals("")) {
            return null;
        }
        String hexString = str.replace(" ", "");
        hexString = hexString.toUpperCase();
        int length = hexString.length() / 2;
        char[] hexChars = hexString.toCharArray();
        byte[] d = new byte[length];
        for (int i = 0; i < length; i++) {
```

```
                int pos =i *2;
                d[i] = (byte) (charToByte(hexChars[pos]) << 4 | charToByte(hexChars
[pos +1]));
            }
            return d;
        }

        private static byte charToByte(char c) {
            return (byte) "0123456789ABCDEF".indexOf(c);
        }

        //byte[]转十六进制字符串
        public static String bytes2HexString(byte[] b, int len) {
            String ret ="";
            for (int i =0; i <len; i++) {
                String hex =Integer.toHexString(b[i] & 0xFF);
                if (hex.length() ==1) {
                    hex ='0' +hex;
                }
                ret +=hex.toUpperCase() +" ";
            }
            return ret;
        }
    }
```

图 7-8 和图 7-9 所示为 UDP Socket 通信示意图。

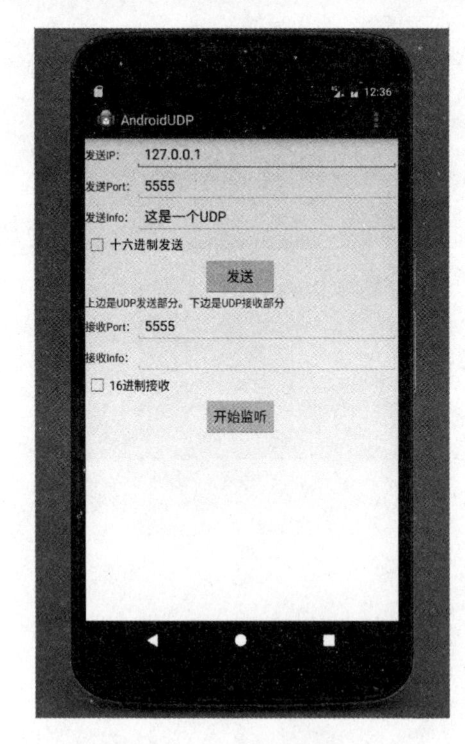

图 7-8 UDP Socket 通信(一)

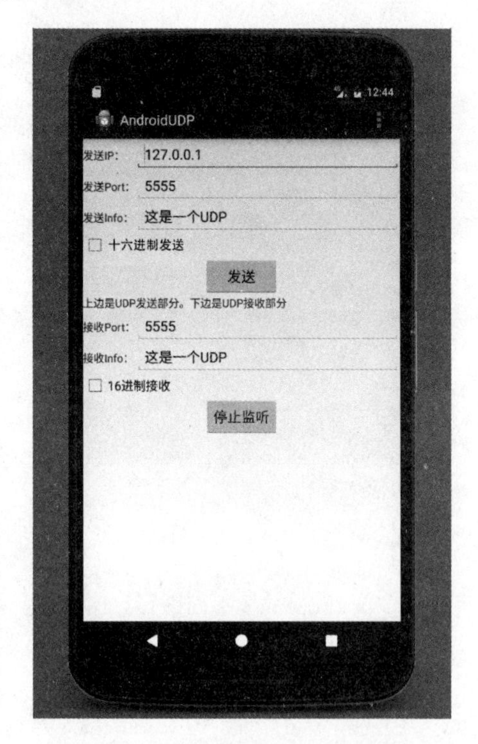

图 7-9 UDP Socket 通信(二)

7.4 网络数据解析

由于移动设备存储的限制，应用程序的数据不能全部存储在本地，许多应用程序数据需要保存在服务器中。这些数据以一定的格式，通过网络传输到Android客户端，客户端再把数据解析出来，显示到应用中。通常，Http通信以XML(eXtensible MarkupLanguage)或者JSON(JavaScript Object Notation)为载体，相互通信数据。

7.4.1 XML数据解析

XML在各种开发中广泛应用，Android也不例外。作为承载数据的一个重要角色，如何读写XML成为Android开发中一项重要的技能。在Android应用开发中，通常可以使用以下3种方式来解析XML文件：

(1) SAX解析(Simple API for XML)；

(2) DOM解析(Document Object Model)；

(3) Android附带的PULL解析器。

本节以解析书籍信息为例来讲解XML文件的解析方式。

首先复制一个book_ info. xml文件到raw目录下，如文件清单7-7所示。

文件清单7-7 book_info. xml

```
<?xml version="1.0" encoding="UTF-8"?>
<books>
    <book id="978-7-115-35407-5">
        <name>Android开发与实践</name>
        <price>59.0</price>
        <introduction>本书作为Android课程的教材,…… </introduction>
    </book>
    <book id="978-7-121-13576-7">
        <name>疯狂Android讲义</name>
        <price>89.0</price>
        <introduction>疯狂源自梦想,技术成就辉煌……</introduction>
    </book>
    <book id="978-7-115-36286-5">
        <name>Android第一行代码</name>
        <price>79.0</price>
        <introduction>全书是Android初学者的最佳人门书……</introduction>
      </book>
    <book id="978-7-121-26773-4">
        <name>Android群英传</name>
        <price>69.0</price>
        <introduction>本书针对具有一定Android开发基础的读者……</introduction>
    </book>
</books>
```

根据 book_info. xml 中的节点信息，创建相应的实体类 Book. java 文件，如文件清单 7-8 所示。

文件清单 7-8 Book. java

```
public class Book {
    public static final String BOOKS ="books";
    public static final String BOOK ="book";
    public static final String ID ="id";
    public static final String NAME ="name";
    public static final String PRICE ="price";
    public static final String INTRODUCTION ="introduction";

    private String id;
    private String name;
    private float price;
    private String introduction;

    public String getId() {
        return id;
    }

    public void setId(String id) {
        this.id =id;
    }

    public String getName() {
        return name;
    }

    public void setName(String name) {
        this.name =name;
    }

    public float getPrice() {
        return price;
    }

    public void setPrice(float price) {
        this.price =price;
    }

    public String getIntroduction() {
        return introduction;
    }

    public void setIntroduction(String introduction) {
        this.introduction =introduction;
```

```
    }

    @Override
    public String toString() {
        return "id" + id + ",name:" + name + ",price:" + price+",introduction:" +
introduction;
    }
}

public abstract class TestXMLFactory {
    //读取指定的 XML 文件
    //参数 inputStream XML 文件输入流
    public abstract void readXML(InputStream inputStream);

    // 获取 Book 对象列表
    public abstract List<Book>getBookList();

    //设置 Book 对象列表
    public abstract void setBookList(List<Book>bookList);
}
```

图 7-10 所示为页面布局设计。图 7-11 所示为 SAX 解析效果。

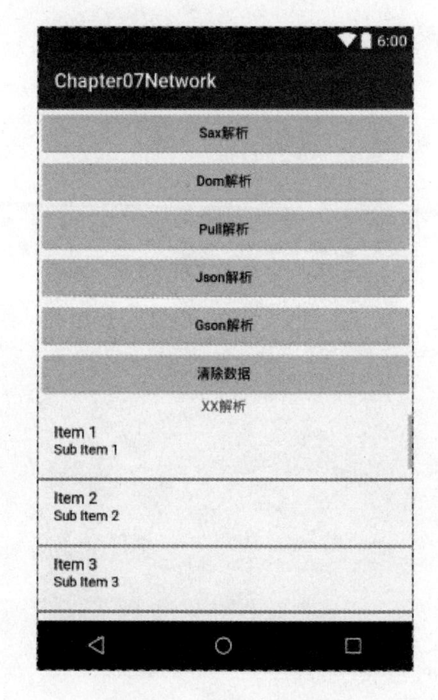

图 7-10　页面布局设计

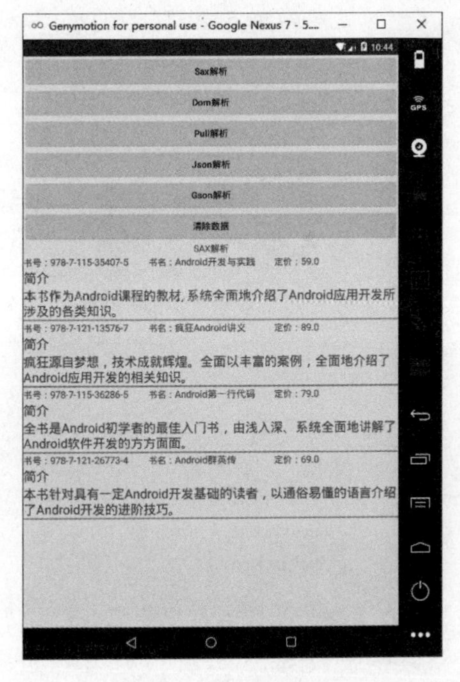

图 7-11　SAX 解析效果

1. SAX 解析

SAX (Simple API XML)解析,是事件驱动型 XML 解析的一个标准接口,对文档进行顺序扫描，当扫描到文档(document)开始与结束、元素(element)开始与结束等地方时通知事件处理函数，由事件处理函数做相应动作，然后继续扫描，直至文档结束，SAX 解析代码如文件清单 7-9 所示。

文件清单 7-9 TestSAX extends. java

```
public class TestSAX extends TestXMLFactory {

    private SAXHandler mHandler =new SAXHandler();
    private List<Book>mBookList;

    @Override
    public void readXML(InputStream inputStream) {
        try {
            //实例化一个 SAXParseFactory 对象
            SAXParseFactory factory =SAXParseFactory.newInstance();
            //实例化 SAXParse 对象，创建 XMLReader 对象，解析器
            SAXParse parse =factory.newSAXParse();
            //解析文件
            parse.parse(inputStream, mHandler);
        } catch (ParseConfigurationException e) {
            e.printStackTrace();
        } catch (SAXException e) {
            e.printStackTrace();
        } catch (IOException e) {
            e.printStackTrace();
        } finally {
            if (inputStream !=null) {
                try {
                    inputStream.close();
                } catch (IOException e) {
                    e.printStackTrace();
                }
            }
        }
    }

    @Override
    public List<Book>getBookList() {
        if (mHandler ==null) {
            return null;
        }
        return mHandler.getBookList();
    }

    @Override
    public void setBookList(List<Book>bookList) {
        mBookList =bookList;
```

```
    }

    //实例化 handler,事件处理器
    class SAXHandler extends DefaultHandler {

        private List<Book>mBookList;
        private Book mBook;
        private String mTargetName;

        public List<Book>getBookList() {
            return mBookList;
        }

        //用于处理文档解析开始事件
        @Override
        public void startDocument() throws SAXException {
            super.startDocument();
            mBookList =new ArrayList<Book> ();
        }

        // 接收文档结束的通知
        @Override
        public void endDocument() throws SAXException {
            super.endDocument();
        }

        //处理元素开始事件
//从参数中可以获得元素所在名称空间的 URI、元素名称、属性类表等信息
        @Override
        public void startElement(String uri, String localName, String qName,
                              Attributes attributes) throws SAXException {
            super.startElement(uri, localName, qName, attributes);
            if (localName.equals(Book.BOOK)) {
                mBook =new Book();
                mBook.setId(attributes.getValue(Book.ID));
            }

            mTargetName =localName;
        }

        //处理元素结束事件
//从参数中可以获得元素所在名称空间的 URI、元素名称等信息
        @Override
        public void endElement(String uri, String localName, String qName)
                throws SAXException {
            super.endElement(uri, localName, qName);
            if (Book.BOOK.equals(localName)) {
                mBookList.add(mBook);
```

```
            }

            mTargetName =null;
        }

        //处理元素的字符内容,从参数中可以获得内容
        @Override
        public void characters(char[] ch, int start, int length) throws SAXException {
            super.characters(ch, start, length);
            if (Book.NAME.equals(mTargetName)) {
                //获得 book 中的 name 属性
                mBook.setName(new String(ch, start, length));
            } else if (Book.PRICE.equals(mTargetName)) {
                //获得 book 中的 price 属性
                mBook.setPrice(Float.valueOf(new String(ch, start, length)));
            }else if (Book.INTRODUCTION.equals(mTargetName)){
                //获得 book 中的 introduction 属性
                mBook.setIntroduction(new String(ch, start, length));
            }
        }
    }
}
```

点击 SAX 解析按钮：

```
case R.id.id_saxread:
      parseFactory =new TestSAX();
      parseFactory.readXML(inputStream);
      showBookList(parseFactory.getBookList());
      tvShow.setText("SAX 解析");
      break;
```

方法 showBookList()用于把 List 数据在 ListView 中显示出来：

```
private void showBookList(List<Book>bookList) {
    List<Map<String, String<<list =new ArrayList<Map<String, String<<();
    for(int i =0; i <bookList.size();i++){
        Map<String, String>map =new HashMap<String, String>();
        Book book =bookList.get(i);
        map.put("id", book.getId());
        map.put("name", book.getName());
        map.put("price", book.getPrice()+"");
        map.put("introduction",book.getIntroduction());
          list.add(map);
    }
    SimpleAdapter adapter =new SimpleAdapter(this, list,R.layout.list_item,
new String[] { "id", "name","price","introduction" },
```

```
new int[] { R.id.tv_id, R.id.tv_name,R.id.tv_price,R.id.tv_introduce });
    listView.setAdapter(adapter);
}
```

关于ListView的使用，在前面章节已经深入讲解了，这里不再赘述。读者可自行完成列表项的布局文件list_item.xml文件。

2. DOM解析

DOM (Document Object Model)，即文档对象模型，定义了访问和操作XML文档的标准方法。它是将整个XML文档载入内存(所以效率较低，不推荐使用)，每一个节点当作一个对象。DOM把XML文档作为树结构来查看。能够通过DOM树访问所有元素，可以修改或删除它们的内容，并创建新的元素。元素、它们的文本，以及它们的属性，都被认为是节点，如文件清单7-10所示。

文件清单7-10　TestDOM.java

```
public class TestDOM extends TestXMLFactory {
    private static final String TAG = "TestDOM";
    private List<Book>mBookList;

    @Override
    public void readXML(InputStream inputStream) {
        //得到 DocumentBuilderFactory 对象,可以得到 DocumentBuilder 对象
        DocumentBuilderFactory factory = DocumentBuilderFactory.newInstance();
        try {
            //得到 DocumentBuilder 对象
            DocumentBuilder builder = factory.newDocumentBuilder();
            //得到代表整个 XML 的 Document 对象
            Document parse =builder.parse(inputStream);
            //得到 "根节点"
            Element root =parse.getDocumentElement();
            //获取根节点的所有 items 的节点
            NodeList nodeList =root.getElementsByTagName(Book.BOOK);

            mBookList =new ArrayList<Book>();
            //遍历根节点所有子节点,books 下所有 book
            for (int i =0; i <nodeList.getLength(); i++) {
                Book book =new Book();
                //获得 book 元素节点
                Element item = (Element) nodeList.item(i);
                book.setId(item.getAttribute(Book.ID));
                NodeList nodes =item.getChildNodes();
                //遍历 book 的所有节点
                for (int j =0; j <nodes.getLength(); j++) {
                    Node node =  nodes.item(j);
                    if (node.getNodeType() ==Node.ELEMENT_NODE) {
```

```
                        if (Book.NAME.equals(node.getNodeName())) {
                            //获得 book 中的 name 属性
                            String content =node.getTextContent();
                            book.setName(content);
                        } else if (Book.PRICE.equals(node.getNodeName())) {
                            //获得 book 中的 price 属性
                            String content =node.getTextContent();
                            book.setPrice(Float.valueOf(content));
                        }else if (Book.INTRODUCTION.
equals(node.getNodeName())){
                            //获得 book 中的 introduction 属性
                            String content =node.getTextContent();
                            book.setIntroduction(content);
                        }
                    }
                }
                mBookList.add(book);
            }

        } catch (ParseConfigurationException e) {
            e.printStackTrace();
        } catch (SAXException e) {
            e.printStackTrace();
        } catch (IOException e) {
            e.printStackTrace();
        } finally {
            if (inputStream !=null) {
                try {
                    inputStream.close();
                } catch (IOException e) {
                    e.printStackTrace();
                }
            }
        }
    }

    @Override
    public List<Book>getBookList() {
        return mBookList;
    }

    @Override
    public void setBookList(List<Book>bookList) {
        mBookList =bookList;
    }
}
```

图 7-12 所示为 DOM 解析效果。图 7-13 所示为 PULL 解析效果。

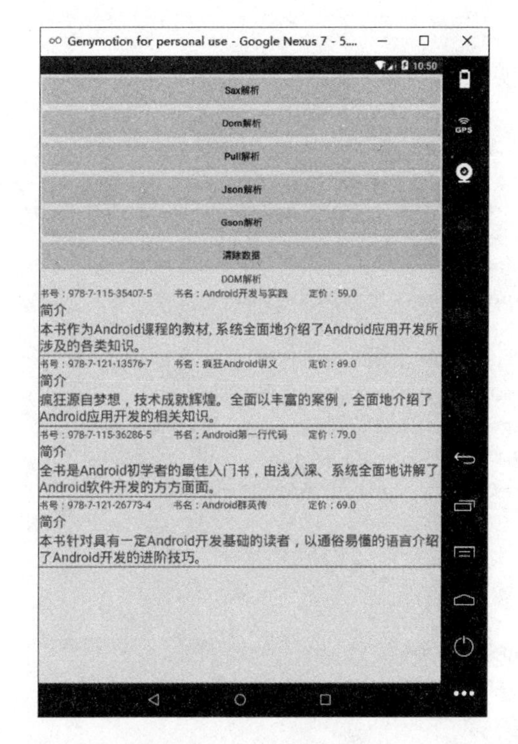

图 7-12　DOM 解析效果

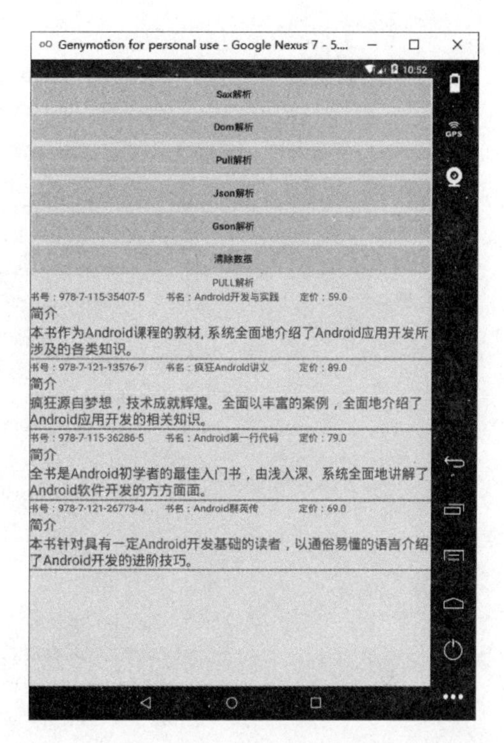

图 7-13　PULL 解析效果

3. PULL 解析

Pull 解析器的运行方式与 SAX 解析器相似。它提供了类似的事件，如开始元素和结束元素事件，使用 parse. next()可以进入下一个元素并触发相应事件。与 SAX 不同的是，Pull 解析器产生的事件是一个数字而非方法，因此可以使用一个 switch 对感兴趣的事件进行处理。当元素开始解析时，调用 parse. nextText()方法可以获取下一个 Text 类型节点的值，如文件清单 7-11 所示。

文件清单 7-11　TestPULL. java

```java
public class TestPULL extends TestXMLFactory {
    private List<Book>mBookList;
    private Book mBook;

    @Override
    public void readXML(InputStream inputStream) {
        try {
            XmlPullParse parse =Xml.newPullParse();
            parse.setInput(inputStream, "UTF-8");

            int eventType =parse.getEventType();
            // 直到文档的结尾处
            while (eventType !=XmlPullParse.END_DOCUMENT) {
                String name =parse.getName();
```

```
                switch (eventType) {
                    case XmlPullParse.START_DOCUMENT:
                        mBookList =new ArrayList<Book> ();
                        break;

                    case XmlPullParse.START_TAG:
                        if (Book.BOOK.equals(name)) {
                            mBook =new Book();
                            mBook.setId(parse.getAttributeValue("", Book.ID));
                        } else if (Book.NAME.equals(name)) {
                            mBook.setName(parse.nextText());
                        } else if (Book.PRICE.equals(name)) {
                            mBook.setPrice(Float.valueOf(parse.nextText()));
                        }else if (Book.INTRODUCTION.equals(name)){
                            mBook.setIntroduction(parse.nextText());
                        }
                        break;

                    case XmlPullParse.END_TAG:
                        if (Book.BOOK.equals(name)) {
                            mBookList.add(mBook);
                        }
                        break;

                    default:
                        break;
                }
                //获取解析下一个事件
                eventType =parse.next();
            }

        } catch (XmlPullParseException e) {
            e.printStackTrace();
        } catch (IOException e) {
            e.printStackTrace();
        } finally {
            if (inputStream !=null) {
                try {
                    inputStream.close();
                } catch (IOException e) {
                    e.printStackTrace();
                }
            }
        }
    }

    @Override
    public List<Book>getBookList() {
```

```
        return mBookList;
    }

    @Override
    public void setBookList(List<Book>bookList) {
        mBookList =bookList;
    }
}
```

7.4.2 JSON数据解析

JSON(JavaScript Object Notation)是一种轻量级的数据交换格式,比XML更轻巧,易于阅读和编写。JSON是JavaScript的原生格式,这意味着在JavaScript中处理JSON数据不需要任何特殊的API或工具包,也易于机器解析和生成(官方网站http://www.json.org/)。

和XML一样,JSON也是基于纯文本的数据格式。由于JSON天生是为JavaScript准备的,所以JSON的数据格式非常简单,可以用JSON传输一个简单的String、Number、Boolean,也可以传输一个数组,或者一个复杂的Object对象。

1. JSON对象与JSON数组

JSON对象以(key/value)对形式存在,key值必须是String类型;而对于value,则可以是String、Number、Object、Array等数据类型。

一个JSON对象以"{"(左花括号)开始,"}"(右花括号)结束。每个"名称"后跟一个":"(冒号);"'名称/值'对"之间使用","(逗号)分隔,如图7-14所示。

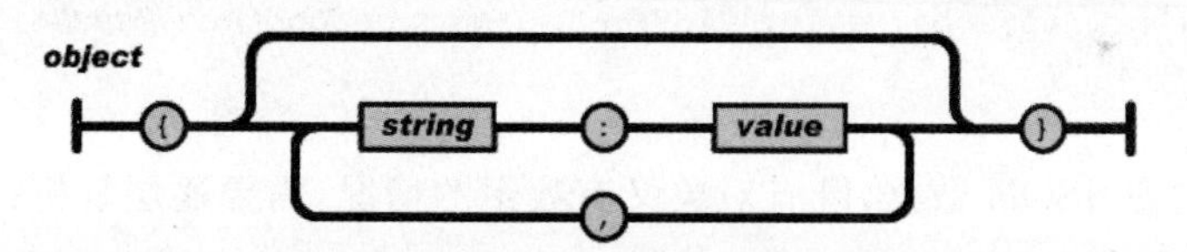

图7-14 AsyncTask模拟下载(一)

例如,{"name": "xiaoluo","phone":"82876598"}

花括号保存对象,数据在名称/值对中,数据之间由逗号分隔。

一个JSON数组以"["(左中括号)开始,"]"(右中括号)结束,值之间使用","(逗号)分隔,如图7-15所示。

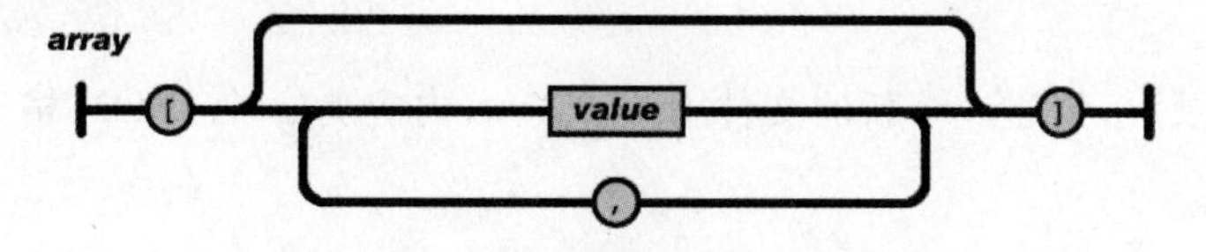

图7-15 AsyncTask模拟下载(二)

例如：

```
[
{"address":"成都","id":1,"name":"张三"},
{"address":"武汉","id":2,"name":"李四"}
]
```

2. JSON 数据解析

在解析 JSON 数据及生成 JSON 数据格式时，需要用到 JSON 解析库，常见的有 JSON-lib、GSON 等。

当使用 JSON-lib 解析时，需要引入第三方的 JSON-lib 包。当生成 JSON 数据时，通过 JSONObject jsonObject = new JSONObject()；得到一个 JSON 对象，然后通过 put() 方法给 JSON 对象添加 key/value 对。读者可自行尝试 JSON 数据生成，这里重点讲解 JSON 数据解析。根据 JSON 数据的不同，可分为以下 3 种情况。

(1) 解析 JSON 对象(JSONObject)：

```
String data ="{\"id\":\"1000\", \"name\":\"王五\"}";
JSONObject json =new JSONObject(data);
System.out.println(json.getString("id") +"," +json.getString("name"));
```

(2) 解析 JSON 数组(JSONArray)：

```
String data2 ="[{\"id\":10, \"sex\":true}, {\"id\":20, \"sex\":false}]";
JSONArray json2 =new JSONArray(data2);
for (int i =0; i <json2.length(); i++) {
JSONObject obj = (JSONObject) json2.get(i);
System.out.println(obj.getInt("id") +"," +obj.getBoolean("sex"));
}
```

(3) 解析复杂 JSON 数据，既有对象又有数组的情况，需要逐层解析：

```
String data3 ="{\"persons\":[{\"name\":\"renhaili\",\"age\":20},
{\"name\":\"zhouxiaodong\",\"age\":21}]}";
JSONObject json3 =new JSONObject(data3);
JSONArray persons =json3.getJSONArray("persons");
for (int i =0; i <persons.length(); i++) {
JSONObject obj = (JSONObject) persons.get(i);
System.out.println(obj.getString("name") +"," +obj.getInt("age"));
}
```

在 raw 目录下，创建一个新的文件 book_json，内容为一个 JSON 格式的数据，具体如下：

```
[
{
```

```
        "id":"978-7-115-35407-5",
        "name":"Android 开发与实践",
        "price":"59.0",
        "introduction":"本书作为 Android 课程的教材……"
    },
    {
        "id":"978-7-121-13576-7",
        "name":"疯狂 Android 讲义",
        "price":"89.0",
        "introduction":"疯狂源自梦想,技术成就辉煌……"
    },
    {
        "id":"978-7-115-36286-5",
        "name":"Android 第一行代码",
        "price":"79.0",
        "introduction":"本书是 Android 初学者的最佳入门书……"
    },
    {
        "id":"978-7-121-26773-4",
        "name":"Android 群英传",
        "price":"69.0",
        "introduction":"本书针对具有一定 Android 开发基础的读者……"
    }
]
```

读取该文件：

```
InputStream jsonInputStream;
jsonInputStream =this.getResources().openRawResource(R.raw.book_json);
```

JSON 解析按钮的点击事件如下：

```
case R.id.id_jsonread:
      tvShow.setText("JSON 解析");
      try {
          byte[] buffer =new byte[jsonInputStream.available()];
          jsonInputStream.read(buffer);
          String json =new String(buffer,"utf-8");
          bookList =new ArrayList<Book>();
          JSONArray jsonArray =new JSONArray(json);
          for (int i =0; i<jsonArray.length();i++){
              JSONObject obj =jsonArray.getJSONObject(i);
              Book book =new Book();
              book.setId(obj.getString("id"));
              book.setName(obj.getString("name"));
              book.setPrice((float) obj.getDouble("price"));
              book.setIntroduction(obj.getString("introduction"));
              bookList.add(book);
          }
              showBookList(bookList);
```

```
        } catch (IOException e) {
            e.printStackTrace();
        } catch (JSONException e) {
            e.printStackTrace();
        }
        break;
```

从创建的JSON文件可以看出读取的JSON数据为一个JsonArray,在JsonArray里面放了4个JsonObject,逐层解析,把每一个JsonObject解析成一个Book对象,添加到Lis<Book>中,并用ListView展示出来。页面运行效果如图7-16和图7-17所示。

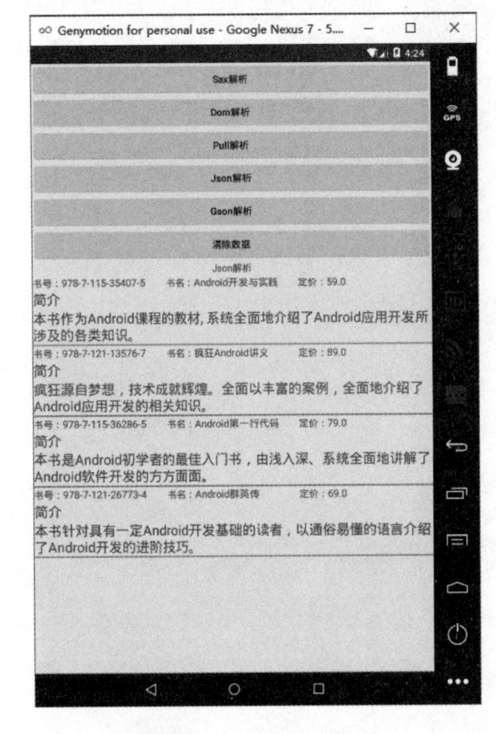

图7-16 JSON解析效果

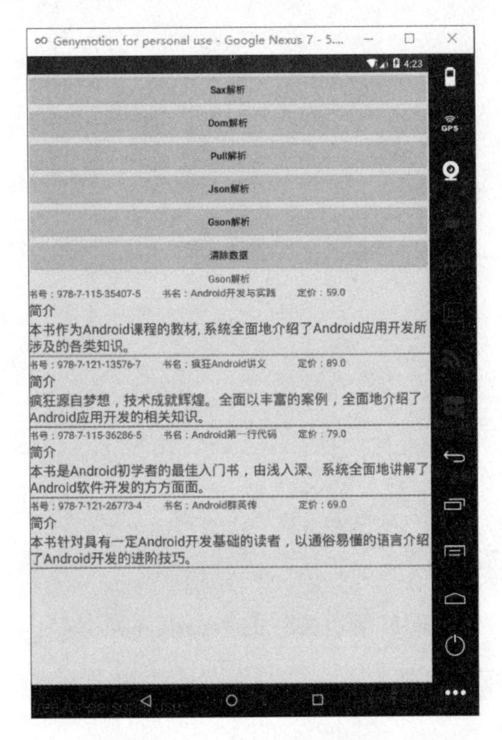

图7-17 GSON解析效果

3. GSON数据解析

除了使用JSON-lib库来解析JSON数据外,我们经常也使用GSON来解析。GSON是Google公司发布的一个开放源代码的Java库,主要用途为序列化Java对象为JSON字符串,或反序列化JSON字符串成Java对象。

Google公司提供的GSON这个JSON解析库,同样需要下载GSON这个jar包,导入到我们的项目中。使用GSON,可以非常轻松地实现数据对象和JSON对象的相互转化,其中最常用的是以下两个方法。

(1) fromJSON(),将JSON对象转换成需要的数据对象:

```
Gson gson =new GsonBuilder().create();
String s =gson.toJson(data);
```

(2) toJSON(),将数据对象转换成 JSON 对象:

```
Gson gson =new GsonBuilder().create();
String success =gson.fromJson(data, String.class);
```

针对前面的例子,用 GSON 解析,按钮点击事件如下:

```
case R.id.id_gsonread:
      tvShow.setText("Gson 解析");
      try {
          byte[] buffer =new byte[jsonInputStream.available()];
          jsonInputStream.read(buffer);
          String json =new String(buffer,"utf-8");
          Type listType =new TypeToken<List<Book<<(){}.getType();
          Gson gson =new Gson();
          bookList =gson.fromJson(json,listType);
          showBookList(bookList);
      } catch (IOException e) {
            e.printStackTrace();
      }
      break;
```

从解析过程来看,使用 GSON 解析 JSON 数据要相对简单一些,代码更加精简。

7.5 WebView

在很多 Android 网络应用中都内置了可以显示 Web 页面的界面,这个界面一般都是由一个称为 WebView 的组件渲染出来的。Android 提供的 WebView 组件是一个浏览器实现,它的内核基于开源的 WebKit 引擎。作为 Android 开发者,掌握 WebView 的用法,对 WebKit 进行一些美化、包装,可以轻松地开发出自己的浏览器,为 APP 开发提升扩展性。

WebView 浏览器操作的常用方法如下:

(1) void goBack():后退一页;

(2) void goForward():前进一页;

(3) void loadUrl(String url):加载指定 URL 对应的网页;

(4) boolean zoomIn():放大网页;

(5) boolean zoomOut():缩小网页。

在 WebView 应用的开发过程中,需要注意以下几点:

(1) 在 Activity 中实例化 WebView 组件:WebView webView = new WebView(this);或者在布局文件中声明 WebView 组件,然后在 Activity 中实例化。

(2) 调用 WebView 的 loadUrl()方法,设置 WebView 要显示的网页:

互联网网页:webView.loadUrl("http://www.google.com");

本地网页文件：webView.loadUrl("file:///android_asset/XX.html")；本地文件存放在 assets 文件中。

(3) 调用 Activity 的 setContentView( )方法来显示网页视图。

(4) 在 WebView 中看了很多页以后，为了让 WebView 支持回退功能，需要覆盖 Activity 类的 onKeyDown()方法，屏蔽系统返回键的作用；如果不做任何处理，点击系统返回键，整个浏览器会调用 finish()而结束自身，而不是回退到上一页面。

(5) 需要在 AndroidManifest.xml 文件中添加网络权限，否则会出现“Web page not available”错误。权限声明：< uses-permission android:name = "android.permission.INTERNET" />。

(6) 针对 WebView 页面中的链接，如果希望点击链接继续在当前的 WebView 中响应，而不是新开 Android 系统的浏览器来响应该链接，则必须覆盖 WebView 的 WebViewClient 对象。

(7) 在 Android 中使用 WebView 时，经常会同时用到 EditText 控件(比如浏览器地址栏)，这样就会出现 EditText 和 WebView 抢占焦点，导致 WebView 中的控件无法输入。可以在触摸 WebView 控件时请求获取焦点，这样就不会出现上述问题。

接下来使用 WebView 控件，开发一个小的浏览器应用。简易浏览布局如图 7-18 所示，包含 3 个控件(EditText，Button，WebView)。最终的运行效果如图 7-19 所示，输入网址，点击“GO”按钮，显示相应的网页。

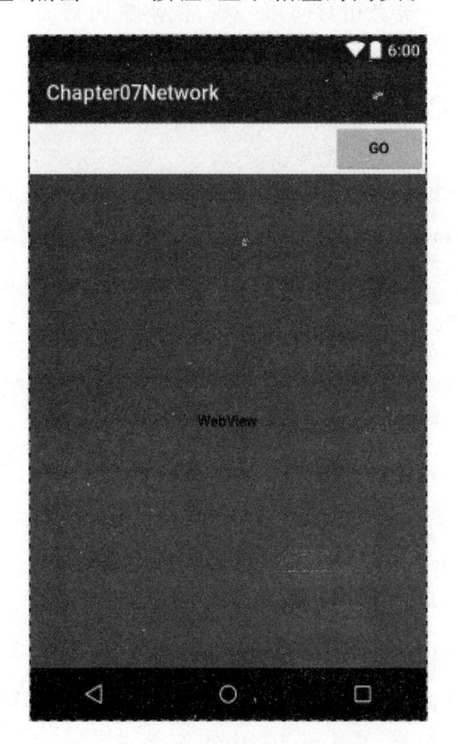

图 7-18 简易浏览器布局

图 7-19 简易浏览器运行效果

核心代码如文件清单 7-12 所示。

文件清单 7-12 SimpleBrowser. java

```
public class SimpleBrowser extends Activity{

    private EditText url;
    private WebView webView;
    private Button search;

    @Override
    protected void onCreate(Bundle savedInstanceState) {
        // TODO Auto-generated method stub
        super.onCreate(savedInstanceState);
        setContentView(R.layout.simple_browser);
        // 获取页面中文本框、WebView 组件
        url = (EditText) findViewById(R.id.url);
        webView = (WebView) findViewById(R.id.show);
        search = (Button) findViewById(R.id.search);

        search.setOnClickListener(new View.OnClickListener() {
            @Override
            public void onClick(View v) {
                // TODO Auto-generated method stub
                //设置 WebView 属性,能够执行 JavaScript 脚本
                webView.getSettings().setJavaScriptEnabled(true);
                String urlStr =url.getText().toString().trim();
                // 加载并显示 urlStr 对应的网页
                webView.loadUrl(urlStr);
                //覆盖 WebView 默认使用第三方或系统默认浏览器打开网页的行为
                //使网页用 WebView 打开
                webView.setWebViewClient(new WebViewClient(){
                    @Override
                    public boolean shouldOverrideUrlLoading(WebView view,
                            String url) {
                        //返回值为 true 时控制 WebView 打开
                        //为 false 则调用系统浏览器或第三方浏览器
                        view.loadUrl(url);
                        return true;
                    }

                });
                //在触摸 WebView 控件时,请求获取焦点
                webView.setOnTouchListener(new View.OnTouchListener() {

                    @Override
                    public boolean onTouch(View v, MotionEvent event) {
                        webView.requestFocus();
                        return false;
                    }
```

```
                    });
                }
            });
        }

        //设置回退,覆盖 Activity 类的 onKeyDown(int keyCoder,KeyEvent event)方法
        @Override
        public boolean onKeyDown(int keyCode, KeyEvent event) {
            // TODO Auto-generated method stub
            if ((keyCode ==KeyEvent.KEYCODE_BACK) && webView.canGoBack()) {
                //goBack()表示返回 WebView 的上一页面
    webView.goBack();
                return true;
            }
            return super.onKeyDown(keyCode, event);
        }
    }
```

需要说明的是，shouldOverrideUrlLoading()方法在 API Level≥24 时被标记 deprecated，官方不建议使用，替代方法是 shouldOverrideUrlLoading(WebView view, WebResourceRequest request)。但是都是向下兼容的，方法 public boolean shouldOverrideUrlLoading(WebView view, String url)支持更广泛的 API，我们这里还是使用它。

7.6 WebService

WebService(Web 服务)是一个用于支持网络间不同机器互操作的软件系统，它是一种自包含、自描述和模块化的应用程序，它可以在网络中被描述、发布和调用，可以将它看作是基于网络的、分布式的模块化组件。

7.6.1 WebService 简介

通常所说的 WebService 都是远程的某个服务器对外公开了某种服务，或者理解为对外公开了某个功能或者方法，而我们可以通过编程来调用该服务以获得所需要的信息。例如，www. webxml. com. cn 对外公开了手机号码归属地查询服务，我们只需要在调用该服务时传入一个手机号段(号码)，就能立即获取该号段的归属地信息。

WebService 建立在 HTTP、SOAP、WSDL 等通用协议的基础之上。WebService 数据通信模型如图 7-20 所示。客户端通过 HTTP 协议向 WebService 服务器发送 XML 数据(内部包含调用的一些方法和相关参数数据)，然后 WebService 服务器给客户端返回一定的 XML 格式的数据，客户端通过解析这些 XML 数据即可得到需要的数据。

SOAP(Simple Object Access Protocol，简单对象访问协议)是一种轻量级的、简单的、基于 XML 的协议，被设计用于在分布式环境中交换格式化和固化信息的简单协议。

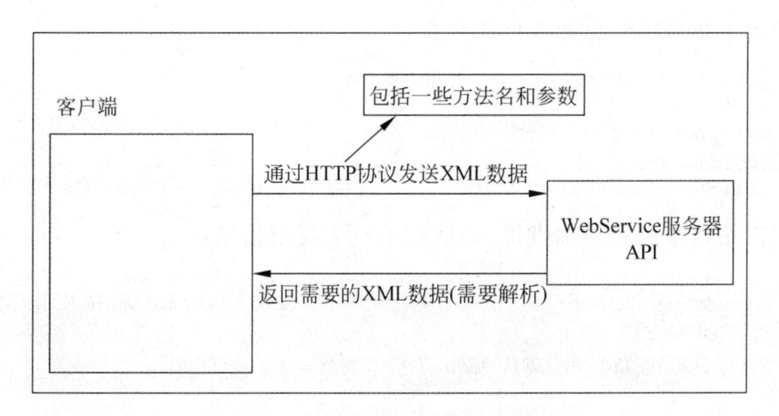

图 7-20 WebService 数据通信模型

也就是说,要进行通信,进行数据访问传输,就必须依赖于一定的协议,而 SOAP 正是 WebService 通信中所依赖的一种协议。目前经常使用的 SOAP 协议有两个版本: SOAP 1.1 和 SOAP 1.2。

WSDL(WebService Description Language,Web 服务描述语言)是一种用来描述 Web 服务的 XML 语言,它描述了 Web 服务的功能、接口、参数、返回值等,便于用户绑定和调用服务。它以一种和具体语言无关的方式定义了给定 Web 服务调用和应答的相关操作和消息。

7.6.2 Android 平台调用 WebService

在 Android 平台调用 WebService,需要依赖第三方类库 KSOAP2,它是一个 SOAP WebService 客户端开发包,主要用于资源受限制的 Java 环境,如 Applets 或 J2ME 应用程序(CLDC/ CDC/MIDP)。KSOAP2 Android 是 Android 平台上一个高效、轻量级的 SOAP 开发包。

Android 平台调用 WebService 步骤如下:

(1) 指定 WebService 的命名空间和调用的方法名。

(2) 如果有其他参数,需要设置调用方法的参数值。

(3) 生成调用 WebService 方法的 SOAP 请求信息。该信息由 SoapSerializationEnvelope 对象描述。

(4) 创建 HttpTransportSE 对象。通过 HttpTransportSE 类的构造方法可以指定 WebService 的 WSDL 文档的 URL。

(5) 使用 call 方法调用 WebService 方法。

(6) 使用 getResponse 方法获得 WebService 方法的返回结果。

接下来,使用 WebService 开发查询手机号码归属地应用。

手机号码归属地数据源:

```
http://ws.webxml.com.cn/WebServices/MobileCodeWS.asmx
getDatabaseInfo
```

获得国内手机号码归属地数据库信息：

```
输入参数：无；
返回数据：一维字符串数组(省份城市记录数量)。
getMobileCodeInfo
```

获得国内手机号码归属地省份、地区和手机卡类型信息：

```
输入参数：mobileCode=字符串(手机号码，最少前7位数字)，userID=字符串(商业用户 ID)
免费用户为空字符串；
返回数据：字符串(手机号码：省份 城市 手机卡类型)。
```

图 7-21 所示为号码查询页面布局。图 7-22 所示为号码查询运行效果。

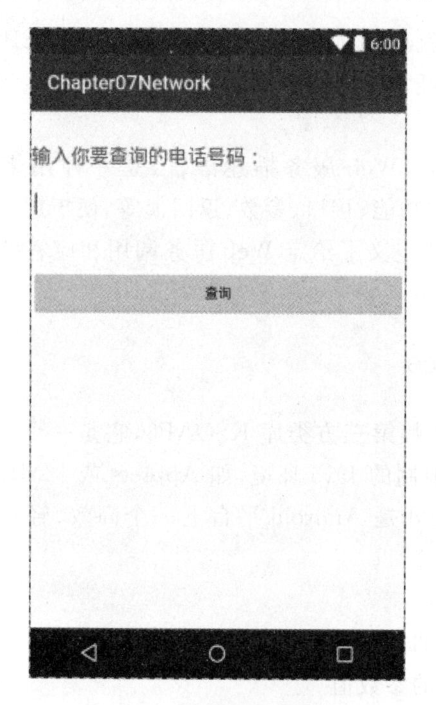

图 7-21 号码查询页面布局

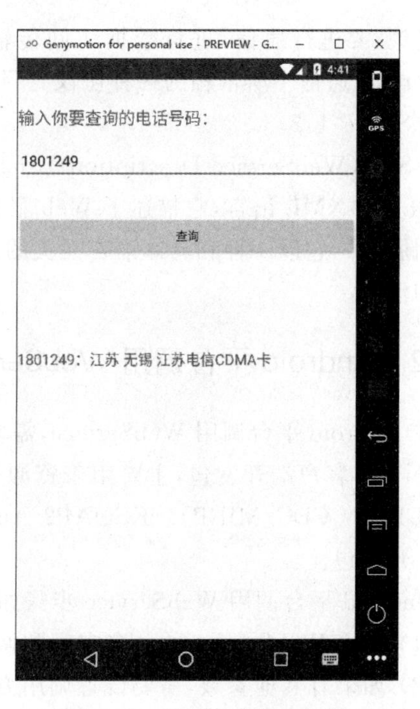

图 7-22 号码查询运行效果

核心代码如文件清单 7-13 所示。

文件清单 7-13 WebServiceActivity.java

```java
public class WebServiceActivity extends Activity {

    private Button searchButton;
    private EditText numEditText;
    private TextView resultTextView;

    @Override
    protected void onCreate(Bundle savedInstanceState) {
```

```
        super.onCreate(savedInstanceState);
        setContentView(R.layout.web_service);

        if (Build.VERSION.SDK_INT >=11) {
            StrictMode.setThreadPolicy(new StrictMode.ThreadPolicy.Builder()
                    .detectDiskReads().detectDiskWrites().detectNetwork()
                    .penaltyLog().build());
            StrictMode.setVmPolicy(new StrictMode.VmPolicy.Builder()
                    .detectLeakedSqlLiteObjects().detectLeakedClosableObjects()
                    .penaltyLog().penaltyDeath().build());
        }

        searchButton = (Button)findViewById(R.id.btnSearch);
        numEditText = (EditText)findViewById(R.id.editTextNum);
        resultTextView = (TextView)findViewById(R.id.textViewResult);

        searchButton.setOnClickListener(new View.OnClickListener() {

            @Override
            public void onClick(View v) {
                // TODO Auto-generated method stub
                String phoneNum =numEditText.getText().toString().trim();
                if ("".equals(phoneNum) || phoneNum.length() <7) {
                    // 输入的手机号不合规范
                    Toast.makeText(getApplicationContext(), "输入的手机号不合规
范!",Toast.LENGTH_LONG).show();
                    numEditText.requestFocus();
                    return;
                }
                resultTextView.setText(getRemoteInfo(phoneNum));
            }
        });

    }

    private String getRemoteInfo(String phoneNum) {
        // 命名空间
        String nameSpace ="http://WebXml.com.cn/";
        // 调用的方法名称
        String methodName ="getMobileCodeInfo";
        // EndPoint
        String endPoint = " http://ws. webxml. com. cn/WebServices/MobileCodeWS.
asmx";
        // SOAP Action
        String soapAction ="http://WebXml.com.cn/getMobileCodeInfo";
        // 指定 WebService 的命名空间和调用的方法名
        SoapObject rpc =new SoapObject(nameSpace, methodName);
        // 设置调用 WebService 接口需要传入的两个参数 mobileCode、userId
```

```
// 不可以随便写,必须和提供的参数名相同
        rpc.addProperty("mobileCode", phoneNum);
        rpc.addProperty("userId", "");
        // 生成调用 WebService 方法的 SOAP 请求信息,并指定 SOAP 的版本
        SoapSerializationEnvelope envelope =new SoapSerializationEnvelope(
                SoapEnvelope.VER11);
        envelope.bodyOut =rpc;
        // 设置是否调用的是 dotNet 开发的 WebService
        envelope.dotNet =true;
        // 等价于 envelope.bodyOut =rpc;
        envelope.setOutputSoapObject(rpc);
        HttpTransportSE transport =new HttpTransportSE(endPoint);
        try {
            // 调用 WebService
            transport.call(soapAction, envelope);
        } catch (Exception e) {
            e.printStackTrace();
        }
        // 获取返回的数据
        SoapObject object = (SoapObject) envelope.bodyIn;
        // 获取返回的结果
        String result =object.getProperty("getMobileCodeInfoResult").toString();
        return result;
    }
}
```

本章小结

本章主要讲解了 Android 应用开发中与网络相关的内容,包括 Android Http 通信、Android Socket 通信、XML 和 JSON 文件解析、WebView 和 WebService 等。在进行 Android 应用开发中,凡是与网络相关的都需要在 AndroidManifest. xml 中添加网络权限:

```
<uses-permission
    android:name="android.permission.INTERNET">
</uses-permission>
```

在实际的 Android 应用开发中,大多数的应用程序都需要联网进行操作,熟练掌握本章 Android 网络编程的相关内容,结合前面多线程的知识,能更有效地开发 Android 网络应用。在本章内容的基础上,感兴趣的读者可以去学习 Android 网络框架,如 OkHttp 等,会对 Android 网络编程有更加娴熟的操作。

习　题

1. 下列关于 Android 网络编程选项中,描述错误的是(　　)。
   A. 可以使用 URL 方式获取网络上的资源
   B. 可以使用 HttpURLConnection 方式获取网络上的资源
   C. 可以使用 HttpClient 的方式获取网络资源
   D. 可以直接在 UI 线程中访问网络并获取对应的网络资源
2. 关于 HttpURLConnection 访问网络的基本用法,描述错误的是(　　)。
   A. HttpURLConnection 对象需要设置请求网络的方式
   B. HttpURLConnection 对象需要设置超时时间
   C. 需要通过 new 关键字来创建 HttpURLConnection 对象
   D. 访问网络完毕需要关闭 HTTP 连接
3. 从网络获取天气预报 API,实现一个天气预报应用程序。

第8章 chapter 8

Android 高级编程

主要内容：Android 多媒体编程，图像处理和 Android 动画

课　　时：8 课时

知识目标：（1）熟悉 Android 多媒体编程；

（2）熟悉图像处理常用工具类；

（3）掌握 Android 动画编程。

能力目标：（1）具备 Android 多媒体开发能力；

（2）具备 Android 动画开发能力。

为了更好地通过示例讲解 Android 多媒体编程、图像处理和 Android 动画相关知识点，我们首先创建 Android 项目 Chapter08Application，然后通过对本章的学习，逐步在该项目中完成本章的示例代码。最初 Chapter08Application 项目结构如图 8-1 所示。

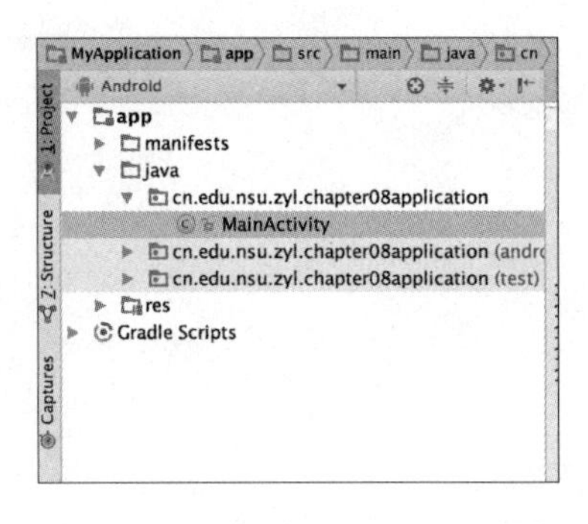

图 8-1　Chapter08Application 项目文件结构图

其中 MainActivity 是 Chapter08Application 项目的入口界面，MainActivity 将以列表的形式展示本章各个知识点的实例，用户通过点击某一个具体的列表项，该应用将跳转至该列表项对应的代码实例。因此我们让 MainActivity 继承 ListActivity，这样整个 MainActivity 就是以列表的样式展示。MainActivity.java 的源代码如文件清单 8-1 所示。

文件清单 8-1　MainActivity. java

```
package cn.edu.nsu.zyl.chapter08application;

import android.app.ListActivity;
import android.os.Bundle;
import android.view.View;
import android.widget.ArrayAdapter;
import android.widget.ListView;

import java.util.ArrayList;
```

```
public class MainActivity extends ListActivity {

    @Override
    protected void onCreate(Bundle savedInstanceState) {
        super.onCreate(savedInstanceState);
        String[] medias=new String[]{"MediaPlayer 播放音频","MediaPlayer 播放视
频","VideoView 播放视频", "Canvas 和 Paint 画图","帧动画示例","补间动画示例","属性
动画示例"};
        ArrayList list=new ArrayList();
        for(String media:medias){
            list.add(media);
        }
         ArrayAdapter adapter= new ArrayAdapter (MainActivity.this, android.R.
layout.simple_list_item_1, list);
        setListAdapter(adapter);
    }

    @Override
    protected void onListItemClick(ListView l, View v, int position, long id) {
        super.onListItemClick(l, v, position, id);

    }
}
```

为了实现点击 MainActivity 页面列表项跳转至某一具体页面的操作，在后面的案例中我们将逐步更新 onListItemClick()方法，从而将各个小节相关案例关联起来。

8.1 Android 多媒体基础

Android 多媒体框架包含了对 MP3、WMA、MP4、AVI 等多种通用媒体类型的支持，借助它可以方便地实现音频和视频的播放功能。播放的资源可以是网络上的多媒体流、本地文件或应用程序资源中(/res/raw 文件夹)获取的各种音频和视频数据。Android 多媒体框架中用于播放音频和视频的类是 MediaPlayer 类和 AudioManager 类，其中 MediaPlayer 类主要用来实现音/视频播放功能，它提供了播放音/视频所需要的所有基础 API，AudioManager 类主要用来管理音频资源和音频输出设备。

8.1.1 使用 MediaPlayer 音频播放

MediaPlayer 支持多种格式的音频播放，并且提供全面的控制方法。MediaPlayer 控制音频播放常用方法如表 8-1 所示。

表 8-1 MediaPlayer 常用方法

方　法	描　述
setDataSource()	设置要播放的音频文件
prepare()	在开始播放之前调用该方法完成准备工作
start()	开始或恢复播放

续表

方　法	描　述
pause()	暂停播放音频
reset()	将 MediaPlayer 对象重置为刚创建的状态
seekTo()	从指定位置开始播放
stop()	停止音频播放，调用该方法后 MediaPlayer 对象无法再播放
release()	释放与 MediaPlayer 相关的资源
isPlaying()	判断当前 MediaPlayer 是否正在播放
getDuration()	获取载入音频文件的时长

在程序中，使用 MediaPlayer 播放音频文件的步骤如下：

1. 创建 MediaPlayer 对象

在程序中可以调用 MediaPlayer 的无参构造方法创建 MediaPlayer 对象，也可以调用 MediaPlayer 类中提供的静态 create()方法创建 MediaPlayer 对象。

```
MediaPlayer mediaPlayer=new MediaPlayer();                    //创建 MediaPlayer 对象
mediaPlayer.setAudioStreamType(AudioManager.STREAM_MUSIC)//设置音频类型
```

上述代码使用构造方法创建 MediaPlayer 对象后，调用 setAudioStreamType()方法设置音频类型。MediaPlayer 可以接收的常用音频类型有如下 4 种：

AudioManager.STREAM_MUSIC：音乐

AudioManager.STREAM_RING：响铃

AudioManager.STREAM_ALARM：闹铃

AudioManager.STREAM_NOTIFICATION：提示音

不同音频类型占用的内存空间是不一样的，音频时间越短，占用的内存空间越小。例如，提示音占用的内存最少，播放音乐占用的内存最多。

MediaPlayer 类提供了多个静态的 create()方法用于创建 MediaPlayer 对象，常用的有如下两种。

static MediaPlayer create(Context context, int resId)：用于从 resId 对应的资源文件中装载音频文件，并返回新建的 MediaPlayer 对象。如下代码用于创建 MediaPlayer 对象，并且装载位于 res/raw/中的音频文件：

```
MediaPlayer mediaPlayer=MediaPlayer.create(context,R.raw.sound_file);
```

static MediaPlayer create(Context context,Uri uri)：用于根据指定的 URI 装载音频文件，并返回新建的 MediaPlayer 对象。如下代码创建了 MediaPlayer 对象，并且加载从 ContentProvider 得到的资源：

```
MediaPlayer mediaPlayer=MediaPlayer.create(context,Uri.parse("content://…"));
```

2. 设置播放数据源

使用 MediaPlayer 的无参构造方法创建 MediaPlayer 对象时，还需要单独指定要装载的音频文件。这时可以调用 MediaPlayer 类的 setDataSource()方法设置数据源，然后调用 prepare()方法完成音频资源的加载。

如下代码用于指定和加载需要播放的音频文件：

```
mediaPlayer.setDataSource("mnt/sdcard/xxx.mp3");     //指定要播放的音频文件
mediaPalyer.parpare();                               //装载要播放的音频文件
```

对于使用静态的 create()方法创建的 MediaPlayer 对象，在创建 MediaPlayer 对象时就已经装载好要播放的音频文件了，因此不需要再次设置数据源。通常，音频文件主要有 3 种来源，分别是当前应用的资源文件、外存储设备的资源文件和网络上的资源文件。需要注意，在使用网络上的资源或外部存储设备上的资源时，需要在 AndroidManifest. xml 中加入相应权限是声明。例如需要播放来自网络的多媒体流时，要在 AndroidManifest. xml 中加入如下权限声明：

```
<uses-permission android:name="android.permission.INTERNET"/>
```

3. 开始播放

获取 MediaPlayer 对象后，可以调用 MediaPlayer 对象的 start()方法开始播放音频文件。

```
mediaPlayer.start();
```

4. 暂停播放

调用 MediaPlayer 对象的 pause()方法可以暂停正在播放的音频。

```
//暂停播放前判断 MediaPlayer 对象是否存在，并且正处于播放状态
if(mediaPlayer!=null&&mediaPlayer.isPlaying()){
currentPosition =mMediaPlayer.getCurrentPosition();
mediaPlayer.pause();
}
```

5. 重新开始播放

重新开始播放使用 seekTo(int msec)方法，该方法将播放时间定位到指定的时间处，从头开始播放就是 seekTo(0)。

```
//播放状态下重播
if(mediaPlayer!=null&&mediaPlayer.isPlaying()){
    mediaPlayer.seekTo(currentPosition);
    mMediaPlayer.start();
}
//暂停状态下重播
if(mediaPlayer!=null){
```

```
    mediaPlayer.seekTo(0);
    mediaPlayer.start();
}
```

6. 停止播放

停止播放使用 stop()方法，停止播放后需要调用 MediaPlayer 的 release()方法将占用的资源释放并将 MediaPlayer 置为空。

```
if(mediaPlayer!=null&&mediaPlayer.isPlaying()){
    mediaPlayer.stop();
    mediaPlayer.release();
    mediaPlayer=null;
}
```

MediaPlayer 状态迁移如图 8-2 所示。

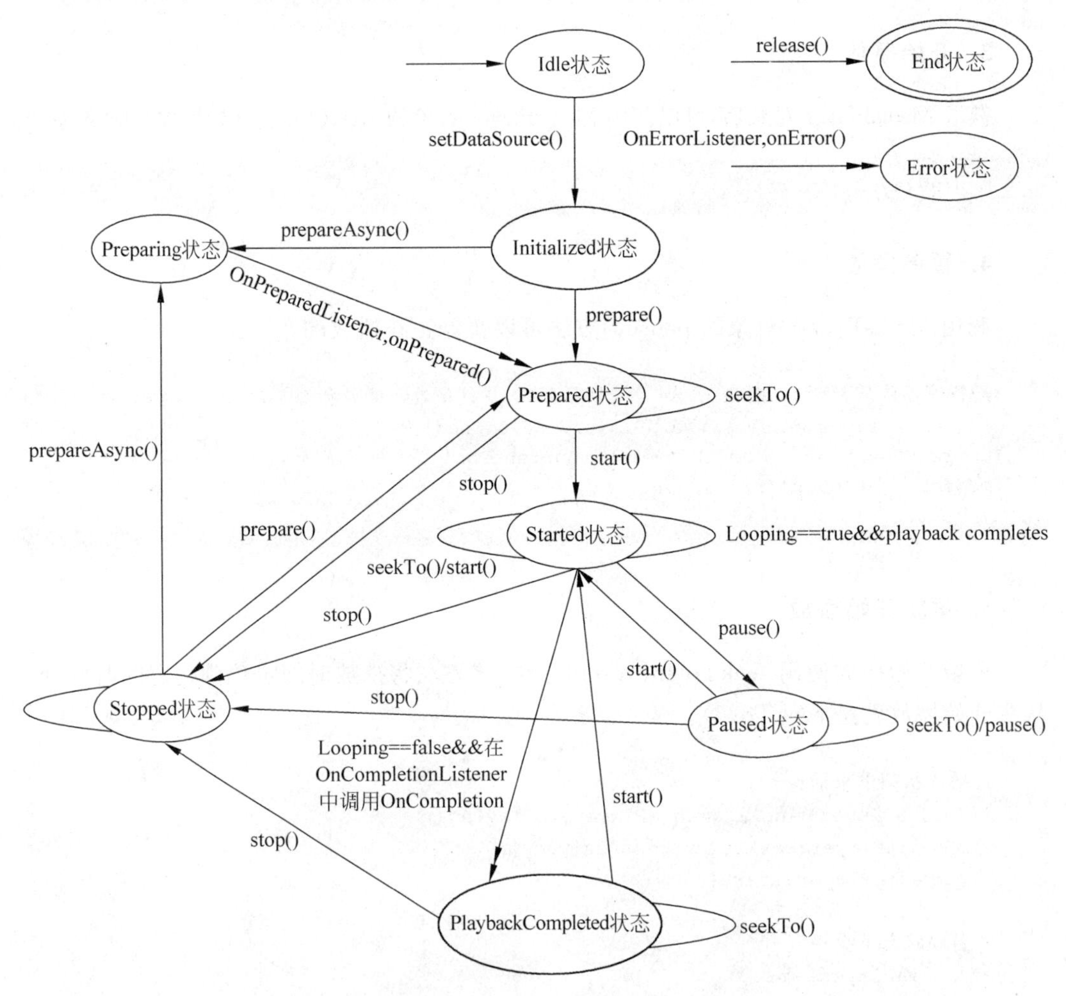

图 8-2 MediaPlayer 状态迁移

8.1.2 音频播放案例

下面通过示例介绍 MediaPlayer 的使用，该示例实现一个具有播放、暂停/继续和停止功能的简易音乐播放器。在 Android 项目中新建 AudioPlayActivity，该音乐播放器界面对应的布局文件 activity_audio_play. xml 代码如文件清单 8-2 所示。

文件清单 8-2 activity_audio_play. xml

```
<?xml version="1.0" encoding="utf-8"?>
<RelativeLayout xmlns:android="http://schemas.android.com/apk/res/android"
    xmlns:tools="http://schemas.android.com/tools"
    android:layout_width="match_parent"
    android:layout_height="match_parent"
    android:paddingBottom="@dimen/activity_vertical_margin"
    android:paddingLeft="@dimen/activity_horizontal_margin"
    android:paddingRight="@dimen/activity_horizontal_margin"
    android:paddingTop="@dimen/activity_vertical_margin"
    tools:context="com.nsu.zyl.chapter08application.AudioPlayActivity">

    <TextView
        android:layout_width="wrap_content"
        android:layout_height="wrap_content"
        android:text="输入音频地址："
        android:id="@+id/textView" />

    <EditText
        android:layout_width="match_parent"
        android:layout_height="wrap_content"
        android:id="@+id/edtMusic"
        android:layout_below="@+id/textView"
        android:layout_alignParentLeft="true"
        android:layout_alignParentStart="true" />

    <Button
        android:layout_width="wrap_content"
        android:layout_height="wrap_content"
        android:id="@+id/btnPlay"
        android:layout_below="@+id/edtMusic"
        android:layout_alignParentLeft="true"
        android:layout_alignParentStart="true"
        android:text="Play" />

    <Button
        android:layout_width="wrap_content"
        android:layout_height="wrap_content"
        android:id="@+id/btnPause"
        android:text="Pause"
```

```
        android:layout_below="@+id/edtMusic"
        android:layout_centerHorizontal="true" />

    <Button
        android:layout_width="wrap_content"
        android:layout_height="wrap_content"
        android:text="STOP"
        android:id="@+id/btnStop"
        android:layout_below="@+id/edtMusic"
        android:layout_alignRight="@+id/edtMusic"
        android:layout_alignEnd="@+id/edtMusic" />
</RelativeLayout>
```

由 activity_audio_play.xml 内容可知，该音乐播放器的界面包括一个用于提示信息的 TextView 控件，一个用于输入音频文件地址的 EditText 控件及三个控制播放、暂停和停止的按钮控件。

在 AudioPlayActivity 类中定义 play()方法播放音乐，实现音乐播放功能。在 play()方法中首先调用 MediaPlayer 对象的 reset()方法重置 MediaPlayer 对象，然后重新设置要播放的音频文件，并预加载该音频，最后调用 start()方法开始播放音频。

play()方法的代码如下：

```
private void play(){
        mediaPlayer.reset();
        try {
            mediaPlayer.setDataSource(file.getAbsolutePath());
            mediaPlayer.prepare();
            mediaPlayer.start();
        } catch (IOException e) {
            e.printStackTrace();
        }
    }
```

当音频文件播放完毕，为了重新开始播放，我们为 MediaPlayer 对象设置完成事件监听器，具体代码如下：

```
mediaPlayer.setOnCompletionListener(new MediaPlayer.OnCompletionListener() {
            @Override
            public void onCompletion(MediaPlayer mediaPlayer) {
                play();
            }
        });
```

为播放按钮 PLAY 添加点击事件监听器，首先获得用户输入音频文件地址，判断该地址是否存在，如果存在，调用 play()方法播放音频并且在提示信息文本控件中给用户提示；如果音频文件不存在，直接在提示信息文本控件中提示用户文件不存在，并且将焦点定位到文件编辑控件。关键代码如下：

```
btnPlay.setOnClickListener(new View.OnClickListener() {
            @Override
            public void onClick(View view) {
                String path=edtMusic.getText().toString();
                file=new File(path);
                if(file.exists()){
                    play();
                    if(isPause){
                        isPause=false;
                        btnPause.setText("continue");
                    }
                    txtMessage.setText("正在播放音乐.....");
                }else{
                    txtMessage.setText("音频文件不存在,请确认后重新输入");
                    edtMusic.setFocusable(true);
                }

            }
        });
```

为暂停按钮PAUSE添加点击事件监听器，如果MediaPlayer处于播放状态并且标记isPause的值为false，则暂停播放音频，并设置相关信息；否则调用MediaPlayer对象的start()方法继续播放音乐，并设置相关信息。关键代码如下：

```
btnPause.setOnClickListener(new View.OnClickListener() {
            @Override
            public void onClick(View view) {
                if(!isPause&&mediaPlayer.isPlaying()){
                    mediaPlayer.pause();
                    isPause=true;
                    btnPause.setText("continue");
                }else{
                    mediaPlayer.start();
                    isPause=false;
                    btnPause.setText("pause");
                }
            }
        });
```

为停止按钮STOP添加点击事件监听器，调用MediaPlayer对象的stop()方法停止播放音频，然后设置相关信息。关键代码如下：

```
btnStop.setOnClickListener(new View.OnClickListener() {
            @Override
            public void onClick(View view) {
```

```
                mediaPlayer.stop();
                txtMessage.setText("音乐停止播放");
            }
        });
```

重写onDestroy()方法，用于在当前Activity销毁时释放MediaPlayer所占用的资源。具体代码如下：

```
@Override
    protected void onDestroy() {
        if(mediaPlayer!=null&&mediaPlayer.isPlaying()){
            mediaPlayer.stop();
            mediaPlayer.release();
            mediaPlayer=null;
        }
        super.onDestroy();
    }
```

AudioPlayActivity具体代码如文件清单8-3所示。

文件清单8-3 AudioPlayActivity.java

```
import android.media.MediaPlayer;
import android.support.v7.app.AppCompatActivity;
import android.os.Bundle;
import android.view.View;
import android.widget.Button;
import android.widget.EditText;
import android.widget.TextView;
import java.io.File;
import java.io.IOException;

public class AudioPlayActivity extends AppCompatActivity {
    private Button btnPlay,btnStop,btnPause;
    private TextView txtMessage;
    private EditText edtMusic;
    private MediaPlayer mediaPlayer;
    private File file;
    private boolean isPause=false;

    @Override
    protected void onCreate(Bundle savedInstanceState) {
        super.onCreate(savedInstanceState);
        setContentView(R.layout.activity_main);
        btnPlay=(Button)findViewById(R.id.btnPlay);
        btnPause=(Button)findViewById(R.id.btnPause);
        btnStop=(Button)findViewById(R.id.btnStop);
        edtMusic=(EditText) findViewById(R.id.edtMusic);
```

```
        txtMessage= (TextView) findViewById(R.id.textView);

        mediaPlayer=new MediaPlayer();
        mediaPlayer.setOnCompletionListener(new MediaPlayer.OnCompletionListener() {
            @Override
            public void onCompletion(MediaPlayer mediaPlayer) {
                play();
            }
        });

        btnPlay.setOnClickListener(new View.OnClickListener() {
            @Override
            public void onClick(View view) {
                String path=edtMusic.getText().toString();
                file=new File(path);
                if(file.exists()){
                    play();
                    if(isPause){
                        isPause=false;
                        btnPause.setText("CONTINUE");
                    }
                    txtMessage.setText("正在播放音乐...");
                }else{
                    txtMessage.setText("输入的音频文件不存在,请确认后重新输入");
                    edtMusic.setFocusable(true);
                }
            }
        });
        btnPause.setOnClickListener(new View.OnClickListener() {
            @Override
            public void onClick(View view) {
                if(!isPause&&mediaPlayer.isPlaying()){
                    mediaPlayer.pause();
                    isPause=true;
                    btnPause.setText("CONTINUE");
                }else if(isPause){
                    mediaPlayer.start();
                    isPause=false;
                    btnPause.setText("PAUSE");
                }
            }
        });
        btnStop.setOnClickListener(new View.OnClickListener() {
            @Override
            public void onClick(View view) {
                mediaPlayer.stop();
                txtMessage.setText("音乐停止播放");
            }
```

```
        });

    }
    @Override
    protected void onDestroy() {
        if(mediaPlayer!=null&&mediaPlayer.isPlaying()){
            mediaPlayer.stop();
            mediaPlayer.release();
            mediaPlayer=null;
        }
        super.onDestroy();
    }
    private void play(){
        mediaPlayer.reset();
        try {
            mediaPlayer.setDataSource(file.getAbsolutePath());
            mediaPlayer.prepare();
            mediaPlayer.start();
        } catch (IOException e) {
            e.printStackTrace();
        }
    }
}
```

重写 Chapter08Applicatin 项目下 MainActivity.java 文件中的 void onListItemClick (ListView l, View v, int position, long id)方法：

```
protected void onListItemClick(ListView l, View v, int position, long id) {
        super.onListItemClick(l, v, position, id);
        Intent intent=new Intent();
        switch (position){
            case 0:

intent.setClassName("cn.edu.nsu.zyl.chapter08application","AudioPlayActivity");
                break;

        }
        startActivity(intent);

    }
```

运行 Chapter08Applicatin 项目，当用户点击列表项第一项“MediaPlayer 播放音频文件”时，显示为一个简易的音乐播放器，用户在文本编辑框中输入播放的音频文件地址；点击播放 PLAY 按钮，将开始播放音乐；点击 PAUSE 按钮，音乐停止播放并且按钮变为 CONTINUE，再次点击该按钮，继续播放音乐；点击 STOP 按钮，音乐停止播放。音乐播放器如图 8-3 所示。

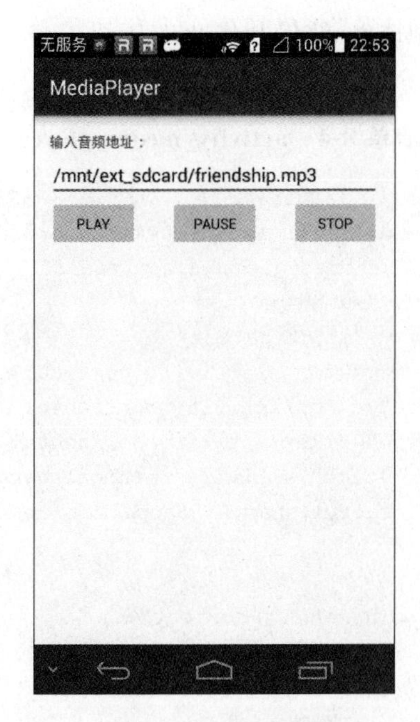

图 8-3 音乐播放器

8.1.3 使用 MediaPlayer 和 SurfaceView 播放视频

MediaPlayer 主要用于播放音频，没有提供输出图像的输出界面，因此播放视频时需要结合 SurfaceView 控件来达到视频输出的效果。

SurfaceView 是继承自 View 用于显示图像的组件，其中内嵌一个专门用于绘制的 Surface。开发者可以控制这个 Surface 的尺寸和格式，并让 SurfaceView 控制这个 Surface 的绘制位置。Surface 对应一块屏幕缓冲区，每个 Window 对应一个 Surface，任何 View 都是画在 Surface 上的。

使用 MediaPlayer 和 SurfaceView 相结合的方式实现视频播放的步骤如下：

(1) 创建 MediaPlyer 的对象，并加载指定的视频文件。

(2) 在界面布局文件中定义 SurfaceView 组件，或在程序中创建 SurfaceView 组件，并为 SurfaceView 的 SurfaceHolder 添加 Callback 监听器。

(3) 调用 MediaPlayer 对象的 setDisplay(Surfaceolder sh)方法将所播放的视频图像输出到指定的 SurfaceView 组件。

(4) 调用 MediaPlayer 对象的 start()、stop()和 pause()方法控制视频的播放。

8.1.4 视频播放案例(一)

下面示例通过使用 MediaPlayer 和 SurfaceView 开发一款简易的视频播放器，实现视频播放的播放、暂停和停止功能。在 Chapter08Application 项目中新建 MediaPlayerActivity 实现视频播放功能，该视频播放器的界面包括一个用于提示信息的 TextView 控件、一个用于输入视频文件地址的 EditText 控件、一个用于显示视频画面的

SurfaceView控件及三个控制播放、暂停和停止的按钮控件。界面对应的布局文件代码如文件清单8-4所示。

文件清单8-4 activity_media_player.xml

```
<?xml version="1.0" encoding="utf-8"?>
<RelativeLayout xmlns:android="http://schemas.android.com/apk/res/android"
    xmlns:tools="http://schemas.android.com/tools"
    android:layout_width="match_parent"
    android:layout_height="match_parent"
    android:paddingBottom="@dimen/activity_vertical_margin"
    android:paddingLeft="@dimen/activity_horizontal_margin"
    android:paddingRight="@dimen/activity_horizontal_margin"
    android:paddingTop="@dimen/activity_vertical_margin"
    tools:context="com.nsu.zyl.chapter08application.MediaPlayerActivity">

    <TextView
        android:layout_width="wrap_content"
        android:layout_height="wrap_content"
        android:text="请输入视频地址："
        android:id="@+id/txtMessage" />

    <EditText
        android:layout_width="match_parent"
        android:layout_height="wrap_content"
        android:id="@+id/edtPath"
        android:layout_below="@+id/txtMessage"
        android:layout_alignParentLeft="true"
        android:layout_alignParentStart="true" />

    <SurfaceView
        android:layout_width="match_parent"
        android:layout_height="360dp"
        android:id="@+id/surfaceView"
        android:layout_below="@+id/edtPath"
        android:layout_alignParentLeft="true"
        android:layout_alignParentStart="true"
        android:keepScreenOn="true"/>

    <Button
        android:layout_width="wrap_content"
        android:layout_height="wrap_content"
        android:text="START"
        android:id="@+id/btnStart"
        android:layout_below="@+id/surfaceView"
        android:layout_alignParentLeft="true"
        android:layout_alignParentStart="true" />
```

```
    <Button
        android:layout_width="wrap_content"
        android:layout_height="wrap_content"
        android:text="PAUSE"
        android:id="@+id/btnPause"
        android:layout_below="@+id/surfaceView"
        android:layout_centerHorizontal="true" />

    <Button
        android:layout_width="wrap_content"
        android:layout_height="wrap_content"
        android:text="STOP"
        android:id="@+id/btnStop"
        android:layout_below="@+id/surfaceView"
        android:layout_alignParentRight="true"
        android:layout_alignParentEnd="true" />

</RelativeLayout>
```

在主界面需要控制视频的播放、暂停/继续和停止，点击主界面相关按钮，实现不同操作，具体代码如文件清单 8-5 所示。

文件清单 8-5 在主界面控制视频的播放、暂停/继续和停止的代码

```
package cn.edu.nsu.zyl.chapter08application;
import android.content.res.AssetManager;
import android.media.MediaPlayer;
import android.content.res.AssetFileDescriptor;
import android.support.v7.app.AppCompatActivity;
import android.os.Bundle;
import android.view.SurfaceView;
import android.view.View;
import android.widget.Button;
import android.widget.TextView;
import android.widget.Toast;
import java.io.IOException;
public class MediaPlayerActivity extends AppCompatActivity {
    private Button btnStart,btnStop,btnPause;
    private EditText edtPath;
    private TextView txtMessage;
    private SurfaceView surfaceView;
    private MediaPlayer mediaPlayer;
    private boolean isPause=false;

    @Override
    protected void onDestroy() {
        if(mediaPlayer.isPlaying()){
            mediaPlayer.stop();
```

```
        }
        mediaPlayer.release();
        super.onDestroy();
    }

    @Override
    protected void onCreate(Bundle savedInstanceState) {
        super.onCreate(savedInstanceState);
        setContentView(R.layout.activity_main);

        btnStart=(Button)findViewById(R.id.btnStart);
        btnPause=(Button)findViewById(R.id.btnPause);
        btnStop=(Button)findViewById(R.id.btnStop);
        edtPath=(EditText)findViewById(R.id.edtPath);
        txtMessage=(TextView) findViewById(R.id.txtMessage);
        surfaceView=(SurfaceView) findViewById(R.id.surfaceView);

        mediaPlayer=new MediaPlayer();
        mediaPlayer.setOnCompletionListener(new MediaPlayer.OnCompletionListener() {
            @Override
            public void onCompletion(MediaPlayer mediaPlayer) {
                Toast.makeText(MediaPlayerActivity.this,"视频播放完毕",
Toast.LENGTH_LONG).show();
            }
        });
        btnStart.setOnClickListener(new View.OnClickListener() {
            @Override
            public void onClick(View view) {
              File file=new File(edtPath.getText().toString());
                if(file.exists()){
                    try {
                        mediaPlayer.reset();
                        mediaPlayer.setDataSource(file.getAbsolutePath());
                        mediaPlayer.setDisplay(surfaceView.getHolder());
                        mediaPlayer.prepare();
                        mediaPlayer.start();
                        btnPause.setText("PAUSE");
                        isPause=false;
                        txtMessage.setText("视频播放中");
                    } catch (IOException e) {
                        e.printStackTrace();
                    }
                }else{
                    txtMessage.setText("输入的视频地址不存在！");
                }
            }
        });
        btnPause.setOnClickListener(new View.OnClickListener() {
```

```
            @Override
            public void onClick(View view) {
                if(mediaPlayer.isPlaying()&&!isPause){
                    mediaPlayer.pause();
                    isPause=true;
                    btnPause.setText("CONTINUE");
                }else if(isPause){
                    mediaPlayer.start();
                    isPause=false;
                    btnPause.setText("PAUSE");
                }
            }
        });
        btnStop.setOnClickListener(new View.OnClickListener() {
            @Override
            public void onClick(View view) {
                if(mediaPlayer.isPlaying()){
                    mediaPlayer.stop();
                }
            }
        });

    }
}
```

同理,需要重写 MainActivity. java 中 onListItemClick()方法,运行 Chapter08Application 项目,在主页面点击“MediaPlayer 播放视频”列表项,跳转至图 8-4 所示界面。

图 8-4 视频播放器

8.1.5 使用 VideoView 播放视频

为了在 Android 中播放视频，Android 提供 VideoView 组件，该组件将视频显示与控制集于一身，可以实现简易的视频播放功能。Android 提供 MediaController 类用于提供图形控制界面，通过该控制界面来控制视频的播放。

VideoView 提供的用于控制视频播放的常用方法如表 8-2 所示。

表 8-2　VideoView 常用方法

方　法	描　述
setVideoPath	设置要播放视频文件位置
start()	开始或恢复播放
pause()	暂停播放
stop()	停止播放
resume()	从头开始播放视频
seekTo()	从指定位置开始播放
isPlaying()	判断当前是否正在播放
getDuration()	获取载入视频文件的时长

使用 VideoView 播放视频文件的步骤如下：

(1) 在布局文件中使用 VideoView 组件或在程序中创建 VideoView 组件。

```
<VideoView android:id="@+id/videoView"
android:background="@drawable/mpbackground"
android:layout_width="match_parent"
android:layout_height="wrap_content"
android:layout_gravity="center"/>
```

(2) 在 Activity 代码中获取布局文件中定义的 VideoView 对象。

```
VideoView videoView=(VideoView)findViewById(R.id.videoView);
```

(3) 播放视频文件，视频文件既可以是本地视频又可以是网络视频。

```
//播放本地视频
videoView.setVideoPath("mnt/sdcard/mis.avi");
//播放网络视频
videoView.setVideoURI("http://www.xxx.xx/film/a.avi");
videoView.start();
```

当播放网络视频时，需要在清单文件中声明权限。

```
<uses-permission android:name="android.permission.INTERNET"/>
```

（4）添加控制器。

可以为 VideoView 添加播放控制器 MediaController，通过控制器来操控视频的播放。

```
MediaController controller=new MediaController(context);
videoView.setMediaController(controller);
```

8.1.6 视频播放案例(二)

在 Chapter08Application 项目中新建 VideoViewActivity.java，通过 VideoView 控件和 MediaPlayer 控件结合使用，实现简单的视频播放功能。

在布局文件中添加 VideoView 组件，布局文件具体代码如下：

```
<?xml version="1.0" encoding="utf-8"?>
<RelativeLayout xmlns:android="http://schemas.android.com/apk/res/android"
    xmlns:tools="http://schemas.android.com/tools"
    android:layout_width="match_parent"
    android:layout_height="match_parent"
    android:paddingBottom="@dimen/activity_vertical_margin"
    android:paddingLeft="@dimen/activity_horizontal_margin"
    android:paddingRight="@dimen/activity_horizontal_margin"
    android:paddingTop="@dimen/activity_vertical_margin"
    tools:context="com.nsu.zyl.chapter08application.VideoViewActivity">

    <VideoView
        android:layout_width="match_parent"
        android:layout_height="wrap_content"
        android:id="@+id/videoView"
        android:layout_centerHorizontal="true" />
</RelativeLayout>
```

界面交互代码如下：

```
public class VideoViewActivity extends AppCompatActivity {
    private VideoView videoView;
    private MediaController mediaController;


    @Override
    protected void onCreate(Bundle savedInstanceState) {
        super.onCreate(savedInstanceState);
        setContentView(R.layout.activity_main);
        videoView=(VideoView) findViewById(R.id.videoView);
        mediaController=new MediaController(this);

        File videoFile=new File("/mnt/ext_sdcard/kb.mp4");
```

```
        if(videoFile.exists()){
            videoView.setVideoPath(videoFile.getAbsolutePath());
            videoView.setMediaController(mediaController);
            mediaController.setMediaPlayer(videoView);
            videoView.requestFocus();
        }
        videoView.setOnCompletionListener(new MediaPlayer.OnCompletionListener() {
            @Override
            public void onCompletion(MediaPlayer mediaPlayer) {
                Toast.makeText(MainActivity.this,"视频播放完毕！",
                                Toast.LENGTH_LONG).show();
            }
        });

    }
}
```

在 onCreate()方法中，首先获取布局文件中添加的 VideoView，然后创建一个 MediaController 对象用于控制视频的播放。最后判断视频文件是否存在，如果存在，则开始视频的播放。图 8-5 所示为 VideoView 视频播放器的界面。

图 8-5 VideoView 视频播放器的界面

8.2 Android 图像处理

Android 提供了多种处理图形图像的工具类，常用的工具类包括 Canvas 类、Paint 类、Bitmap 类和 BitmapFactory 类，其中，Canvas 类代表画布，Paint 类代表画笔，Bitmap

类代表位图，BitmapFactory 类代表位图工厂。

8.2.1 Canvas 类和 Paint 类

Canvas 类代表画布，通过该类提供的方法，可以绘制各种图形。要在 Android 中绘图，需要创建一个继承 View 类的视图，并重写类中的 onDraw()方法。Canvas 类提供的常用方法如表 8-3 所示。

表 8-3 Canvas 类常用方法

方 法	描 述
void drawArc (RectF oval, float startAngles, float void sweepAngle, boolean useCenter, Paint paint)	绘制弧
void drawBitmap(Bitmap bitmap, Rect src, Rect dst, Paint paint)	在指定点绘制从源位图挖取的一块区域
void drawBitmap(Bitmap bitmap, float left, float top, Paint paint)	在指定点绘制位图
void drawCircle(float x, float y, float radius, Paint paint)	在指定点绘制一个指定半径的圆形
void drawLine(float startX, float startY, float stopX, float stopY, Paint paint)	在指定起始点绘制一条线
void drawLines(float[] pts, int offset, int count, Paint paint)	绘制多条线
void drawRoundRect(RectF rect, float rx, float ry, Paint paint)	绘制指定的圆角矩形，其中 rx 表示 X 轴圆角半径，ry 表示 Y 轴圆角半径
void drawOval(RectF oval, Paint paint)	绘制椭圆
void drawPath(Path path, Paint paint)	沿着指定 Path 绘制任意形状
drawText(String text, int start, int end, Paint paint)	绘制字符串
void rotate(float degrees, float x, float y)	对 Canvas 进行旋转操作
void scale(float sx, float sy, float px, float py)	对 Canvas 进行缩放操作
void skew(float x, float y)	对 Canvas 进行倾斜变化
void translate(float x, float y)	移动 Canvas，向下移动 x 距离，向右移动 y 距离

Paint 类代表画笔，用来设置绘图风格，包括颜色、线宽、透明度等。使用 Paint 类时，需要首先创建 Paint 对象，然后利用 Paint 类提供的方法更改默认设置。

例如，定义一个颜色为红色并带有阴影的画笔，可以使用如下代码：

```
Paint paint=new Paint();
paint.setColor(Color.red);
paint.setShadowLayer(2,3,3,Color.rgb(180,180,180));
```

Paint 类常用的方法如表 8-4 所示。

表 8-4　Paint 类常用方法

方　　法	描　　述
Paint()	创建一个 Paint 对象,并使用默认值
Paint(int flags)	创建一个 Paint 对象,并使用指定属性
void setARGB(int a,int r int g,int b)	设置 ARGB
void setColor(int color)	设置颜色
void setAlpha(int a)	设置透明度
void setAntiAlias(boolean aa)	设置是否抗锯齿
void setDither(boolean dither)	设置是否使用图像抖动处理
void setShadowLayer(float radius, float dx, float dy, int color)	设置阴影,radius 为阴影角度,dx 和 dy 是阴影在 x 轴和 y 轴上的距离,color 为阴影的颜色
void setTextAlign(Align align)	设置文本的对齐方式,参数值为 Align. CENTER, Align. LEFT, Align. RIGHT
void setTextSize(float textSize)	设置绘制文本时文字的大小
void setStrokeWidth(float width)	设置笔触的宽度
void setStrokeJoin(Paint. Join join)	设置笔画转弯处的风格

8.2.2　绘图案例

下面通过程序展示如何在 Android 中利用 Canvas 类和 Paint 类绘制基本的图形。首先定义一个继承 View 的 MyView 类；然后重写 onDraw()方法,在 onDraw()方法中绘制奥林匹克五环标志。MyView 类的具体代码如下：

```
public class MyView extends View {

    public MyView(Context context) {
        super(context);
    }

    public void onDraw(Canvas canvas) {

        Paint paint_blue =new Paint();                          //绘制蓝色的环
        paint_blue.setColor(Color.BLUE);
        paint_blue.setStyle(Paint.Style.STROKE);
        paint_blue.setStrokeWidth(10);
        canvas.drawCircle(160,150,60,paint_blue);

    Paint paint_black =new Paint();                             //绘制黑色的环
```

```
        paint_black.setColor(Color.BLACK);
        paint_black.setStyle(Paint.Style.STROKE);
        paint_black.setStrokeWidth(10);
        canvas.drawCircle(295, 150, 60, paint_black);

    Paint paint_red =new Paint();                          //绘制红色的环
        paint_red.setColor(Color.RED);
        paint_red.setStyle(Paint.Style.STROKE);
        paint_red.setStrokeWidth(10);
        canvas.drawCircle(430, 150, 60, paint_red);

        Paint paint_yellow =new Paint();                   //绘制黄色的环
        paint_yellow.setColor(Color.YELLOW);
        paint_yellow.setStyle(Paint.Style.STROKE);
        paint_yellow.setStrokeWidth(10);
        canvas.drawCircle((float)225.5, 210, 60, paint_yellow);


        Paint paint_green =new Paint();                    //绘制绿色的环
        paint_green.setColor(Color.GREEN);
        paint_green.setStyle(Paint.Style.STROKE);
        paint_green.setStrokeWidth(10);
        canvas.drawCircle(361, 210, 60, paint_green);

        Paint paint_string =new Paint();                   //绘制字符串
        paint_string.setColor(Color.BLUE);
        paint_string.setTextSize(20);
        canvas.drawText("Faster,Higher,Stronger", 245, 310, paint_string);
    }
}
```

在上述代码中，五环通过调用 Canvas 类中的 drawCircle()方法绘制，五环颜色通过设置 Paint 的颜色实现。在五环标志下方，调用 canvas. drawText()方法绘制字符串到 Canvas 类中。在 MainActivity 中加载 MyView，运行效果如图 8-6 所示。

```
public class MainActivity extends Activity {

    @Override
    public void onCreate(Bundle savedInstanceState) {
        super.onCreate(savedInstanceState);
        setContentView(new MyView(this));       //加载 MyView
    }
}
```

图 8-6 奥运五环运行效果

8.2.3 Bitmap 类和 BitmapFactory 类

Bitmap 类是 Android 中一个非常重要的图像处理类。当需要对已有的图片资源进行操作时，通常不会对原始图片文件直接操作，而是将文件加载到 Bitmap 对象中，然后再对 Bitmap 实例对象进行操作。利用 Bitmap 对象可以获取图像文件信息，进行图像的剪切、旋转、缩放等操作，并可以指定格式保存图像文件。Bitmap 类常用的方法如表 8-5 所示。

表 8-5 Bitmap 类常用方法

方　法	描　述
static Bitmap createBitmap(Bitmap sources, int x, int y, int width, int height)	从源位图 source 的指定坐标点(给定 x,y)开始，从中"挖取"宽度 width,高度 height 的一块区域，创建新的 Bitmap
static Bitmap createScaleBitmap(Bitmap src, int src, int dstWidth, int dstHeight, boolean filter)	对源位图 src 进行缩放，缩放成宽度 dstWidth、高度 dstWidth 的新位图
static Bitmap createBitmap(int width, int height, Confit config)	创建指定格式、大小的位图
static Bitmap createBitmap(Bitmap source,int x, int y,int width,int height)	以 source 为源图，创建新的图片，指定起始坐标以及新图像的高度
static Bitmap createBitmap(Bitmap source, int x, int y, int width, int height, Matrix m, boolean filter)	从源位图 source 指定的坐标点(给定 x,y)开始，从中"挖取"宽 width、高 height 的一块区域，创建新的 Bitmap 对象，并按照 Matrix 指定的规则变换
boolean isRecycled()	判断 Bitmap 对象是否被收回
void recycle()	回收 Bitmap 对象

续表

方　法	描　述
boolean compress(Bitmap. CompressFormat format, int quality, OutputStream stream)	用于将 Bitmap 对象压缩为指定格式并保存到指定文件输出流中，其中，format 参数值可以是 Bitmap. CompressFormat. PNG， Bitmap. ComressFormat. JPEG，Bitmap. CompressFormat. WEBP

BitmapFactory 类是一个工具类，主要用于从不同数据源解析和创建 Bitmap 对象。BitmapFactory 类提供的创建 Bitmap 类的常用方法如表 8-6 所示。

表 8-6　BitmapFactory 类的常用方法

方　法	描　述
static Bitmap decodeFile(String pathName)	从 pathName 指定的文件中解析、创建 Bitmap 对象
static Bitmap decodeFileDescriptor(FileDescriptor fd)	从 fd 对应的文件中解析、创建 Bitmap 对象
static Bitmap decodeStream(InputStream is)	从指定输入流中解析、创建 Bitmap 对象
static Bitmap decodeResource (Resources res,int id)	根据给定的 ID，从给定的资源中解析和创建 Bitmap 对象
static Bitmap decodeByteArray (byte [ ] data, int offset, int length)	从指定的字节数组 offset 位置开始读取 length 长字节解析成 Bitmap 对象

解析 Drawable 文件夹下图片文件并创建相应 Bitmap 对象，相关代码如下：

```
Bitmap bitmap = BitmapFactory. decodeResource (getResources (), R. drawable. ic_
launcher);
```

解析指定路径下(/sdcard/images/)图像文件并创建图像对象，相关代码如下：

```
Bitmap bitmap=BitmapFacory.decodeFile("/sdcard/images/meinv.jpg");
```

8.3 Android 动画

在 Android 应用开发中经常涉及动画，Android 3.0 版本之前，Android 支持两种动画模式，即帧动画(Frame Animation)和补间动画(Tween Animation)，Android 3.0 版本中引入了属性动画(Property Animation)。

8.3.1 帧动画

帧动画(Frame Animation)：通过一系列图像依次显示来模拟动画效果。实现帧动画可以分为以下几步。

(1) 在 XML 资源文件中定义一组用于生成动画的图片资源。

在 XML 中使用< animation-list >包含一系列的< item >标记来定义一组用于生成动画的图片资源，其中< item >表示要轮换显示的图片。

```
<animation-list xmlns:android="http://schemas.android.com/apk/res/android"
    android:oneshot="true">
    <item android:drawable="@drawable/flower1" android:duration="200" />
    <item android:drawable="@drawable/flower2" android:duration="200" />
    <item android:drawable="@drawable/flower3" android:duration="200" />
</animation-list>
```

上述<item>标记中的 duration 属性表示图像显示的时间。定义好的 XML 文件要存放在/res/drawable/目录下。

(2) 将步骤(1)中定义的动画资源作为组件的背景使用。例如：

```
protected void onCreate(Bundle savedInstanceState) {
        // TODO Auto-generated method stub
        super.onCreate(savedInstanceState);
        setContentView(R.layout.main);
        imageView = (ImageView) findViewById(R.id.imageView1);
        imageView.setBackgroundResource(R.drawable.drawable_anim);
        anim = (AnimationDrawable) imageView.getBackground();
    }
```

(3) 在程序中启动动画播放。

由于 AnimationDrawable 对象是默认不播放的，所以需要在程序中启动动画播放。AnimationDrawable 提供 start()方法和 stop()方法，用于开始和结束动画。

除了可以在 XML 中控制帧动画外，还可以直接通过 Java 程序实现帧动画。帧动画主要用到的类是 AnimationDrawable，通过它的 addFrame(Drawable frame, int duration)方法，将 Drawable 对象放到 AnimationDrawable 中，并设定跳转时间，然后将 AnimationDrawable 对象设置为 ImageView 的背景，最后调用 start()方法启动。

8.3.2 帧动画案例

下面通过示例介绍帧动画。在 Chapter08Application 项目中，创建一个 FrameAnimationActivity.java 用于展示帧动画的创建和使用。首先定义动画对应的 XML 文件，该文件共有 6 帧。如文件清单 8-6 所示。

文件清单 8-6 FrameAnimationActivity

```
<?xml version="1.0" encoding="utf-8"?>
<animation-list android:oneshot="false"
    xmlns:android="http://schemas.android.com/apk/res/android">

    <item
        android:drawable="@drawable/animation_1"
        android:duration="200"/>
    <item
        android:drawable="@drawable/animation_2"
        android:duration="200"/>
```

```
    <item
        android:drawable="@drawable/animation_3"
        android:duration="200"/>
    <item
        android:drawable="@drawable/animation_4"
        android:duration="200"/>
    <item
        android:drawable="@drawable/animation_5"
        android:duration="200"/>
    <item
        android:drawable="@drawable/animation_6"
        android:duration="200"/>
</animation-list>
```

然后将动画资源作为 ImageView 控件的背景资源,并且启动该动画。

```
public class FrameAnimationActivity extends AppCompatActivity {

    protected void onCreate(Bundle savedInstanceState) {
        super.onCreate(savedInstanceState);
        setContentView(R.layout.activity_frame_animation);
        ImageView imageView = (ImageView) findViewById(R.id.imageView);
          AnimationDrawable drawable = (AnimationDrawable)   getResources ().
getDrawable(R.drawable.frame_animation);
        imageView.setImageDrawable(drawable);
        drawable.start();
    }
}
```

该示例运行效果如图 8-7 所示。

图 8-7 盛开的花朵运行效果

8.3.3 补间动画

补间动画(Tween Animation)是指开发者只需要指定动画开始、动画结束“关键帧”，动画变化的“中间帧”由系统计算补齐。补间动画通过对View中的内容进行一系列的图形变换来实现动画效果。补间动画只能应用在View对象，并且只支持部分属性，例如支持旋转缩放但不支持背景颜色改变等，因此补间动画又称为View Animation。补间动画既可以使用代码定义也可以用XML定义。

Android系统目前支持5种补间动画：透明度渐变动画(AlphaAnimation)、旋转渐变动画(RotateAnimation)、缩放渐变动画(ScaleAnimation)、位移渐变动画(TranslateAnimation)和组合动画(AnimationSet)。

1. 透明度渐变动画

创建时允许指定开始以及结束透明度，还有动画的持续时间。透明度的变化范围(0,1)，0代表完全透明，1代表完全不透明。透明的渐变对应<alpha/>标签。在XML中定义透明度渐变动画的基本语法格式如下：

```
<?xml version="1.0" encoding="utf-8">
<set xmlns:android=http://schema.android.com/apk/res/android
    android:interpolate="@android:anim/linear_interpolator">
<alpha android:repeatMode="restart"
      android:repeatCount="infinite"
      android:duration="1000"
      android:fromAlpha="0.0"
      android:toAlpha="1.0"/>
</set>
```

上述XML定义了一个让View从完全透明到不透明，持续时间为1s的动画。

上述定义的动画中使用到的属性说明如下：

(1) android:interpolate:用于控制动画变化速度，使得动画效果可以匀速、加速、减速或者抛物线速度等各种速度，其取值可为：

① @android:anim/linear_interpolator:匀速；

② @android:anim/accelerate_interpolator:加速；

③ @android:anim/decelerate_interpolator:减速；

④ @android:anim/accelerate_decelerate_interpolator:开始和结束减速，中间加速；

⑤ @android:anim/cycle_interpolator:动画循环播放特定次数，变化速度按正弦曲线改变；

⑥ @android:anim/anticipate_interpolator:先向相反方向改变一段再加速播放；

⑦ @android:anim/anticipate_overshoot_interpolator:开始时向后然后向前甩一定值后返回最后的值；

⑧ @android:anim/bounce_interpolator:跳跃，快到目的值时值会跳跃，如目的值为

100,后面的值可能依次为85,77,70,80,90,100;

⑨ @android:anim/overshoot_interpolator:回弹,最后超出目的值然后缓慢改变到目的值等。

(2) android:repeatMode:用于设置动画重复的方式,可选值为reverse(反向)、restart(重新开始)。

(3) android:repeatCount:用于设置动画重复次数,属性值可以为正整数,也可以为infinite(无限循环)。

(4) android:duration:用于指定动画播放时长。

(5) android:fromAlpha:用于指定动画开始时的透明的,0.0代表完全透明,1.0代表不透明。

2. 缩放渐变动画

缩放渐变动画通过为动画指定开始时的缩放系数、结束时的缩放系数以及持续时间来创建动画。在XML中缩放渐变动画对应< scale/>标签。

在XML中定义缩放渐变动画基本格式如下:

```
<?xml version="1.0" encoding="utf-8">
<set xmlns:android=http://schema.android.com/apk/res/android
    android:interpolate="@android:anim/linear_interpolator">
<scale android:repeatMode="restart"
       android:repeatCount="infinite"
       android:duration="2000"
       android:fromXScale="1.0"
       android:fromYScale="1.0"
       android:toXScale="2.0"
       android:toYScale="0.5"
       android:pivotX="50% "
       android:pivotY="50% "/>
</set>
```

上述代码定义了一个在X轴上放大2倍,在Y轴上缩小1/2的缩放动画。

缩放渐变动画常用属性如下:

(1) android:fromXScale: 指定动画开始时在X轴上的缩放系数,值为1.0表示不变化。

(2) android:fromYScale: 指定动画开始时在Y轴上的缩放系数,值为1.0表示不变化。

(3) android:toXScale: 指定动画结束时在X轴上的缩放系数,值为1.0表示不变化。

(4) android:toYScale: 指定动画结束时在Y轴上的缩放系数,值为1.0表示不变化。

3. 位移渐变动画

位移渐变动画是指通过为动画指定起始以及结束位置，以及持续时间来创建的动画；在 XML 中位移渐变动画对应< translate/>标签。

```
<?xml version="1.0" encoding="utf-8">
<set xmlns:android=http://schema.android.com/apk/res/android
    android:interpolate="@android:anim/linear_interpolator">
<translate android:fromXDelta="50"
      android:fromYDelta="50"
      android:toXDelta="200"
      android:toYDelta="200"
      android:repeatMode="reverse"
      android:repeatCount="infinite"
      android:duration="2000"/>
</set>
```

上述代码实现 View 对象的平移，从(50,50)平移到(200,200)，持续时间 2s。

位移渐变动画常用属性如下：

(1) android:fromXDelta：指定动画开始时 View 的 X 坐标。

(2) android:fromYDelta：指定动画开始时 View 的 Y 坐标。

(3) android:toXDelta：指定动画结束时 View 的 X 坐标。

(4) android:toYDelta：指定动画结束时 View 的 Y 坐标。

4. 旋转渐变动画

旋转渐变动画是指通过指定动画起始以及结束的旋转角度，以及动画的持续时间和旋转的轴心来创建的动画；在 XML 中旋转渐变动画对应< rotate/>标签。

```
<?xml version="1.0" encoding="utf-8">
<set xmlns:android=http://schema.android.com/apk/res/android
    android:interpolate="@android:anim/linear_interpolator">
<rotate android:fromDegree="0"
      android:toDegrees="180"
      android:pivotX="50% "
      android:pivotY="50% "
      android:repeatMode="reverse"
      android:repeatCount="infinite"
      android:duration="2000"/>
</set>
```

上述定义一个让 View 从 0°旋转到 180°持续时间 2s 的旋转动画。

旋转动画常用属性为：

(1) android:fromDegrees：指定动画开始时的角度。

(2) android:toDegrees：指定动画结束时的角度。

(3) android:pivotX:指定轴心X的坐标。

(4) android:pivotY:指定轴心Y的坐标。

(5) android:repeatMode:用于设置动画的重复方式,可以选reverse(反向)或restart(重新开始)

(6) android:repeatCount:用于设置动画的重复次数。

5. 组合动画

组合渐变,就是前面多种渐变的组合,对应<set/>标签

用XML定义的动画放在/res/anim/文件夹内,XML文件的根元素可以为<alpha>,<scale>,<translate>,<rotate>,interpolator元素或<set>。默认情况下,所有动画是同时进行的,可以通过startOffset属性设置各个动画的开始偏移来达到动画顺序播放的效果。可以通过设置interpolator属性改变动画渐变的方式,如AccelerateInterpolator,开始时慢,然后逐渐加快。默认为AccelerateDecelerateInterpolator。

定义动画XML资源后,就可以利用AnimationUtils工具类加载指定的动画资源,加载成功后会返回一个Animation对象,利用该对象就可以控制图片播放动画。

```
ImageView spaceshipImage= (ImageView) findViewById(R.id.spaceshipImage);
Animation hAnimation = AnimationUtils. loadAnimation (this, R. anim. hyperspace_
jump);
spaceshipImage.startAnimation(hAnimation);
```

8.3.4 补间动画案例

在Chapter08Application项目中新建TweenAnimationActivity,由该Activity展示补间动画的使用。首先新建一个动画资源,该动画资源如文件清单8-7所示。

文件清单8-7 anim_set.xml

```
<?xml version="1.0" encoding="utf-8"?>
<set xmlns:android="http://schemas.android.com/apk/res/android">
    <!--定义缩放动画 -->
    <scale android:fromXScale="1.0"
        android:fromYScale="1.0"
        android:toXScale="0.01"
        android:toYScale="0.01"
        android:pivotX="50% "
        android:pivotY="50% "
        android:duration="3000"/>
    <!--定义动画透明度 -->
    <alpha android:fromAlpha="1"
        android:toAlpha="0.05"
        android:duration="3000"/>
    <!--定义旋转动画 -->
    <rotate android:fromDegrees="0"
```

```
        android:toDegrees="1800"
        android:pivotX="50% "
        android:pivotY="50% "
        android:duration="3000"/>
</set>
```

上述动画资源指定动画匀速变化，同时进行缩放、透明度和旋转 3 种改变，动画持续时间为 3s。

在 Activity 加载上述动画资源，并使用 Animation 控制动画的播放。

```
public class TweenAnimationActivity extends AppCompatActivity {
    private ImageView imageView;
    private Animation animation;

    @Override
    protected void onCreate(Bundle savedInstanceState) {
        super.onCreate(savedInstanceState);
        setContentView(R.layout.activity_main);
        imageView=(ImageView)findViewById(R.id.imageView);
        animation=AnimationUtils.loadAnimation(this,R.anim.anim_set);
                                              //加载动画资源
        animation.setFillAfter(true);         //设置动画结束后保留结束状态
        imageView.startAnimation(animation);

    }
}
```

8.3.5 属性动画

补间动画(Tween Animation)改变的是 View 的绘制效果，但无法改变 View 的属性，比如将 Button 位置移动后，其位置属性并没有改变，再次点击 Button 时没有任何效果；又如无论如何缩放 Button 的大小，其大小的属性都不改变，缩放后有效的点击区域还是初始的 Button 大小的点击区域。属性动画(Property Animator)是为了解决补间动画存在的问题而引入的，属性动画不但可以实现补间动画中的 4 种动画效果，还可以定义任何属性的变化，例如当 Button 缩放时，Button 位置和大小的属性值都会发生改变。另外，属性动画不仅可以应用于 View，还可以应用于任何对象。

属性动画的实现机制是通过对目标对象进行赋值并修改其属性来实现的，创建属性动画常用方式有如下几种。

1. ValueAnimator 类

属性动画的运行机制是通过不断地对属性值进行操作来实现的，初始值和结束值之间的动画过渡就是由 ValueAnimator 类来负责计算的。它的内部使用一种时间循环的机制计算值与值之间的动画过渡，只需要将初始值和结束值提供给 ValueAnimator，并且告诉它动画所需运行的时长，那么 ValueAnimator 就会自动完成从初始值到结束值之间属性值的计算。对于指定对象属性值的更新，需要通过为 ValueAnimator 设置监听器实现。此外，ValueAnimator 还负责管理动画的播放次数、播放模式等。

使用 ValueAnimator 创建动画的步骤如下：

(1) 调用 ValueAnimator 的 ofInt()，ofFloat() 或 ofObject() 静态方法创建 ValueAnimator 对象。

(2) 调用 ValueAnimator 相应的 setXxx()方法设置动画的持续时间、插值方式、重复次数等。

(3) 为 ValueAnimator 注册监听器，监听 ValueAnimator 计算出的值的改变，并将这些值应用到指定对象。

(4) 调用 ValueAnimator 的 start()方法启动动画。

```
ValueAnimator anim =ValueAnimator.ofFloat(0f, 1f);
anim.setDuration(300);
anim.addUpdateListener(new AnimatorUpdateListener() {
    @Override
    public void onAnimationUpdate(ValueAnimator animation) {
        Log.i("update", ((Float) animation.getAnimatedValue()).toString());
        imageView.setAlpha((float)animator.getAnimatedValue());
    }
});
anim.setInterpolator(new CycleInterpolator(3));
anim.start();
```

上述代码中，调用 ValueAnimator 的 ofFloat()方法就可以构建出一个 ValueAnimator 的实例对象。ofFloat()方法当中允许传入多个 float 类型的参数，这里传入 0 和 1 就表示将值从 0 平滑过渡到 1，然后调用 ValueAnimator 的 setDuration()方法设置动画运行的时长，接着为 ValueAnimation 值的变化添加监听器，最后调用 start()方法启动动画。

2. ObjectAnimator 类

ObjectAnimator 类继承自 ValueAnimator，用于对指定对象的一个属性执行动画。相比 ValueAnimator，ObjectAnimator 会在属性值计算完成时自动设置对象的相应属性，因此使用 ObjectAnimator 就不需要注册 AnimatorUpdateListener 监听器，这样用起来更加简单。实际应用中一般都会用 ObjectAnimator 来改变某一对象的某一属性。使用 ObjectAnimator 的 ofInt()、ofFloat()或 ofObject()静态方法创建 ObjectAnimator 时，需要知道具体的对象以及对象的属性名。第一个参数为对象名，第二个为属性名，后面的参数为可变参数。

例如，下面是把一个 TextView 的透明度在 3s 内从 0 变至 1 的代码段：

```
tv=(TextView)findViewById(R.id.textview1);
btn=(Button)findViewById(R.id.button1);
```

```
btn.setOnClickListener(new OnClickListener() {
    @Override
    public void onClick(View v) {
        ObjectAnimator oa=ObjectAnimator.ofFloat(tv, "alpha", 0f, 1f);
        oa.setDuration(3000);
        oa. .setInterpolator(new AccelerateDecelerateInterpolator());
        oa.start();
    }
});
```

需要注意的是，使用 ObjectAnimator 有一定的限制。要想使用 ObjectAnimator，应该满足以下条件：

(1) 对象对应的属性应该有一个 set 方法，set<PropertyName>。

如上面的例子中，tv 需要具有 setAlpha(float value)方法。

(2) 当调用 ObjectAnimator 的 ofInt()、ofFloat()或 ofObject()之类的工厂方法时，如果 values…参数只设置了一个值，那么会假定为目的值，属性值的变化范围为当前值到目的值。为了获得当前值，该对象要有相应属性的 get 方法 get<PropertyName>。如果有 get 方法，则应返回值类型应与相应的 set 方法的参数类型一致。

如果不满足上述条件，则不能用 ObjectAnimator，应用 ValueAnimator 代替。

3. AnimatorSet 类

AnimatorSet 是 Animator 的子类，用于组合多个 Animator 的执行，并制定多个 Animator 按照次序播放还是同时播放。这个类提供了一个 play()方法，如果向这个方法中传入一个 Animator 对象（ValueAnimator 或 ObjectAnimator）将会返回一个 AnimatorSet. Builder 的实例，AnimatorSet. Builder 中包括以下 4 种方法：

(1) after(Animator anim)：将现有动画插入到传入的动画之后执行。

(2) after(long delay)：将现有动画延迟指定毫秒后执行。

(3) before(Animator anim)：将现有动画插入到传入的动画之前执行。

(4) with(Animator anim)：将现有动画和传入的动画同时执行。

下列代码段实现了让 TextView 先从屏幕外移动进屏幕，然后旋转 360°，旋转的同时进行淡入/淡出操作的动画效果：

```
ObjectAnimator moveIn =ObjectAnimator.ofFloat(textview, "translationX", -500f, 0f);
ObjectAnimator rotate =ObjectAnimator.ofFloat(textview, "rotation", 0f, 360f);
ObjectAnimator fadeInOut =ObjectAnimator.ofFloat(textview, "alpha", 1f, 0f, 1f);
AnimatorSet animSet =new AnimatorSet();
animSet.play(rotate).with(fadeInOut).after(moveIn);
animSet.setDuration(5000);
animSet.start();
```

8.3.6 属性动画案例

以下通过示例程序展示属性动画的使用。在 Chapter08Application 项目中新建 PropertyAnimationActivity，其界面展示如图 8-8 所示。

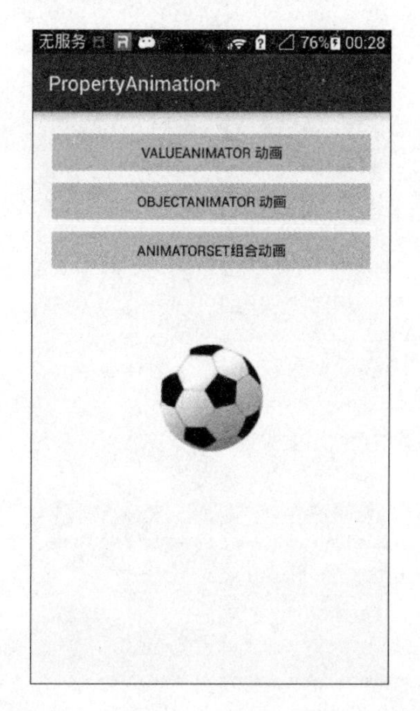

图 8-8 PropertyAnimationActivity 展示

PropertyAnimationActivity 对应的布局文件如文件清单 8-8 所示。

文件清单 8-8 activity_property_animation.xml

```
<?xml version="1.0" encoding="utf-8"?>
<RelativeLayout xmlns:android="http://schemas.android.com/apk/res/android"
    xmlns:tools="http://schemas.android.com/tools"
    android:layout_width="match_parent"
    android:layout_height="match_parent"
    android:paddingBottom="@dimen/activity_vertical_margin"
    android:paddingLeft="@dimen/activity_horizontal_margin"
    android:paddingRight="@dimen/activity_horizontal_margin"
    android:paddingTop="@dimen/activity_vertical_margin"
    tools:context="com.nsu.zyl.chapter08application.
                                    PropertyAnimationActivity">

    <Button
        android:layout_width="match_parent"
        android:layout_height="wrap_content"
        android:text="ValueAnimator 动画"
        android:id="@+id/btnVA"
        android:layout_alignParentTop="true"
```

```
        android:layout_centerHorizontal="true" />

    <Button
        android:layout_width="fill_parent"
        android:layout_height="wrap_content"
        android:text="ObjectAnimator 动画"
        android:id="@+id/btnOA"
        android:layout_below="@+id/btnVA"
        android:layout_alignRight="@+id/btnVA"
        android:layout_alignEnd="@+id/btnVA" />

    <Button
        android:layout_width="match_parent"
        android:layout_height="wrap_content"
        android:text="AnimatorSet 组合动画"
        android:id="@+id/btnAS"
        android:layout_below="@+id/btnOA"
        android:layout_centerHorizontal="true" />

    <ImageView
        android:layout_width="wrap_content"
        android:layout_height="wrap_content"
        android:id="@+id/imageView"
        android:layout_centerVertical="true"
        android:layout_centerHorizontal="true"
        android:src="@drawable/soccer_icon" />
</RelativeLayout>
```

在 Activity 上监听 3 个按钮，分别实现 3 种属性动画效果。

```
public class PropertyAnimationActivity extends AppCompatActivity {
    private Button btnVA,btnAS,btnOA;
    private ImageView imageView;

    @Override
    protected void onCreate(Bundle savedInstanceState) {
        super.onCreate(savedInstanceState);
        setContentView(R.layout. activity_property_animation);
        btnVA=(Button)findViewById(R.id.btnVA);
        btnOA=(Button)findViewById(R.id.btnOA);
        btnAS=(Button)findViewById(R.id.btnAS);
        imageView=(ImageView)findViewById(R.id.imageView);

        btnVA.setOnClickListener(new View.OnClickListener() {
            @Override
            public void onClick(View view) {
                ValueAnimator valueAnimator=ValueAnimator.ofFloat(1,0,1);
                valueAnimator.setDuration(5000);
```

```
                valueAnimator.addUpdateListener(new
ValueAnimator.AnimatorUpdateListener() {
                    @Override
                    public void onAnimationUpdate(ValueAnimator valueAnimator) {

imageView.setAlpha((float)valueAnimator.getAnimatedValue());
                    }
                });
                valueAnimator.start();

            }
        });
        btnOA.setOnClickListener(new View.OnClickListener() {
            @Override
            public void onClick(View view) {
                ObjectAnimator
objectAnimator=ObjectAnimator.ofFloat(imageView,"rotation",0f,360f,0f);
                objectAnimator.setDuration(5000);
                objectAnimator.setInterpolator(new
AccelerateDecelerateInterpolator());
                objectAnimator.start();
            }
        });
        btnAS.setOnClickListener(new View.OnClickListener(){
            @Override
            public void onClick(View view) {
                Rect outRect =new Rect();

MainActivity. this. getWindow(). findViewById (Window. ID _ ANDROID _ CONTENT ).
getDrawingRect(outRect);
                ObjectAnimator  animator1  =  new  ObjectAnimator().  ofFloat
(imageView,"y",(outRect.height()-imageView.getHeight())/2,outRect.height()-
imageView.getHeight());
                animator1.setInterpolator(new BounceInterpolator());

                ObjectAnimator  animator2 = ObjectAnimator. ofFloat ( imageView,
"rotation",0f,360f,0f);
                animator2.setInterpolator(new AccelerateDecelerateInterpolator());

                AnimatorSet animatorSet=new AnimatorSet();
                animatorSet.play(animator1).with(animator2);
                animatorSet.setDuration(5000);
                animatorSet.start();

            }
        });
    }
}
```

本章小结

本章主要讲解 Android 多媒体、图像处理和动画相关的知识点。首先介绍 Android 音频播放和视频播放的处理方式；然后介绍图像处理中常用的 Canvas、Paint、Bitmap 和 BitmapFactory 几个工具类；最后介绍 Android 中常用的逐帧动画、补间动画和属性动画的使用。

习　题

1. MediaPlayer 接收的常见声音类型有________、________、________和________4 种。

2. Android 提供多种处理图形图像的工具类，常用的工具类包括________、________、________和________。

3. Android 三种常用的动画形式是________，________和________。

4. 补间动画作用于 View 对象，主要包括对 View 对象________、________、________和透明度的变化。

5. 补间动画和属性动画的区别是什么？

6. 编写程序，实现小球的自由落体动画效果。

第 9 章 chapter 9

Android 综合案例

主要内容：Android 综合案例的实现及服务器交互

建议课时：4 课时

知识目标：(1) 掌握 Android 应用的开发实现过程；

(2) 了解 Android 快速开发框架的使用；

(3) 了解 Web 后台服务器的配置与交互。

能力目标：(1) 具备开发 Android 应用程序的能力；

(2) 初步具备开发 Android 应用框架的能力。

本书前面章节对 Android 平台移动应用开发的知识点进行了具体介绍，为了让读者体会如何将零散的知识点应用到具体的产品研发中，我们开发了“吃都”App。本章将重点讲解“吃都”App 的实现过程，希望读者能在理解本案例的基础上，去修改、完善该 App。

“吃都”App 包括 4 大基础模块的内容：首页、附近、预约、我的。思维导图如图 9-1 所示，具体功能以最终实现为准。

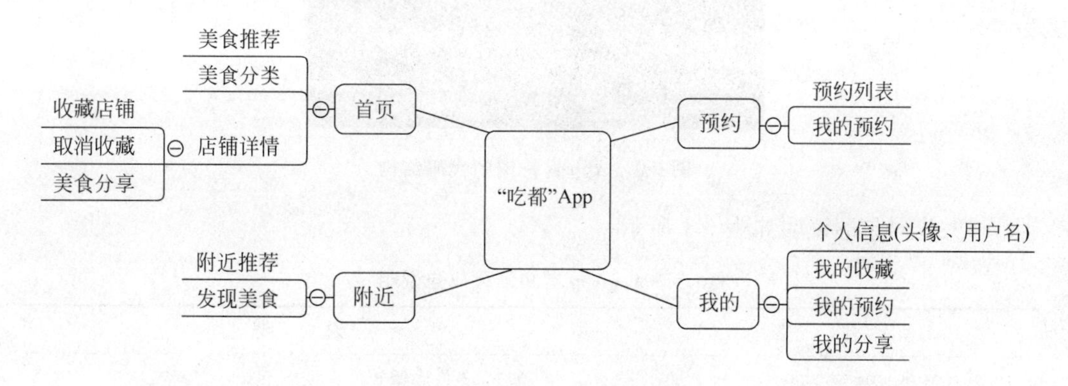

图 9-1 “吃都”App 思维导图

“首页”模块用于展示一些美食及店铺信息，细分为热门、火锅、西餐、甜品、饮品等类型。

“附近”模块用于展示附近的店铺。由于没有做 GPS 定位处理，系统采用随机显示。

感兴趣的读者可将地图定位功能添加进来。

“预约”模块用于展示可供用户进行预约操作的店铺,并能查看“我的预约”信息。

“我的”模块用于展示“头像”“用户名”“我的收藏”“我的预约”“我的分享”等信息。“我的收藏”用于展示登录用户所收藏的店铺信息,“我的预约”用于展示登录用户所预约的店铺信息及预约状态,“我的分享”用于展示登录用户所分享的店铺信息。

9.1 Android客户端开发

9.1.1 客户端程序整体说明

“吃都美食”App客户端的代码结构如图9-2所示。

图9-2 App客户端的代码结构

相关代码包的说明如表9-1所示。

表9-1 App客户端代码包说明

包名	说明
cn. sharesdk. onekeyshare	社会化分享依赖包
food. neusoft. com. food	总包
food. neusoft. com. food. activity	存放所有的Activity类
food. neusoft. com. food. adapter	存放适配器类
food. neusoft. com. food. domain	存放相关的实体类,如商品信息、订单信息等

续表

包　　名	说　　明
food. neusoft. com. food. Fragment. main	存放 Fragment 类,Fragment 主要用于页面的导航
food. neusoft. com. food. thread	线程,网络相关的工具类
food. neusoft. com. food. utils	通用工具包
food. neusoft. com. food. view	自定义控件,自定义 ViewPager 顶部轮播图
food. neusoft. com. food. widget	自定义组件
food. neusoft. com. food. wxapi	微信 API,项目集成微信时采用

9.1.2 Android框架使用

在“吃都美食”App的开发中使用了一些Android快速开发框架来提高开发效率,如AsyncHttpClient框架、xUtils框架、Universal-Image-Loader框架和ShareSDK等。要在程序中使用这些框架,需要在build. gradle中添加相关的依赖:

```
dependencies {
    ...
    compile 'com.jiechic.library:xUtils:2.6.14'
    compile 'com.nostra13.universalimageloader:universal-image-loader:1.9.5'
    compile 'com.loopj.android:android-async-http:1.4.9'
    compile files('libs/mta-sdk-1.6.2.jar')
    compile files('libs/open_sdk.jar')
    compile files('libs/ShareSDK-Core-2.7.10.jar')
}
```

为了让读者能在阅读案例的过程中更好地理解代码,先对这些框架作简要说明。

1. AsyncHttpClient框架

Android中网络请求一般使用ApacheHttp Client或者采用HttpURLConnect,但是直接使用这两个类库需要写大量的代码才能完成网络post和get请求,而使用android-async-http库可以大大简化操作。AsyncHttpClient基于Apache HttpClient,所有的请求都独立在UI主线程之外,通过回调方法处理请求结果,采用Handler机制传递信息。

使用方法:

(1) 在官网(http://loopj. com/android-async-http/)下载最新AsyncHttpClient的Jar包,将Jar包添加进Android应用程序的libs文件夹;

(2) 在代码中通过import引入相关类;

(3) 创建异步请求,一般使用静态的HttpClient对象,调用相应的方法,通常涉及AsyncHttpClient,RequestParams,AsyncHttpResponseHandler 3个类的使用。

AsyncHttpClient类通常用于在Android应用程序中创建异步请求,如GET、

POST、PUT 和 DELETE 等，请求参数通过 RequestParams 实例创建，响应通过重写匿名内部类 ResponseHandlerInterface 的方法处理。使用 AsyncHttpClient 执行网络请求时，最终都会调用 sendRequest( ) 方法，在这个方法内部将请求参数封装成 AsyncHttpRequest 交由内部的线程池执行。

RequestParams 类用于创建 AsyncHttpClient 实例中请求参数的集合，参数可以是 String、File 和 InputStream 等。

AsyncHttpResponseHandler 继承自 ResponseHandlerInterface 类，主要用于拦截和处理由 AsyncHttpClient 创建的请求。在匿名类 AsyncHttpResponseHandler 中重写 onSuccess(int, org. apache. http. Header[], byte[])方法来处理响应成功的请求。此外，也可以重写 onFailure(int, org. apache. http. Header[], byte[], Throwable)，onStart()，onFinish()，onRetry()和 onProgress(int, int)等方法。

本案例中有大量的网络操作，为了更好地使用网络请求功能，我们对 AsyncHttpClient 框架进行了封装，提供了 HttpUtils. java 工具类。代码如下：

```
package food.neusoft.com.food.thread;

import android.content.Context;
import android.widget.Toast;
import com.loopj.android.http.AsyncHttpClient;
import com.loopj.android.http.AsyncHttpResponseHandler;
import com.loopj.android.http.RequestParams;
import cz.msebera.android.httpclient.entity.StringEntity;
import food.neusoft.com.food.R;

public class HttpUtils {

    private static AsyncHttpClient client =new AsyncHttpClient();

    public static void get(Context context, String actionUrl, RequestParams params,
                           AsyncHttpResponseHandler responseHandler) {
        client.get(context, actionUrl, params, responseHandler);
    }

    public static void post(Context context, String actionUrl, RequestParams params,
                            AsyncHttpResponseHandler responseHandler) {
        client.post(context, actionUrl, params, responseHandler);
    }

    public static void postJson(Context context, String actionUrl, String json,
                           AsyncHttpResponseHandler responseHandler){
        try {
            client.post(context, actionUrl, new StringEntity(json, "UTF-8"),
                           "application/json", responseHandler);
        } catch (Exception e) {
            Toast.makeText(context, R.string.toast_network_error4,
```

```
                        Toast.LENGTH_SHORT).show();
            e.printStackTrace();
        }
    }
}
```

2. xUtils 框架

xUtils是基于Afinal开发的目前功能比较完善的一个Android开源框架。xUtils一共有4大模块：ViewUtils模块、HttpUtils模块、BitmapUtils模块和DbUtils模块。

ViewUtils模块：以注解方式进行UI、资源和事件的绑定。

HttpUtils模块：支持同步、异步方式的请求，支持GET、POST等请求。

BitmapUtils模块：支持加载网络图片和本地图片。

DbUtils模块：更直观地查询语义，一行代码就可以进行增删改查。

本案例中主要使用ViewUtils模块。在Activity和Fragment中，以注解的方式完成UI、资源和事件的绑定。ViewUtils的使用机制如下：

(1) 在Application的onCreate()方法中加入下面代码：

```
x.Ext.init(this);
x.Ext.setDebug(BuildConfig.DEBUG);
```

(2) 在Activity的onCreate()方法中加入下面代码：

```
x. view().inject(this);
```

(3) 加载当前的Activity布局需要如下注解：

```
@ContentView
```

(4) 给View进行初始化需要如下注解：

```
@InjectView
```

(5) 处理控件的各种响应事件需要如下注解：

```
@Envent
```

案例中大量采用了xUtils框架来进行UI资源的绑定，以LoginActivity为例，具体代码如下：

```
public class LoginActivity extends AppCompatActivity {

    ...
    //xUtils的view注解要求必须提供ID,以免代码混淆不受影响
```

```
    @ViewInject(R.id.et_number)
    private EditText et_number;
    @ViewInject(R.id.et_password)
    private EditText et_password;
    @ViewInject(R.id.bt_login)
    private Button bt_login;
    @ViewInject(R.id.rl_register)
    private RelativeLayout rl_register;
    @ViewInject(R.id.iv_weixin)
    private ImageView iv_weixin;
    @ViewInject(R.id.iv_qq)
    private ImageView iv_qq;
    @ViewInject(R.id.iv_weibo)
    private ImageView iv_weibo;

    @Override
    protected void onCreate(Bundle savedInstanceState) {
        super.onCreate(savedInstanceState);

        setContentView(R.layout.activity_login);
        context =this;
        ViewUtils.inject(this);       //注入 view
        doHandler();
        Init();
    }
    ……
}
```

3. Universal-Image-Loader 框架

Universal-Image-Loader 开源框架主要用于解决异步加载图片，或者加载大量图片的问题，可以实现多线程下载图片，并对线程、图片缓存、下载过程进行管理。

showImageForEmptyUri：设置图片 Uri 为空或是错误的时候显示的图片。

showImageOnFail：设置图片加载或解码过程中发生错误显示的图片。

```
public static void getImage(Context context, int drawble) {
        DisplayImageOptions options =new DisplayImageOptions.Builder()
                .showImageForEmptyUri(drawble)
                .showImageOnFail(drawble)
                .cacheInMemory(true)            // 设置下载的图片是否缓存在内存中
                .cacheOnDisk(true)              // 设置下载的图片是否缓存在 SD 卡中
                .showImageOnLoading(drawble)  // 创建配置过的 DisplayImageOption 对象
                .build();

        ImageLoaderConfiguration config =new ImageLoaderConfiguration.Builder(context)
                .defaultDisplayImageOptions(options)
                .threadPriority(Thread.NORM_PRIORITY -2)
```

```
                .denyCacheImageMultipleSizesInMemory()
                .imageDownloader(new BaseImageDownloader(context))
                .tasksProcessingOrder(QueueProcessingType.LIFO)
                .build();
        ImageLoader.getInstance().init(config);
    }
```

4. ShareSDK

ShareSDK是一种社会化分享组件,为iOS、Android、WP8的App提供社会化功能,集成了一些常用的类库和接口,可以缩短开发者的开发时间,还有社会化统计分析管理后台。支持包括QQ、微信、新浪微博、腾讯微博、开心网、人人网、豆瓣、网易微博、搜狐微博、Facebook、Twitter、Google+等国内外40多家主流社交平台。

本案例中,需要实现店铺分享功能,使用ShareSDK,在各大社交媒体上将本案例App注册,实现社会化分享功能。案例中提供了一个工具方法showShare()来完成部分社交媒体的分享,感兴趣的读者可自行添加其他分享。

```
    public static void showShare(Context context,String title,String text,
    String imageUrl,String APPurl) {
            ShareSDK.initSDK(context);
            OnekeyShare oks =new OnekeyShare();
            //关闭 SSO 授权
            oks.disableSSOWhenAuthorize();

            // title 标题,印象笔记、邮箱、信息、微信、人人网和 QQ 空间等使用
            oks.setTitle(title);
            // titleUrl 是标题的网络链接,QQ 和 QQ空间等使用
            oks.setTitleUrl(APPurl);                          //把它设置为 App 的下载地址
            // text 是分享文本,所有平台都需要这个字段
            oks.setText(text);
            // imagePath 是图片的本地路径,Linked-In 以外的平台都支持此参数
            //oks.setImagePath("/sdcard/test.jpg");        //确保 SDcard 下面存在此张图片
            // URL 仅在微信(包括好友和朋友圈)中使用
            oks.setUrl("http:www.bidu.com");
            // comment 是我对这条分享的评论,仅在人人网和 QQ 空间使用
            // oks.setComment("");
            oks.setImageUrl(imageUrl);
            // Site 是分享此内容的网站名称,仅在 QQ 空间使用
            oks.setSite(context.getString(R.string.app_name));
            // SiteUrl 是分享此内容的网站地址,仅在 QQ空间使用
            oks.setSiteUrl(Appurl);
            // 启动分享 GUI
            oks.show(context);
    }
```

9.1.3 核心功能实现

1. 注册与登录

启动应用程序，首先进入登录页面，如图 9-3 所示。

初次使用，可先注册账号。点击“立即注册”，进入注册页面，如图 9-4 所示。

```
rl_register.setOnClickListener(new View.OnClickListener() {
            @Override
            public void onClick(View view) {
                Toast.makeText(context, "注册", Toast.LENGTH_SHORT).show();
                   startActivity(new Intent(LoginActivity.this, RegisterActivity.
class));
            }
        });
```

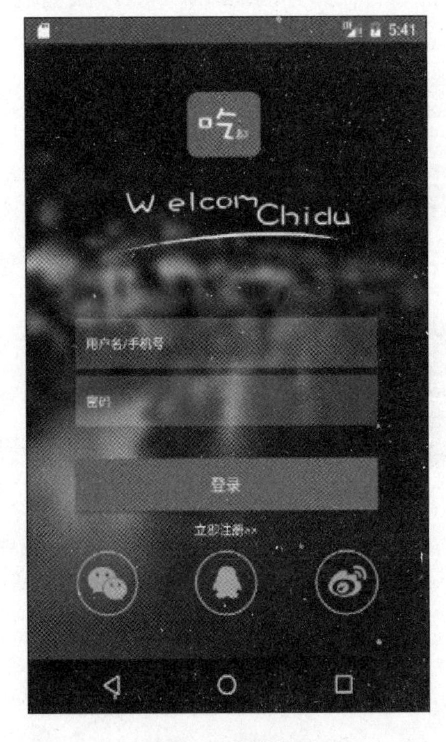

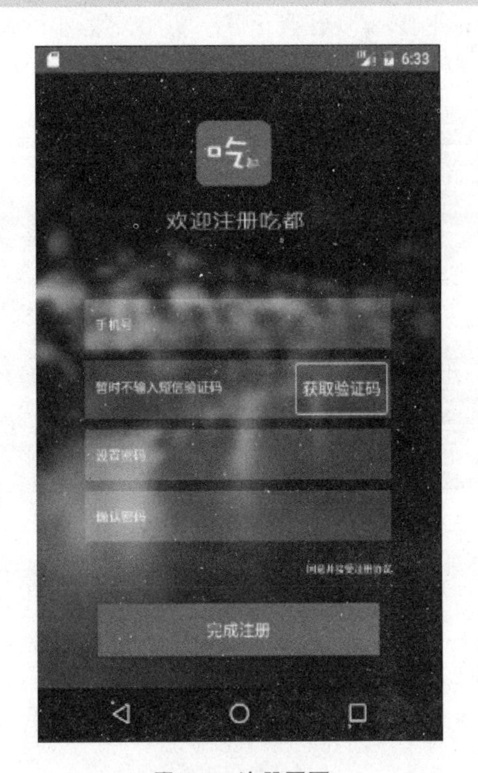

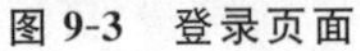
图 9-3　登录页面

图 9-4　注册页面

在注册页面，用户填写相关信息，并同意注册协议，提交注册信息给服务器。注册页面暂时未进行短信验证，只需输入账号名(手机号)、密码和确认密码即可。

```
注册 API：http://100.0.101.18:8080/CDFood/signIn
请求参数：userId,userPassword
返回值：SUCCESS 表示注册成功
```

客户端提交网络请求代码：

```
RequestParams params =new RequestParams();
params.put("userId", et_number.getText().toString());
params.put("userPassword", et_password1.getText().toString());
HttpUtils.post(context, Url.signIn, params, signIn_handler);
```

解析服务器传回的数据，如果为 SUCCESS，提示注册成功，并返回登录页面。数据解析过程使用 AsyncHttpClient 框架，在 onSuccess()方法中解析。

```
private void dohandler() {
    signIn_handler =new AsyncHttpResponseHandler() {

        @Override
        public void onSuccess (int statusCode, Header[] headers, byte[]
responseBody) {
            try {
                String result =new String(responseBody);
                if (result.equals("SUCCESS")) {
                    Toast.makeText(context, "注册成功！",
                            Toast.LENGTH_SHORT).show();
                    startActivity(new Intent(RegisterActivity.this,
                                LoginActivity.class));
                    finish();
                }else{
                    Toast.makeText(context, "用户名已存在",
                            Toast.LENGTH_SHORT).show();
                }
            } catch (Exception e) {
                e.printStackTrace();
            }
        }

        @Override
        public void onFailure (int statusCode, Header[] headers, byte[]
responseBody,
                    Throwable error) {
            Toast.makeText(context, "网络连接错误，请检查网络设置后重试。",
                    Toast.LENGTH_SHORT).show();
        }
    };
}
```

在登录页面，用户输入账号名(手机号)及密码，点击“登录”按钮进行登录。登录过程需要对用户进行验证。把账号名和密码提交给服务器，由服务器进行验证处理。

```
登录 API：http://100.0.101.18:8080/CDFood/login
请求参数：userId，userPassword
```

```
返回值：{  "userAdress":"xxxx",
           "userIconPath":"xxxx",
           "userId":"xxxx",
           "userName":"xxxx",
           "userNo":xxxx,
           "userPassword":"xxxx"}
         "ERROR"表示登录失败
```

服务器验证通过后会以 JSON 格式返回数据。如果验证失败，则传回 ERROR；如果验证通过，则传回该登录用户的相关信息。Android 客户端收到服务器返回数据后，对数据进行解析，如果验证通过，则从 JSON 对象中取出 userId 保存，并进入主页面。

```
private void doHandler() {
      login_handler =new AsyncHttpResponseHandler() {
          @Override
          public void onSuccess (int statusCode, Header[] headers, byte[]
responseBody) {
              String result =new String(responseBody);
              System.out.println(result);
              if (result.equals("ERROR")) {
                  Toast.makeText(context, "登录失败",
                          Toast.LENGTH_SHORT).show();
              } else {
                  try {
                      JSONObject jsonObject=new JSONObject(result);
                      String userId =jsonObject.getString("userId");
                      User user =new User(context);
                      user.saveUserNumber(userId);
                      NApplication.user_number=userId;
                      Toast.makeText(context, "登录成功！",
                              Toast.LENGTH_SHORT).show();
                      startActivity(new Intent(context, MainActivity.class));
                      finish();
                  } catch (JSONException e) {
                      e.printStackTrace();
                  }
              }
          }

          @Override
          public void onFailure(int statusCode, Header[] headers, byte[] responseBody,
                              Throwable error) {
              Toast.makeText(context, "网络连接错误,请检查网络设置后重试。",
                          Toast.LENGTH_SHORT).show();
          }
      };
  }
```

本案例中也允许第三方登录，如QQ、微信、微博等。在登录页面上提供了第三方登录的入口。以QQ登录为例，通过QQAuth进行QQ登录授权认证，如果onComplete()方法被回调，则表示授权成功，引导用户进入系统。相关代码如下：

```
iv_qq.setOnClickListener(new View.OnClickListener() {
    @Override
    public void onClick(View view) {
        Toast.makeText(context, "qq", Toast.LENGTH_SHORT).show();
        onClickQQLogin();
    }
});

private void onClickQQLogin() {
    if (!mQQAuth.isSessionValid()) {
        IUiListener listener =new BaseUiListener() {
            @Override
            protected void doComplete(JSONObject values) {
                updateUserInfo();
            }
        };
        mQQAuth.login(this, "all", listener);
        mTencent.login(this, "all", listener);
    }
}

private void updateUserInfo() {
if (mQQAuth !=null && mQQAuth.isSessionValid()) {
    IUiListener listener =new IUiListener() {

        @Override
        public void onError(UiError e) {
        }

        @Override
        public void onComplete(final Object response) {
            new Thread() {

                @Override
                public void run() {
                    JSONObject json = (JSONObject) response;
                    try {
                        ImageUrl =json.getString("figureurl_qq_2");
                        nickname =json.getString("nickname");
                    } catch (JSONException e) {
                        e.printStackTrace();
                    }
                }
            }.start();
            startActivity(new Intent(LoginActivity.this, MainActivity.class));
```

```
                finish();
            }

            @Override
            public void onCancel() {
            }
        };
        mInfo = new UserInfo(this, mQQAuth.getQQToken());
        mInfo.getUserInfo(listener);
    }
}
```

2. 主页面设计

用户登录成功后,将进入应用程序主页面。主页面采用 Fragment 和 ViewPager 实现“首页”“附近”“预约”“我的”这几个功能页面的切换。

主页面设计如图 9-5 所示,屏幕底部用 4 个 RadioButton 控制“首页”“附近”“预约”“我的”这 4 个页面的切换,而这 4 个页面使用 ViewPager 装载。

图 9-5 主页面设计

主页面布局文件如文件清单 9-1 所示。

文件清单 9-1 activity_main. xml

```
<?xml version="1.0" encoding="utf-8"?>
<RelativeLayout xmlns:android="http://schemas.android.com/apk/res/android"
```

```
xmlns:tools="http://schemas.android.com/tools"
android:id="@+id/activity_main"
android:layout_width="match_parent"
android:layout_height="match_parent"
tools:context="food.neusoft.com.food.MainActivity">

<RadioGroup
    android:id="@+id/radioGroup"
    android:layout_width="match_parent"
    android:layout_height="@dimen/bar_height"
    android:layout_alignParentBottom="true"
    android:layout_alignParentLeft="true"
    android:orientation="horizontal"
    android:paddingBottom="7dp"
    android:paddingTop="6dp" >

<RadioButton
    android:id="@+id/radio0"
    android:layout_width="0dp"
    android:layout_height="wrap_content"
    android:layout_weight="1"
    android:button="@null"
    android:drawableTop="@drawable/main_home"
    android:gravity="center"
    android:text="@string/main_home"
    android:textColor="@drawable/main_textcolor"
    android:textSize="@dimen/bar_text" />

<RadioButton
    android:id="@+id/radio1"
    android:layout_width="0dp"
    android:layout_height="wrap_content"
    android:layout_weight="1"
    android:button="@null"
    android:drawableTop="@drawable/main_nearby"
    android:gravity="center"
    android:text="@string/main_attachment"
    android:textColor="@drawable/main_textcolor"
    android:textSize="@dimen/bar_text" />

<RadioButton
    android:id="@+id/radio2"
    android:layout_width="0dp"
    android:layout_height="wrap_content"
    android:layout_weight="1"
    android:button="@null"
    android:drawableTop="@drawable/main_order"
    android:gravity="center"
```

```
        android:text="@string/main_order"
        android:textColor="@drawable/main_textcolor"
        android:textSize="@dimen/bar_text" />

    <RadioButton
        android:id="@+id/radio3"
        android:layout_width="0dp"
        android:layout_height="wrap_content"
        android:layout_weight="1"
        android:button="@null"
        android:drawableTop="@drawable/main_mine"
        android:gravity="center"
        android:text="@string/main_mine"
        android:textColor="@drawable/main_textcolor"
        android:textSize="@dimen/bar_text" />
    </RadioGroup>

    <food.neusoft.com.food.widget.NoScrollViewPager
        android:id="@+id/viewPager"
        android:layout_width="match_parent"
        android:layout_height="wrap_content"
        android:layout_above="@+id/radioGroup"
        android:layout_alignParentLeft="true" />
</RelativeLayout>
```

布局中使用了自定义的 ViewPager 控件 NoScrollViewPager，继承 ViewPager，覆盖 ViewPager 的 onInterceptTouchEvent(MotionEvent ev)方法和 onTouchEvent(MotionEvent ev)方法，将这两个方法的返回值改为 false，那么 ViewPager 就不会响应滑动的事件。

自定义 NoScrollViewPager 代码如文件清单 9-2 所示。

文件清单 9-2 NoScrollViewPager

```
public class NoScrollViewPager extends ViewPager{
    public NoScrollViewPager(Context context) {
        super(context);
    }

    public NoScrollViewPager(Context context, AttributeSet attrs) {
        super(context, attrs);
    }

    @Override
    public boolean onInterceptTouchEvent(MotionEvent ev) {
        return false;
    }

    @Override
    public boolean onTouchEvent(MotionEvent ev) {
        return false;
```

```
    }
}
```

在主页面 MainActivity 中加载 Fragment：

```
private void setupFragment(){
    FragmentManager fm=getSupportFragmentManager();
    List<BaseFragment>fs=new ArrayList<>();
    fs.add(new HomeFragment());
    fs.add(new NearbyFragment());
    fs.add(new OrderFragment());
    fs.add(new MineFragment());

    MyFragmentAdapter adapter=new MyFragmentAdapter(fm,fs);
    viewPager.setAdapter(adapter);

    //初始选中第1页→首页
    viewPager.setCurrentItem(0,false);
    radio0.setChecked(true);

    viewPager.setOnPageChangeListener(new ViewPager.OnPageChangeListener() {
        @Override
        public void onPageSelected(int position) {
            switch (position){
                  case 0:
                    radio0.setChecked(true);
                    break;
                case 1:
                    radio1.setChecked(true);
                    break;
                case 2:
                    radio2.setChecked(true);
                    break;
                case 3:
                    radio3.setChecked(true);
                    break;
                default:
                    break;
            }
        }
        @Override
        public void onPageScrolled(int position, float positionOffset,
                                        int positionOffsetPixels) {
        }
        @Override
        public void onPageScrollStateChanged(int state) {
        }
    });
}
```

在主页面上，设置连按两次 back 键，退出程序。

```
@Override
public void onBackPressed() {
    if (isExit) {
        super.onBackPressed();
        ((NApplication) this.getApplication()).destoryAllActivity();
        finish();
    } else {
        isExit =true;
        new Timer().schedule(new TimerTask() {
            @Override
            public void run() {
                isExit =false;
            }
        }, 2000);
        Toast.makeText(getApplicationContext(), "再点击一次退出程序",
                Toast.LENGTH_SHORT).show();
    }
}
```

3. “首页”功能

“首页”主要用于向用户展示一些美食及店铺信息，分为热门(图 9-6)、火锅(图 9-7)、西餐、甜品、饮品几种类型。页面依然采用 Fragment 和 ViewPager 的方式实现。

图 9-6 “热门”页面

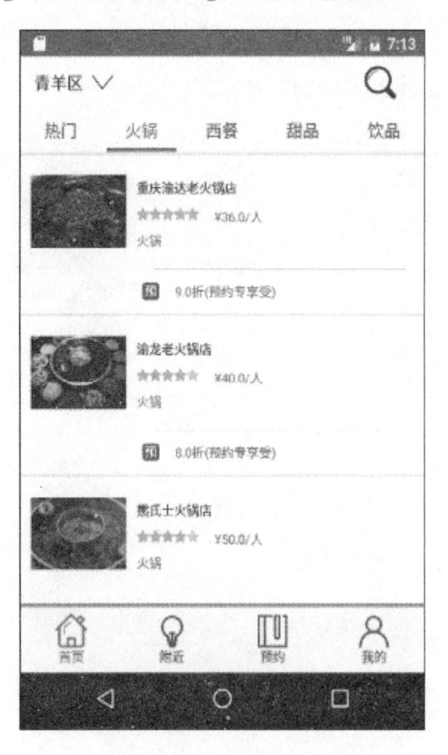

图 9-7 “火锅”页面

“热门”页面顶部图片轮播,如文件清单 9-3 所示。

文件清单 9-3 实现顶部图片轮播的代码

```
class MyPagerAdapter extends PagerAdapter {

    @Override
    public int getCount() {
        return imageId.size();
    }

    @Override
    public boolean isViewFromObject(View view, Object object) {
        return view ==object;
    }

    @Override
    public Object instantiateItem(ViewGroup container, int position) {
        ImageView image =new ImageView(getContext());
        image.setScaleType(ImageView.ScaleType.FIT_XY);
        image.setImageResource(imageId.get(position));
        container.addView(image);
        image.setOnTouchListener(new TopNewsTouchListener());
        return image;
    }

    @Override
    public void destroyItem(ViewGroup container, int position, Object object) {
        container.removeView((View) object);
    }
}
```

使用 Handler 控制图片的轮播,如文件清单 9-4 所示。

文件清单 9-4 使用 Handler 控制图片轮播的代码

```
private void setupFragment() {

    myimpager.setAdapter(new MyPagerAdapter());
    myimpager.addOnPageChangeListener(new ViewPager.OnPageChangeListener() {
        @Override
        public void onPageScrolled(int position, float positionOffset,
                        int positionOffsetPixels) {

        }

        @Override
        public void onPageSelected(int position) {
            ChoosePhoto(position);
        }

        @Override
        public void onPageScrollStateChanged(int state) {
```

```
            }
        });

        //设置图片轮播
        if (handler ==null) {
            handler =new Handler() {
                @Override
                public void handleMessage(Message msg) {
                    super.handleMessage(msg);
                    int current =myimpager.getCurrentItem();
                    if (current <imageId.size() -1) {
                        current++;
                    } else {
                        current =0;
                    }
                    myimpager.setCurrentItem(current);
                    handler.sendEmptyMessageDelayed(0, 2000);
                }
            };
            handler.sendEmptyMessageDelayed(0, 2000);
        }
    }
```

“火锅”“西餐”“甜品”“饮品”4 个页面均以列表形式显示火锅、西餐(图 9-8)、甜品(图 9-9)、饮品等店铺信息。“饮品”页面与“甜品”页面类似,此处不再列出。

图 9-8 “西餐”页面

图 9-9 “甜品”页面

由于实际中列表数据会很多，这些页面在实现过程中均使用了自定义的列表类控件，如 PullToRefreshLayout、PullableListView 等。

PullToRefreshLayout 自定义的布局，用来管理 3 个子控件：一是下拉头；二是包含内容的 pullableView（可以是实现 Pullable 接口的任何 View）；三是上拉头。

PullableListView 是实现了 Pullable 接口的 ListView，实现下拉刷新，上拉加载。

Pullable 是定义的一个接口，用于判断能否进行上拉和下拉操作。

Pullable.java 源代码如文件清单 9-5 所示。

文件清单 9-5 Pullable.java

```
public interface Pullable{
    //判断是否可以下拉,如果不需要下拉功能可以直接 return false
    //如果可以下拉,返回 true,否则返回 false
    boolean canPullDown();

    //判断是否可以上拉,如果不需要上拉功能可以直接 return false
    //如果可以上拉,返回 true,否则返回 false
    boolean canPullUp();
}
```

"首页"页面的 5 个 Fragment 页面均采用这种自定义的列表实现下拉刷新和上拉加载的功能，本书后面在"附近""预约"页面实现中会详细讲解实现细节，读者也可查阅源码了解更多内容。

4. 店铺详情页面

用户在首页浏览中点击"店铺"或"商家"，进入"店铺"页面，如图 9-10 所示。"店铺"页面用于展示当前店铺的美食信息，并提供店铺收藏和分享的功能。

图 9-10 "店铺"页面

“店铺”页面以列表形式展示“每日推出”和“本店热卖”美食信息。通过店铺编号从服务器可获取该店铺的美食信息。

```
店铺美食 API：http://100.0.101.18:8080/CDFood/getFoods
请求参数：marketNo
返回值：[{"foodNo":xxxx,
        "foodName":"xxxx",
        "foodIntroduce":"xxxx",
        "foodDiscount":xxxx,
        "foodHot":true,
        "foodIconPath":"xxxx",
        "foodPrice":"xxxx"},{……},{……},{……}]
"ERROR"表示数据获取失败
```

解析服务器传回的数据，如果为“ERROR”，表示获取数据失败。正常情况下返回的数据为 JSON 数组字符串，从 JSON 数组中取出每个 JSON 对象，解析出相应的数据信息。解析过程如下：

```
try {
    JSONArray jsonarray=new JSONArray(result);
    for(int i=0;i<jsonarray.length();i++){
        JSONObject jsonobject=jsonarray.getJSONObject(i);
        String foodDiscount=jsonobject.getString("foodDiscount");
        String foodHot=jsonobject.getString("foodHot");
        String foodIconPath=Url.getImgURL(jsonobject.getString("foodIconPath"));
        String foodIntroduce=jsonobject.getString("foodIntroduce");
        String foodName=jsonobject.getString("foodName");
        long foodNo=jsonobject.getLong("foodNo");
        double foodPrice=jsonobject.getDouble("foodPrice");
        FoodInfo foodInfo=new FoodInfo(foodName,foodIntroduce,foodHot,
                        foodIconPath,foodDiscount,foodNo,foodPrice);
        if(foodHot=="null"||foodHot=="true"){
              //是热卖,则加入"本店热卖"列表
              hotinfos.add(foodInfo);
        }else{
              //不是热卖,则加入"本店推出"列表
              putinfos.add(foodInfo);
        }
    }
    putadapter.notifyDataSetChanged();
    hotadapter.notifyDataSetChanged();
} catch (JSONException e) {
    e.printStackTrace();
    Toast.makeText(StroeActivity.this,"解析数据失败",Toast.LENGTH_SHORT).show();
}
```

在“店铺”页面，还可以对店铺进行收藏操作。对“桃心”进行监听，选中则为收藏，没选中则说明用户没收藏。“桃心”状态切换时进行收藏或取消收藏处理。

```
iv_heart.setOnCheckedChangeListener(new CompoundButton.OnCheckedChangeListener() {
        @Override
        public void onCheckedChanged(CompoundButton compoundButton, boolean b) {
            if(b){//收藏
                Collect(NApplication.user_number,marketNo);
                MineCollects.put(marketNo,true);
            }else{//取消收藏
                RemoveCollect(NApplication.user_number,marketNo);
                MineCollects.put(marketNo,false);
            }
        }
    });

收藏店铺 API：http://100.0.101.18:8080/CDFood/ saveCollect
请求参数：userId,marketNo
返回值："SUCCESS"表示收藏成功
```

用户收藏店铺时，根据 userId 和 marketNo 提交网络请求，如果服务器返回 SUCCESS 则表示店铺收藏成功。程序代码如文件清单 9-6 所示。

文件清单 9-6　用户收藏店铺时的代码

```
private void Collect(String userId,long marketNo){
    RequestParams params=new RequestParams();
    params.put("userId",userId);
    params.put("marketNo",marketNo);
    HttpUtils.post(this,Url.saveCollect,params,collect_handler);
}

collect_handler=new AsyncHttpResponseHandler() {
        @Override
          public void onSuccess(int statusCode, Header[] headers,
                                                 byte[] responseBody) {
            String result=new String(responseBody);
            if(result.equals("SUCCESS")){
                Toast.makeText(StroeActivity.this,"收藏成功！",
                                                Toast.LENGTH_SHORT).show();
            }else{
                Toast.makeText(StroeActivity.this,"收藏失败",
                                                Toast.LENGTH_SHORT).show();
            }
          }

          @Override
          public void onFailure(int statusCode, Header[] headers,
                                      byte[] responseBody, Throwable error) {
               Toast.makeText(StroeActivity.this, R.string.toast_network_error1,
                                             Toast.LENGTH_SHORT).show();
          }
};
```

```
取消收藏店铺 API：http://100.0.101.18:8080/CDFood/ removeCollect
请求参数：userId,marketNo
返回值：SUCCESS 表示取消成功
```

用户取消收藏时，仍然根据 userId 和 marketNo 提交网络请求，如果服务器返回“SUCCESS”则表示店铺取消成功。

```
private void RemoveCollect(String userId,long marketNo){
    RequestParams params=new RequestParams();
    params.put("userId",userId);
    params.put("marketNo",marketNo);
    HttpUtils.post(this,Url.removeCollect,params,removecollect_handler);
}
```

数据解析部分和收藏店铺类似，此处不再赘述。完整的代码可参考案例源代码 StoreActivity.java 文件。

点击“分享”图标时，将对该店铺进行社会化分享，页面效果如图 9-11 所示。具体实现细节可参考“我的”模块中的实现。

5. “附近”功能

“附近”页面用于展示附近的店铺，如图 9-12 所示。案例中由于没有做 GPS 定位处理，系统采用随机显示。感兴趣的读者可将地图定位功能添加进来。

图 9-11　店铺分享

图 9-12　“附近”页面

在实际应用中，经常会遇到ListView列表项过多，加载缓慢的问题。因此在装载ListView时，一般采取动态加载的方式。即预先加载一定条目的数据，当用户滑动时进行数据更新或者加载更多条目数据。

自定义上下拉刷新的ListView源代码如文件清单9-7所示。

文件清单9-7　自定义上下拉刷新的ListView源代码

```
public class PullableListView extends ListView implements Pullable {

    public PullableListView(Context context) {
        super(context);
    }

    public PullableListView(Context context, AttributeSet attrs) {
        super(context, attrs);
    }

    public PullableListView(Context context, AttributeSet attrs, int defStyle) {
        super(context, attrs, defStyle);
    }

    @Override
    public boolean canPullDown(){
        if (getCount() ==0) {
            // 没有 item 时也可以下拉刷新
            return true;
        } else if (getFirstVisiblePosition() ==0&& getChildAt(0).getTop() >=0) {
            // 滑到 ListView 的顶部了
            return true;
        } else
            return false;
    }

    @Override
    public boolean canPullUp(){
        if (getCount() ==0) {
            // 没有 item 时也可以上拉加载
            return true;
        } else if (getLastVisiblePosition() == (getCount() -1)) {
            // 滑到底部了
            if (getChildAt(getLastVisiblePosition() -getFirstVisiblePosition()) !=null
                && getChildAt(getLastVisiblePosition()-getFirstVisiblePosition())
                        .getBottom() <=getMeasuredHeight())
                return true;
        }
        return false;
    }
}
```

“附件”页面的 Fragment 代码如文件清单 9-8 所示。

文件清单 9-8 “附件”页面的 Fragment 代码

```
public class NearbytFragment extends BaseFragment {

    private Context context;
    private AsyncHttpResponseHandler attach_handler;
    private View view;

    @ViewInject(R.id.ls_show)
    private ListView ls_show;
    @ViewInject(R.id.refresh_view)
    private PullToRefreshLayout refresh_view;

    private List<AttachInfo> attachInfos;
    public static String LOCAL;
    private boolean isLoadmore;

    private int count=10;                //请求量
    private int firstIndex;              //请求页数
    private AttachAdapter attachAdapter;

    @Override
    public void onCreate(Bundle savedInstanceState) {
        super.onCreate(savedInstanceState);
        context=getContext();
    }

    @Override
    public View onCreateView(LayoutInflater inflater, ViewGroup container,
                                                Bundle savedInstanceState) {
        view =inflater.inflate(R.layout.fragment_attach, container, false);
        ViewUtils.inject(this, view);
        dohandler();
        Init();
        return view;
    }

    private void Init() {
        attachInfos=new ArrayList<>();
        refresh_view.setOnRefreshListener(new MyListener());
        setupFragment();
    }

    private void setupFragment() {
        getFirst();
    }

    private void getFirst(){
        //没做定位功能,这里就随机选择区域了
        Random random=new Random();
```

```
        int index=Math.abs(random.nextInt()% 10);
        LOCAL=local.get(index);

        firstIndex=0;
        isLoadmore=false;
        RequestParams params=new RequestParams();
        params.put("count",count);
        params.put("firstIndex",firstIndex);
        params.put("marketAdress",LOCAL);
        HttpUtils.get(context,Url.getNearMarket,params,attach_handler);

        attachAdapter =new AttachAdapter(context,attachInfos);
        ls_show.setAdapter(attachAdapter);

        //设置点击事件
        ls_show.setOnItemClickListener(new AdapterView.OnItemClickListener() {
            @Override
            public void onItemClick(AdapterView<?>adapterView, View view,
                                    int i, long l) {
                Intent intent=new Intent(context, StroeActivity.class);
                AttachInfo attachInfo=attachInfos.get(i);
                intent.putExtra("marketNo",attachInfo.getMarketNo());
                intent.putExtra("type","附近");
                intent.putExtra("storename",attachInfo.getMarketName());
                intent.putExtra("introduce",attachInfo.getMarketIntroduce());
                intent.putExtra("imagepath",attachInfo.getMarketIconPath());
                startActivity(intent);
            }
        });
    }

    private void getNext(){
        isLoadmore=true;
        RequestParams params=new RequestParams();
        params.put("count",count);
        params.put("firstIndex",firstIndex);
        params.put("marketAdress",LOCAL);
        HttpUtils.get(context,Url.getNearMarket,params,attach_handler);
    }

    private void dohandler(){
        attach_handler=new AsyncHttpResponseHandler() {
            @Override
            public void onSuccess(int statusCode, Header[] headers,
                                  byte[] responseBody) {
                String result=new String(responseBody);
                if(result.equals("ERROR")){
                    Toast.makeText(context,"获取数据失败",
                               Toast.LENGTH_SHORT).show();
                    if(isLoadmore){
                        refresh_view.loadmoreFinish(PullToRefreshLayout.FAIL);
```

```
                }else{
                    refresh_view.refreshFinish(PullToRefreshLayout.FAIL);
                }
            }else{
                //解析附近店铺数据
            }
        }

        @Override
        public void onFailure(int statusCode, Header[] headers,
                        byte[] responseBody, Throwable error) {
            Toast.makeText(context, R.string.toast_network_error1,
                    Toast.LENGTH_SHORT).show();
            if(isLoadmore){
                refresh_view.loadmoreFinish(PullToRefreshLayout.FAIL);
            }else{
                refresh_view.refreshFinish(PullToRefreshLayout.FAIL);
            }
        }
    };
}
```

客户端根据 count,firstIndex 和 marketAdress 提交网络请求,从服务器获取附近店铺信息。

```
附近 API: http://100.0.101.18:8080/CDFood/getBookMarket
请求参数: count,firstIndex,marketAdress
返回值: [{  "marketNo":1,
           "marketName":"重庆渝达老火锅店",
           "marketIntroduce":"来这儿,体验四川的美味~ ",
           "marketAdress":"青羊区",
           "marketIconPath":"HG_pic_one.png",
           "marketBigPicture":"SY_pic.png",
           "marketPrice":36.0,
           "marketDiscount":9.0,
           "marketHotLevel":5,
           "marketDistance":1.0,
           "typeName":"火锅",
           "erectLineIconPath":null,
           "bookIconPath":"view_yu.png",
           "newIconPath":"view_new.png",
           "discountIconPath":"view_hui.png",
           "foodType":null},{……},{……}]
   "ERROR"表示数据获取失败
```

解析服务器传回的数据,如果为 ERROR,表示获取数据失败。正常情况下返回的数据为 JSON 数组字符串,从 JSON 数组中取出每个 JSON 对象,解析出相应的数据信息。解析过程的代码如文件清单 9-9 所示。

文件清单 9-9 解析过程的代码

```
try {
    JSONArray jsonArray=new JSONArray(result);
    if(! isLoadmore){
        attachInfos.clear();
    }
    for(int i=0;i<jsonArray.length();i++){
        JSONObject jsonObject=jsonArray.getJSONObject(i);
        String bookIconPath=Url.getImgURL(jsonObject
                .getString("bookIconPath"));
        String discountIconPath=Url.getImgURL(jsonObject
                .getString("discountIconPath"));
        String marketAdress=jsonObject.getString("marketAdress");
        String marketBigPicture=Url.getImgURL(jsonObject
                .getString("marketBigPicture"));
        double marketDiscount=jsonObject.getDouble("marketDiscount");
        double marketDistance=jsonObject.getDouble("marketDistance");
        double marketHotLevel=jsonObject.getDouble("marketHotLevel");
        String marketIconPath=Url.getImgURL(jsonObject
                .getString("marketIconPath"));
        String marketIntroduce=jsonObject.getString("marketIntroduce");
        String marketName=jsonObject.getString("marketName");
        long marketNo=jsonObject.getLong("marketNo");
        double marketPrice=jsonObject.getDouble("marketPrice");
        String newIconPath=Url.getImgURL(jsonObject
                .getString("newIconPath"));
        String typeName=jsonObject.getString("typeName");
        AttachInfo attachInfo=new AttachInfo(
            bookIconPath,newIconPath,discountIconPath,marketAdress,
            marketBigPicture,marketDiscount,marketDistance,
            marketHotLevel,marketIntroduce,marketIconPath,
            marketName,marketNo,marketPrice,typeName);
        attachInfos.add(attachInfo);
    }
    attachAdapter.notifyDataSetChanged();
    firstIndex+=count;
    if(isLoadmore){
        refresh_view.loadmoreFinish(PullToRefreshLayout.SUCCEED);
    }else{
        refresh_view.loadmoreFinish(PullToRefreshLayout.SUCCEED);
    }
} catch (JSONException e) {
    e.printStackTrace();
    if(isLoadmore){
        refresh_view.loadmoreFinish(PullToRefreshLayout.FAIL);
    }else{
        refresh_view.refreshFinish(PullToRefreshLayout.FAIL);
    }
}
```

自定义监听器,对页面下拉和上拉监听,下拉刷新页面数据,上拉加载更多数据。监听器 MyListener 代码如文件清单 9-10 所示。

文件清单 9-10 监听器 MyListener 代码

```
class MyListener implements PullToRefreshLayout.OnRefreshListener {

    @Override
    public void onRefresh(PullToRefreshLayout pullToRefreshLayout) {
        getFirst();
    }

    @Override
    public void onLoadMore(PullToRefreshLayout pullToRefreshLayout) {
        getNext();
    }
}
```

6. “预约”功能

“预约”页面用于展示可供用户进行预约操作的店铺,如图 9-13 所示。

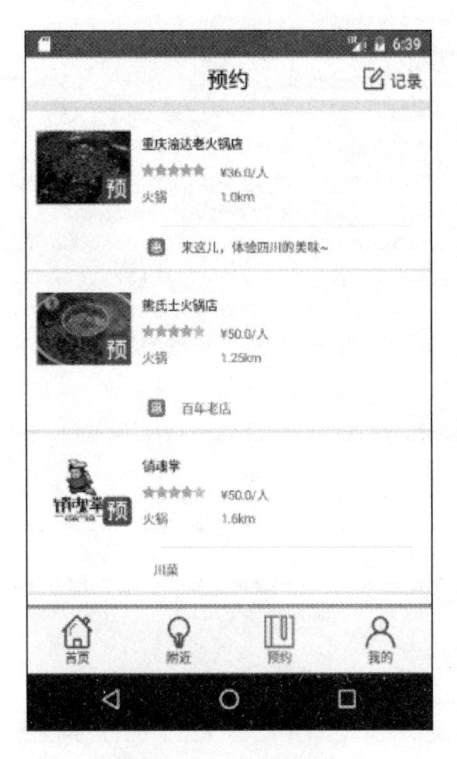

图 9-13 “预约”页面

其实现原理基本同“附近”功能。客户端根据 count、firstIndex 和 marketAdress 提交网络请求,从服务器获取可预约的店铺信息。

```
预约 API: http://100.0.101.18:8080/CDFood/getBookMarket
请求参数: count,firstIndex,marketAdress
返回值: [{ "marketNo":453,
          "marketName":"销魂掌",
          "marketIntroduce":"川菜",
          "marketAdress":"龙泉驿",
          "marketIconPath":"FJ_pic_two.png",
          "marketBigPicture":"SY_pic.png",
          "marketPrice":50.0,
          "marketDiscount":8.0,
          "marketHotLevel":5,
          "marketDistance":1.6,
          "typeName":"火锅",
          "erectLineIconPath":null,
          "bookIconPath":"view_yu.png",
          "newIconPath":"",
          "discountIconPath":null,
          "foodType":null},{……},{……}]
    ERROR 表示数据获取失败。
```

服务器返回的数据仍然为JSON数组字符串，从JSON数组中取出每个JSON对象，解析出相应的数据信息，构造出可预约的店铺列表，展示到页面上。列表依然采用自定义上下拉刷新的PullableListView和PullToRefreshLayout。解析过程如下：

```
try {
    JSONArray jsonArray=new JSONArray(result);
    if(! isLoadmore){
        orderInfos.clear();
    }
    for(int i=0;i<jsonArray.length();i++){
        JSONObject jsonObject=jsonArray.getJSONObject(i);
        String bookIconPath=Url.getImgURL(jsonObject.getString("bookIconPath"));
        String discountIconPath=Url.getImgURL(jsonObject
                    .getString("discountIconPath"));
        String marketAdress=jsonObject.getString("marketAdress");
        String marketBigPicture=Url.getImgURL(jsonObject
                    .getString("marketBigPicture"));
        double marketDiscount=jsonObject.getDouble("marketDiscount");
        double marketDistance=jsonObject.getDouble("marketDistance");
        double marketHotLevel=jsonObject.getDouble("marketHotLevel");
        String marketIconPath=Url.getImgURL(jsonObject
                    .getString("marketIconPath"));
        String marketIntroduce=jsonObject.getString("marketIntroduce");
        String marketName=jsonObject.getString("marketName");
        long marketNo=jsonObject.getLong("marketNo");
        double marketPrice=jsonObject.getDouble("marketPrice");
        String newIconPath=Url.getImgURL(jsonObject.getString("newIconPath"));
```

```
            String typeName=jsonObject.getString("typeName");
            OrderInfo orderInfo=new OrderInfo(
                    bookIconPath,typeName,marketPrice,newIconPath,marketNo,
                    marketName,marketIntroduce,marketIconPath,marketHotLevel,
                    marketDistance,marketDiscount,marketBigPicture,
                    marketAdress,discountIconPath);
            orderInfos.add(orderInfo);
        }
        orderAdapter.notifyDataSetChanged();
        firstIndex+=count;
        if(isLoadmore){
            refresh_view.loadmoreFinish(PullToRefreshLayout.SUCCEED);
        }else{
            refresh_view.loadmoreFinish(PullToRefreshLayout.SUCCEED);
        }
} catch (JSONException e) {
        e.printStackTrace();
        if(isLoadmore){
            refresh_view.loadmoreFinish(PullToRefreshLayout.FAIL);
        }else{
            refresh_view.refreshFinish(PullToRefreshLayout.FAIL);
        }
}
```

在预约列表点击列表项后，跳转至对应的店铺页面。

```
ls_show.setOnItemClickListener(new AdapterView.OnItemClickListener() {
        @Override
        public void onItemClick(AdapterView<?>adapterView, View view, int i, long l) {
                Intent intent=new Intent(context, StroeActivity.class);
                OrderInfo orderInfo=orderInfos.get(i);
                intent.putExtra("marketNo",orderInfo.getMarketNo());
                intent.putExtra("type","预约");
                intent.putExtra("storename",orderInfo.getMarketName());
                intent.putExtra("introduce",orderInfo.getMarketIntroduce());
                intent.putExtra("imagepath",orderInfo.getMarketIconPath());
                startActivity(intent);
        }
});
```

在预约页面点击“记录”按钮，可跳转至“我的预约”页面。

```
iv_history.setOnClickListener(new View.OnClickListener() {
        @Override
        public void onClick(View view) {
                startActivity(new Intent(context, MineOrderActivity.class));
        }
});
```

可预约的店铺会出现"预"字图标，如图 9-14 所示。在预约列表的适配器 OrderAdapter 中进行设置，点击"预"字图标，会弹出图 9-15 所示的对话框，供用户选择。

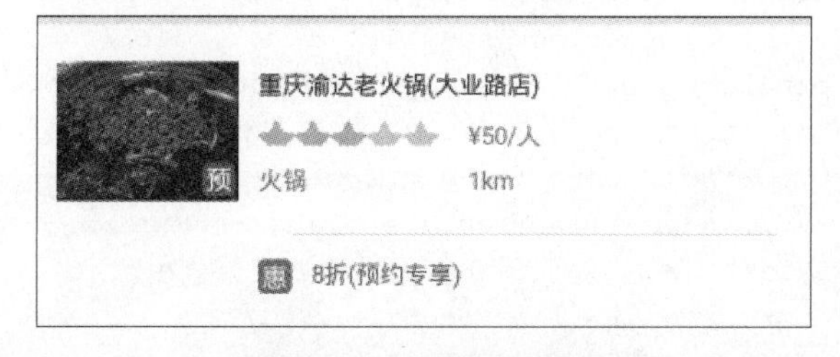

图 9-14 "预约"列表项

图 9-15 预约选择

OrderAdapter.java 源代码如文件清单 9-11 所示。

文件清单 9-11 OrderAdapter.java

```
public class OrderAdapter extends BaseAdapter {

    private Context context;
    private List<OrderInfo>orderInfos;
    private BitmapUtils butils;
    private BitmapUtils yutils;

    public OrderAdapter(Context context, List<OrderInfo>orderInfos) {
        this.context=context;
        this.orderInfos=orderInfos;
        butils=new BitmapUtils(context);
        butils.configDefaultLoadingImage(R.drawable.fj_loading);
        yutils=new BitmapUtils(context);
    }

    @Override
    public int getCount() {
        return orderInfos.size();
    }

    @Override
```

```
public OrderInfo getItem(int i) {
    return orderInfos.get(i);
}

@Override
public long getItemId(int i) {
    return i;
}

@Override
public View getView(int i, View view, ViewGroup viewGroup) {
    ViewHolder holder;
    if(view==null){
        view=View.inflate(context, R.layout.item_order,null);
        holder=new ViewHolder();
        holder.item_icon=(ImageView) view.findViewById(R.id.item_icon);
        holder.iv_order=(ImageView) view.findViewById(R.id.iv_order);
        holder.item_title=(TextView) view.findViewById(R.id.item_title);
        holder.item_rating=(RatingBar) view.findViewById(R.id.item_rating);
        holder.tv_money=(TextView) view.findViewById(R.id.tv_money);
        holder.item_name=(TextView) view.findViewById(R.id.item_name);
        holder.tv_price=(TextView) view.findViewById(R.id.tv_price);
        holder.tv_distance=(TextView) view.findViewById(R.id.tv_distance);
        holder.iv_hui=(ImageView) view.findViewById(R.id.iv_hui);
        view.setTag(holder);
    }else{
        holder=(ViewHolder) view.getTag();
    }
    OrderInfo orderInfo=getItem(i);
    butils.display(holder.item_icon,orderInfo.getMarketIconPath());
    yutils.display(holder.iv_order,orderInfo.getBookIconPath());
    holder.item_title.setText(orderInfo.getMarketName());
    holder.item_rating.setRating((float) orderInfo.getMarketHotLevel());
    holder.tv_money.setText("￥"+orderInfo.getMarketPrice()+"/人");
    holder.item_name.setText(orderInfo.getTypeName());
    holder.tv_distance.setText(orderInfo.getMarketDistance()+"km");
    yutils.display(holder.iv_hui,orderInfo.getDiscountIconPath());
    holder.tv_price.setText(orderInfo.getMarketIntroduce());
    OnClick(holder,i);
    return view;
}

static class ViewHolder{
    public ImageView item_icon;
    public ImageView iv_order;
    public TextView item_title;
    public RatingBar item_rating;
    public TextView tv_money;
    public TextView item_name;
```

```
        public TextView tv_price;
        public TextView tv_distance;
        public ImageView iv_hui;
    }

    private void OnClick(ViewHolder holder,final  int position){
        holder.iv_order.setOnClickListener(new View.OnClickListener() {
            @Override
            public void onClick(View view) {
                OrderInfo orderInfo=orderInfos.get(position);
                TimeChoose timeChoose=new TimeChoose(context, (Activity) context);
                timeChoose.DiaLog(orderInfo.getMarketNo());
            }
        });
    }
}
```

7. “我的”功能

在“我的”功能页面，目前已完成的有三部分内容，即我的收藏、我的预约、我的分享，如图 9-16 所示。点击“我的收藏”，可进入“我的收藏”页面；点击“我的预约”，可进入“我的预约”页面；点击“我的分享”，可进入“我的分享”页面。

“我的收藏”页面如图 9-17 所示，以列表的形式显示用户收藏的店铺信息，包括店铺图片、店铺热度、店铺描述、收藏时间等信息。

图 9-16 “我的”页面

图 9-17 “我的收藏”页面

列表项的设计如图 9-18 所示，对应的 MineCollectAdapter. java 代码如文件清单 9-12 所示。

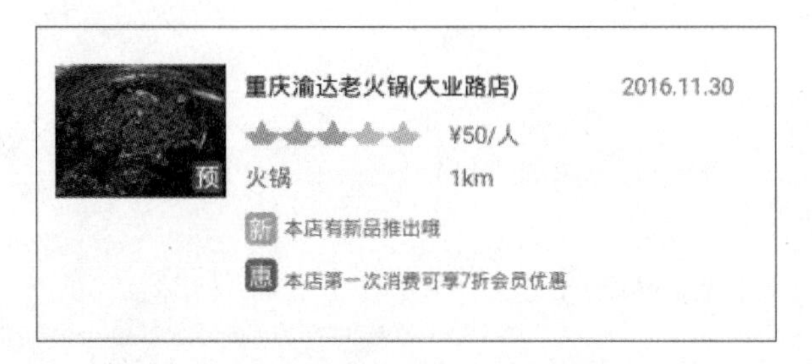

图 9-18 “我的收藏”列表项设计

文件清单 9-12 MineCollectAdapter. java

```
public class MineCollectAdapter extends BaseAdapter{

    private Context context;
    private List<MineCollectInfo>mineCollectInfos;
    private BitmapUtils butils;
    private BitmapUtils yutils;
    //定义构造器
    public MineCollectAdapter (Context context, List < MineCollectInfo >
mineCollectInfos){
        this.context=context;
        this.mineCollectInfos=mineCollectInfos;
        butils=new BitmapUtils(context);
        butils.configDefaultLoadingImage(R.drawable.fj_loading);
        yutils=new BitmapUtils(context);
    }

    @Override
    public int getCount() {
        return mineCollectInfos.size();
    }

    @Override
    public MineCollectInfo getItem(int i) {
        return mineCollectInfos.get(i);
    }

    @Override
    public long getItemId(int i) {
        return i;
    }

    @Override
    public View getView(int i, View view, ViewGroup viewGroup) {
        ViewHolder holder;
```

```
        if(view==null){
            view=View.inflate(context, R.layout.item_minecollect,null);
            holder=new ViewHolder();
            holder.item_icon=(ImageView) view.findViewById(R.id.item_icon);
            holder.iv_order=(ImageView) view.findViewById(R.id.iv_order);
            holder.item_title=(TextView) view.findViewById(R.id.item_title);
            holder.item_rating=(RatingBar) view.findViewById(R.id.item_rating);
            holder.tv_money=(TextView) view.findViewById(R.id.tv_money);
            holder.item_name=(TextView) view.findViewById(R.id.item_name);
            holder.tv_distance=(TextView) view.findViewById(R.id.tv_distance);
            holder.iv_new=(ImageView) view.findViewById(R.id.iv_new);
            holder.tv_new=(TextView) view.findViewById(R.id.tv_new);
            holder.iv_hui=(ImageView) view.findViewById(R.id.iv_hui);
            holder.tv_hui=(TextView) view.findViewById(R.id.tv_hui);
            holder.tv_time=(TextView) view.findViewById(R.id.tv_time);
            view.setTag(holder);
        }else{
            holder=(ViewHolder) view.getTag();
        }
        MineCollectInfo info=getItem(i);
        butils.display(holder.item_icon,info.getMarketIconPath());
        yutils.display(holder.iv_order,info.getBookIconPath());
        holder.item_title.setText(info.getMarketName());
        holder.item_rating.setRating((float) info.getMarketHotLevel());
        holder.tv_money.setText("￥"+info.getMarketPrice()+"/人");
        holder.item_name.setText(info.getTypeName());
        holder.tv_distance.setText(info.getMarketDistance()+"km");
        yutils.display(holder.iv_new,info.getNewIconPath());
        holder.tv_new.setText(info.getMarketIntroduce());
        yutils.display(holder.iv_hui,info.getDiscountIconPath());
        holder.tv_hui.setText(info.getMarketIntroduce());
        holder.tv_time.setText(info.getDate());
        return view;
    }

    static class ViewHolder{
        public ImageView item_icon;
        public ImageView iv_order;
        public TextView item_title;
        public RatingBar item_rating;
        public TextView tv_money;
        public TextView item_name;
        public TextView tv_distance;
        public ImageView iv_new;
        public TextView tv_new;
        public ImageView iv_hui;
        public TextView tv_hui;
        public TextView tv_time;
    }
}
```

客户端根据 userId 提交网络请求，从服务器获取“我的收藏”的数据信息。服务器以 JSON 数组字符串返回信息。

```
我的收藏 API: http://100.0.101.18:8080/CDFood/getMyCollect
请求参数: userId
返回值: [{  "id":1,
          "user":{"userNo":1,
                  "userId":"1",
                  "userPassword":"1",
                  "userAdress":null,
                  "userIconPath":"09b5f265 - 6771 - 485e - be8a - 7082b528c6ad.
jpeg",
                  "userName":null},
          "market":{"marketNo":935,
                  "marketName":"自助汤锅",
                  "marketIntroduce":"自助餐",
                  "marketAdress":"青羊区",
                  "marketIconPath":"Re_pic_three.png.png",
                  "marketBigPicture":"SY_pic.png",
                  "marketPrice":50.0,
                  "marketDiscount":9.9,
                  "marketHotLevel":5,
                  "marketDistance":6.4,
                  "typeName":null,
                  "erectLineIconPath":null,
                  "bookIconPath":"",
                  "newIconPath":"",
                  "discountIconPath":"",
                  "foodType":{"typeNo":1,"typeName":"火锅"}},
          "type":"1",
          "date":"2016-12-01"},{……},{……}]
    "ERROR"表示数据获取失败。
```

案例中“我的收藏”页面主要由 MineCollectActivity 负责实现。MineCollectActivity 的核心代码如文件清单 9-13 所示。

文件清单 9-13　MineCollectActivity 的核心代码

```
public class MineCollectActivity extends BaseActivity {

    private AsyncHttpResponseHandler minecollect_handler;
    @ViewInject(R.id.iv_back)
    private ImageView iv_back;
    @ViewInject(R.id.ls_show)
    private ListView ls_show;

    private List<MineCollectInfo>mineCollectInfos;
    private MineCollectAdapter mineCollectAdapter;
```

```
@Override
protected void onCreate(Bundle savedInstanceState) {
    super.onCreate(savedInstanceState);
    setContentView(R.layout.activity_minecollect);
    ViewUtils.inject(this);
    dohandler();
    Init();
}

private void Init() {
    iv_back.setOnClickListener(new View.OnClickListener() {
        @Override
        public void onClick(View view) {
            finish();
        }
    });
    mineCollectInfos=new ArrayList<>();
    RequestParams params=new RequestParams();
    params.put("userId", NApplication.user_number);
    HttpUtils.get(this, Url.getMyCollect,params,minecollect_handler);
    mineCollectAdapter =new MineCollectAdapter(this,mineCollectInfos);
    ls_show.setAdapter(mineCollectAdapter);
}

private void dohandler() {

    minecollect_handler=new AsyncHttpResponseHandler() {
        @Override
        public void onSuccess(int statusCode, Header[] headers,
                                          byte[] responseBody) {
            String result=new String(responseBody);
            if(result.equals("ERROR")){
                Toast.makeText(MineCollectActivity.this,"获取数据失败",
                    Toast.LENGTH_SHORT).show();
            }else{
                try {
                    JSONArray jsonArray=new JSONArray(result);
                    for(int i=0;i<jsonArray.length();i++){
                        JSONObject jsonObject=jsonArray.getJSONObject(i);
                        String date=jsonObject.getString("date");
                        JSONObject jsonObject1=jsonObject
                                .getJSONObject("market");
                        String bookIconPath=Url.getImgURL(jsonObject1
                                .getString("bookIconPath"));
                        String discountIconPath=Url.getImgURL(jsonObject1
                                .getString("discountIconPath"));
                        double marketDiscount=jsonObject1
```

```
                        .getDouble("marketDiscount");
                double marketDistance=jsonObject1
                        .getDouble("marketDistance");
                double marketHotLevel=jsonObject1
                        .getDouble("marketHotLevel");
                String marketIconPath=Url.getImgURL(jsonObject
                        1.getString("marketIconPath"));
                String marketIntroduce=jsonObject1
                        .getString("marketIntroduce");
                String marketName=jsonObject1
                        .getString("marketName");
                long marketNo=jsonObject1.getLong("marketNo");
                double marketPrice=jsonObject1
                        .getDouble("marketPrice");
                String newIconPath=Url.getImgURL(jsonObject1
                        .getString("newIconPath"));
                JSONObject jsonObject2=jsonObject1
                        .getJSONObject("foodType");
                String typeName=jsonObject2.getString("typeName");
                MineCollectInfo mineCollectInfo=new MineCollectInfo(
                    bookIconPath,typeName,newIconPath,marketPrice,
                    marketNo,marketName,marketIntroduce,
                    marketIconPath,marketHotLevel,marketDistance,
                    marketDiscount,discountIconPath,date);
                mineCollectInfos.add(mineCollectInfo);
            }
            mineCollectAdapter.notifyDataSetChanged();

        } catch (JSONException e) {
            e.printStackTrace();
            Toast.makeText(MineCollectActivity.this,"获取数据失败",
                Toast.LENGTH_SHORT).show();
        }
    }
}

@Override
public void onFailure(int statusCode, Header[] headers,
                byte[] responseBody, Throwable error) {
    Toast.makeText(MineCollectActivity.this,
        "网络连接错误,请检查网络设置后重试。",
            Toast.LENGTH_SHORT).show();
}
};
}
}
```

“我的预约”页面如图 9-19 所示，以列表的形式显示用户预约的店铺信息，包括店铺价格、店铺热度、预约时间及预约状态等信息。

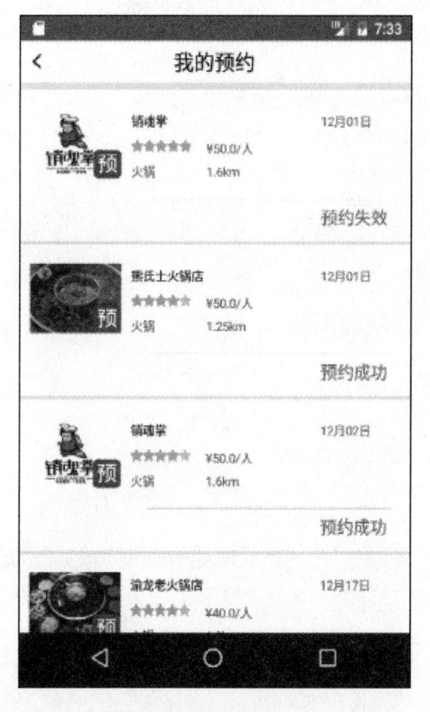

图 9-19 “我的预约”页面

客户端根据 userId、每次加载的条数 count 和当前索引 firstIndex 提交网络请求，从服务器获取“我的预约”的数据信息。

```
RequestParams params=new RequestParams();
params.put("userId", NApplication.user_number);
params.put("count",count);
params.put("firstIndex",firstIndex);
HttpUtils.get(this, Url.getMyOrder,params,mineorder_handler);
```

从服务器获取“我的预约”的数据信息：

```
我的预约 API：http://100.0.101.18:8080/CDFood/getMyOrder
请求参数：userId,count,firstIndex
返回值：[{"orderNo":45,
        "orderDay":"12 月 01 日\n 明天",
        "orderTime":"0:30",
        "orderPeopleCount":3,
        "orderState":"预约失效",
        "orderTotalPrice":150.0,
        "user":{"userNo":1,
                "userId":"1",
```

```
                "userPassword":"1",
                "userAdress":null,
                "userIconPath":"09b5f265-6771-485e-be8a-7082b528c6ad.jpeg",
                "userName":null},
    "market":{"marketNo":186,
                "marketName":"销魂掌",
                "marketIntroduce":"川菜",
                "marketAdress":"温江区",
                "marketIconPath":"FJ_pic_two.png",
                "marketBigPicture":"SY_pic.png",
                "marketPrice":50.0,
                "marketDiscount":8.0,
                "marketHotLevel":5,
                "marketDistance":1.6,
                "typeName":null,
                "erectLineIconPath":null,
                "bookIconPath":"view_yu.png",
                "newIconPath":"view_new.png",
                "discountIconPath":null,
                "foodType":{"typeNo":1,"typeName":"火锅"}}},{……},{……}]
  ERROR 表示数据获取失败。
```

从服务器返回的JSON数据中解析出“我的预约”页面所需要的店铺数据信息。

```
try {
    JSONArray jsonArray=new JSONArray(result);
    for(int i=0;i<jsonArray.length();i++){
        JSONObject jsonObject=jsonArray.getJSONObject(i);
        String orderDay=jsonObject.getString("orderDay");
        String orderState=jsonObject.getString("orderState");
        JSONObject jsonObject1=jsonObject.getJSONObject("market");
        String bookIconPath=Url.getImgURL(jsonObject1.getString("bookIconPath"));
        double marketDistance=jsonObject1.getDouble("marketDistance");
        double marketHotLevel=jsonObject1.getDouble("marketHotLevel");
        double marketPrice=jsonObject1.getDouble("marketPrice");
        String marketIconPath=Url.getImgURL(jsonObject1
                            .getString("marketIconPath"));
        String marketName=jsonObject1.getString("marketName");
        JSONObject jsonObject2=jsonObject1.getJSONObject("foodType");
        String typeName=jsonObject2.getString("typeName");
        //根据获取的数据,构造一个订单
        MineOrderInfo mineOrderInfo=new MineOrderInfo(
            bookIconPath,typeName,orderState,orderDay,marketName,
            marketIconPath,marketHotLevel,marketDistance,marketPrice);
        mineOrderInfos.add(mineOrderInfo);
    }
      mineOrderAdapter.notifyDataSetChanged();
} catch (JSONException e) {
```

```
        e.printStackTrace();
        Toast.makeText(MineOrderActivity.this,"获取数据失败",
                                    Toast.LENGTH_SHORT).show();
    }
```

“我的预约”列表中，如果预约成功，则以红色字体显示“预约成功”；如果预约失效，则以黑色字体显示“预约失效”。在 MineOrderAdapter 中根据预约的状态进行区分。

MineOrderAdapter 代码如下：

```
public class MineOrderAdapter extends BaseAdapter {

    private Context context;
    private List<MineOrderInfo>mineOrderInfos;
    private BitmapUtils butils;
    private BitmapUtils yutils;

    //定义构造器
    public MineOrderAdapter(Context context,List<MineOrderInfo>mineOrderInfos){
        this.context=context;
        this.mineOrderInfos=mineOrderInfos;
        butils=new BitmapUtils(context);
        butils.configDefaultLoadingImage(R.drawable.fj_loading);
        yutils=new BitmapUtils(context);
    }

    @Override
    public int getCount() {
        return mineOrderInfos.size();
    }

    @Override
    public MineOrderInfo getItem(int i) {
        return mineOrderInfos.get(i);
    }

    @Override
    public long getItemId(int i) {
        return i;
    }

    @Override
    public View getView(int i, View view, ViewGroup viewGroup) {
        ViewHolder holder;
        if(view==null){
            view=View.inflate(context, R.layout.item_mineorder,null);
            holder=new ViewHolder();
            holder.item_icon=(ImageView) view.findViewById(R.id.item_icon);
```

```
            holder.iv_order= (ImageView) view.findViewById(R.id.iv_order);
            holder.item_title= (TextView) view.findViewById(R.id.item_title);
            holder.tv_time= (TextView) view.findViewById(R.id.tv_time);
            holder.item_rating= (RatingBar) view.findViewById(R.id.item_rating);
            holder.tv_money= (TextView) view.findViewById(R.id.tv_money);
            holder.item_name= (TextView) view.findViewById(R.id.item_name);
            holder.tv_distance= (TextView) view.findViewById(R.id.tv_distance);
            holder.tv_orderinfo= (TextView) view.findViewById(R.id.tv_orderinfo);
            view.setTag(holder);
        }else{
            holder= (ViewHolder) view.getTag();
        }
        MineOrderInfo mineOrderInfo=getItem(i);
        butils.display(holder.item_icon,mineOrderInfo.getMarketIconPath());
        yutils.display(holder.iv_order,mineOrderInfo.getBookIconPath());
        holder.item_title.setText(mineOrderInfo.getMarketName());
        String[] splie=mineOrderInfo.getOrderDay().split("日");
        holder.tv_time.setText(splie[0]+"日");
        holder.item_rating.setRating((float) mineOrderInfo.getMarketHotLevel());
        holder.tv_money.setText("¥"+mineOrderInfo.getMarketPrice()+"/人");
        holder.item_name.setText(mineOrderInfo.getTypeName());
        holder.tv_distance.setText(mineOrderInfo.getMarketDistance()+"km");
        //如果预约成功,字体为红色,不成功则为黑色
        if(mineOrderInfo.getOrderState().equals(context.getResources()
                    .getString(R.string.order_success))){
            holder.tv_orderinfo.setTextColor(context.getResources()
                        .getColor(R.color.ordersuccess));
        }else{
            holder.tv_orderinfo.setTextColor(context.getResources()
                        .getColor(R.color.orderfailure));
        }
        holder.tv_orderinfo.setText(mineOrderInfo.getOrderState());
        return view;
    }

    static class ViewHolder{
        public ImageView item_icon;
        public ImageView iv_order;
        public TextView item_title;
        public TextView tv_time;
        public RatingBar item_rating;
        public TextView tv_money;
        public TextView item_name;
        public TextView tv_distance;
        public TextView tv_orderinfo;
    }
}
```

“我的分享”页面如图 9-20 所示,以列表的形式显示用户已分享的店铺信息,包括店铺价格、店铺热度、分享时间等信息。

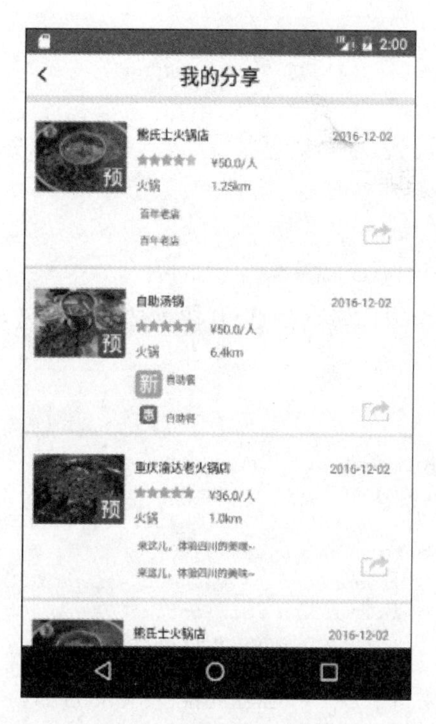

图 9-20 “我的分享”页面

客户端根据 userId 提交网络请求，从服务器获取“我的分享”的数据信息。

```
RequestParams params =new RequestParams();
params.put("userId", NApplication.user_number);
HttpUtils.get(this, Url.getMyShare, params, share_handler);

我的分享 API：http://100.0.101.18:8080/CDFood/getMyShare
请求参数：userId
返回值：[{"id":45,
        "user":{"userNo":1,
                "userId":"1",
                "userPassword":"1",
                "userAdress":null,
                "userIconPath":"09b5f265-6771-485e-be8a-7082b528c6ad.jpeg",
                "userName":null},
        "market":{"marketNo":308,
                "marketName":"熊氏土火锅店",
                "marketIntroduce":"百年老店",
                "marketAdress":"武侯区",
                "marketIconPath":"HG_pic_three.png",
                "marketBigPicture":"SY_pic.png",
                "marketPrice":50.0,
                "marketDiscount":8.8,
                "marketHotLevel":4,
                "marketDistance":1.25,
                "typeName":null,
```

```
                "erectLineIconPath":null,
                "bookIconPath":"view_yu.png",
                "newIconPath":null,
                "discountIconPath":"",
                "foodType":{"typeNo":1,"typeName":"火锅"}},
                "type":"0",
                "date":"2016-12-02"},{……},{……}]
ERROR 表示数据获取失败。
```

服务器以JSON字符串返回数据信息，从服务器返回的数据中解析出“我的分享”页面所需要的店铺数据信息。

```
try {
    JSONArray jsonArray = new JSONArray(result);
    for (int i = 0; i < jsonArray.length(); i++) {
        JSONObject jsonObject = jsonArray.getJSONObject(i);
        String date = jsonObject.getString("date");
        JSONObject jsonObject1 = jsonObject.getJSONObject("market");
        String bookIconPath = Url.getImgURL(jsonObject1
                    .getString("bookIconPath"));
        String discountIconPath = Url.getImgURL(jsonObject1
                    .getString("discountIconPath"));
        double marketDiscount = jsonObject1.getDouble("marketDiscount");
        double marketDistance = jsonObject1.getDouble("marketDistance");
        double marketHotLevel = jsonObject1.getDouble("marketHotLevel");
        String marketIconPath = Url.getImgURL(jsonObject1
                    .getString("marketIconPath"));
        String marketIntroduce = jsonObject1.getString("marketIntroduce");
        String marketName = jsonObject1.getString("marketName");
        long marketNo = jsonObject1.getLong("marketNo");
        double marketPrice = jsonObject1.getDouble("marketPrice");
        String newIconPath = Url.getImgURL(jsonObject1
                    .getString("newIconPath"));
        JSONObject jsonObject2 = jsonObject1.getJSONObject("foodType");
        String typeName = jsonObject2.getString("typeName");
        MineShareInfo mineshareinfo = new MineShareInfo(
            bookIconPath, typeName, newIconPath, marketPrice,
            marketNo, marketName, marketIntroduce, marketIconPath,
            marketHotLevel, marketDistance, marketDiscount,
                discountIconPath, date);
            mineshareinfos.add(mineshareinfo);
    }
    mineshareadapter.notifyDataSetChanged();
} catch (JSONException e) {
    e.printStackTrace();
    Toast.makeText(MineShareActivity.this, "获取数据失败",
                Toast.LENGTH_SHORT).show();
}
```

“我的分享”列表中有一个“分享”图标，如图9-21右下角的小图标所示。

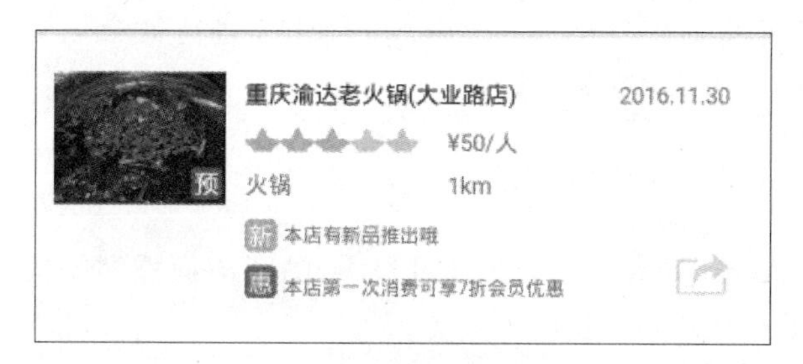

图 9-21 “我的分享”列表项

点击该“分享”图标，弹出当前支持的社会化分享列表，如图 9-22 所示。

```
private void OnClick(ViewHolder holder, final int postion){
        holder.iv_share.setOnClickListener(new View.OnClickListener() {
            @Override
            public void onClick(View view) {
                Tools.showShare(context,"好吃","好吃",
                    mineShareInfos.get(postion).getMarketIconPath(),
                    "http://www.baidu.com");
            }
        });
}
```

选择分享到 Sina Weibo，进入分享页面，如图 9-23 所示。

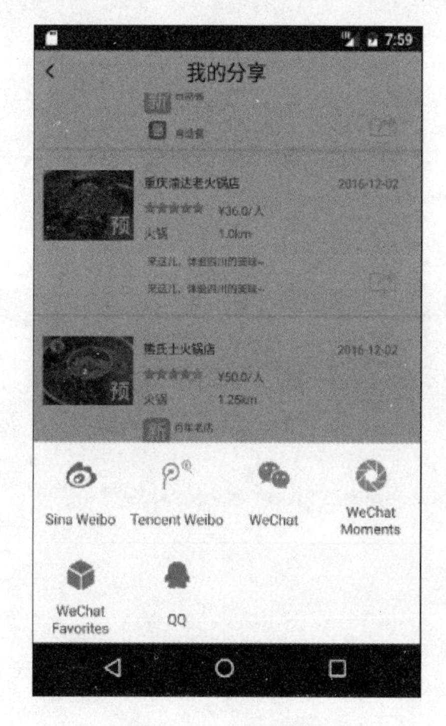

图 9-22 分享操作

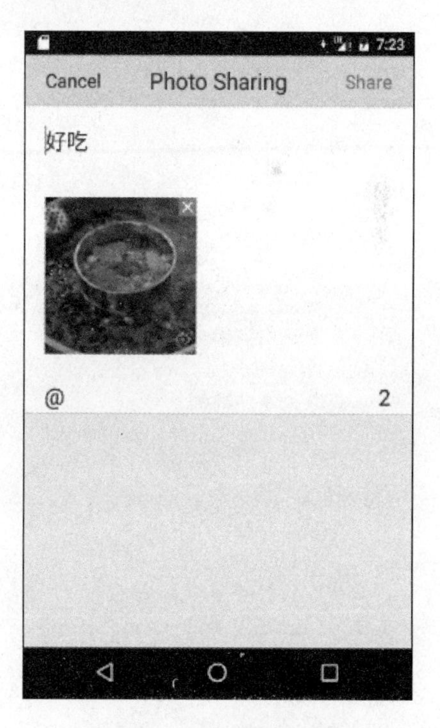

图 9-23 分享到新浪微博

关于分享的实现，9.1.2节介绍了使用ShareSDK进行社会化分享的基本操作。读者可参考综合案例的完整源代码来理解。

9.1.4 辅助工具类

在“吃都”App Android客户端的实现过程中，除了前面提到的核心功能所涉及的类与方法外，还用到了一些常用的辅助工具类。

URL类存放全局的URL相关信息。如文件清单9-14所示。

文件清单9-14 URL类存放全局的URL相关信息的代码

```
public class URL {

    //主机地址 100.0.101.18
    public final static String URL="http://100.0.101.18:8080/CDFood/";

    public final static String ImgURL="http://100.0.101.18:8080/CDFood/Images/";
    private static String getURL(String activity){
        String url=null;
        url=Url+activity;
        return url;
    }

    public static String getImgURL(String activity) {
        String url =null;
        url =ImgURL+activity;
        return url;
    }
    /**用户注册**/
    public static String signIn =getUrl("signIn");
    /**用户登录**/
    public static String login=getUrl("login");
    /**获取热门美食**/
    public static String getHotFood=getUrl("getHotFood");
    /**获取热门商铺**/
    public static String getHotMarket=getUrl("getHotMarket");
    /**获取火锅的商铺**/
    public static String getHotpotMarket=getUrl("getHotpotMarket");
    /**获取西餐的商铺**/
    public static String getWesternMarket=getUrl("getWesternMarket");
    /**获取甜品的商铺**/
    public static String getSweetMarket=getUrl("getSweetMarket");
    /**获取饮料的商铺**/
    public static String getDrinkMarket=getUrl("getDrinkMarket");
    /**获取附近的商铺**/
    public static String getNearMarket=getUrl("getNearMarket");
    /**获取某个店铺里的商品**/
    public static String getFoods=getUrl("getFoods");
```

```
    /**保存预约单号**/
    public static String saveFoodOrder=getUrl("saveFoodOrder");
    /**得到我的预约订单**/
    public static String getBookMarket=getUrl("getBookMarket");
    /**获取可预约的商铺**/
    public static String getMyOrder=getUrl("getMyOrder");
    /**收藏店铺(保存)**/
    public static String saveCollect=getUrl("saveCollect");
    /**得到我的收藏**/
    public static String getMyCollect=getUrl("getMyCollect");
    /**取消收藏**/
    public static String removeCollect=getUrl("removeCollect");
    /**分享店铺(保存)**/
    public static String saveShare=getUrl("saveShare");
    /**分享店铺(保存)**/
    public static String getMyShare=getUrl("getMyShare");
    /**上传我的头像**/
    public static String uploadPhotos=getUrl("uploadPhotos");
    /**获取用户资料**/
    public static String getUserInfo=getUrl("getUserInfo");
}
```

Tools 工具类定义了一些常用的工具方法：

```
public class Tools {
    //列表显示地区名字
    public static void showLocalList(Context context, final View anchor,
                                        List<String>info, int size){
        PopupWindow popupWindow=null;
        View layout=View.inflate(context, R.layout.localinfo_group,null);
        popupWindow=new PopupWindow(layout,size,
                WindowManager.LayoutParams.WRAP_CONTENT,true);
        popupWindow.setBackgroundDrawable(new BitmapDrawable());
        ListView list=(ListView) layout.findViewById(R.id.list);
        list.setAdapter(new LocalInfoAdapter(context,info));

        final PopupWindow finalPopupWindow =popupWindow;
        list.setOnItemClickListener(new AdapterView.OnItemClickListener() {
            @Override
            public void onItemClick(AdapterView<?>adapterView, View view,
int i, long l) {
                ((TextView) anchor).setText(local.get(i));
                finalPopupWindow.dismiss();
            }
        });
        popupWindow.showAsDropDown(anchor,0,0);
    }

    //社会化分享
    public static void showShare(Context context,String title,String text,
```

```
                                String imageUrl,String Appurl) {
        //源码在 ShareSDK 分享中已说明
    }

    //加载图片
    // showImageForEmptyUri 设置图片 URI 为空或是错误时显示的图片
    // showImageOnFail 设置图片加载或解码过程中发生错误显示的图片
    // showImageOnLoading 创建配置过的 DisplayImageOption 对象
    public static void getImage(Context context) {
        DisplayImageOptions options =new DisplayImageOptions.Builder()
                .showImageForEmptyUri(R.drawable.ic_main_errorpicture)
                .showImageOnFail(R.drawable.ic_main_errorpicture)
                .cacheInMemory(true)      // 设置下载的图片是否缓存在内存中
                .cacheOnDisk(true)        // 设置下载的图片是否缓存在 SD 卡中
                .showImageOnLoading(R.drawable.ic_main_nopicture)
                .build();

        ImageLoaderConfiguration config =new
                ImageLoaderConfiguration.Builder(context)
                .defaultDisplayImageOptions(options)
                .threadPriority(Thread.NORM_PRIORITY -2)
                .denyCacheImageMultipleSizesInMemory()
                .imageDownloader(new BaseImageDownloader(context))
                .tasksProcessingOrder(QueueProcessingType.LIFO)
                .build();
        ImageLoader.getInstance().init(config);
    }

    //加载图片
    public static void getImage(Context context, int drawable) {
        DisplayImageOptions options =new DisplayImageOptions.Builder()
                .showImageForEmptyUri(drawable)
                .showImageOnFail(drawable)
                .cacheInMemory(true)
                .cacheOnDisk(true)
                .showImageOnLoading(drawable).build();

        ImageLoaderConfiguration config =new
                ImageLoaderConfiguration.Builder(context)
                .defaultDisplayImageOptions(options)
                .threadPriority(Thread.NORM_PRIORITY -2)
                .denyCacheImageMultipleSizesInMemory()
                .imageDownloader(new BaseImageDownloader(context))
                .tasksProcessingOrder(QueueProcessingType.LIFO)
                .build();
        ImageLoader.getInstance().init(config);
}

    //获取控件的高度或者宽度
    //isHeight=true 则为测量该控件的高度,isHeight=false 则为测量该控件的宽度
    public static int getViewHeight(View view, boolean isHeight) {
```

```
        int result;
        if (view ==null)
            return 0;
        if (isHeight) {
            int h =View.MeasureSpec
                    .makeMeasureSpec(0, View.MeasureSpec.UNSPECIFIED);
            view.measure(h, 0);
            result =view.getMeasuredHeight();
        } else {
            int w =View.MeasureSpec
                    .makeMeasureSpec(0, View.MeasureSpec.UNSPECIFIED);
            view.measure(0, w);
            result =view.getMeasuredWidth();
        }
        return result;
    }

    //得到指定大小,压缩后的图片
    //mcContext 上下文
    //image 原图片
    //return 压缩后的图片
    public static Bitmap getThumbnails(Context mcContext, Bitmap image) {

        Bitmap bitmap;
        int width =mcContext.getResources()
                .getDimensionPixelSize(R.dimen.regist_image_width);
        int height =mcContext.getResources()
                .getDimensionPixelSize(R.dimen.regist_image_height);
        int h =image.getHeight();
        int w =image.getWidth();
        if (h >width || w >height) {
            bitmap =ThumbnailUtils.extractThumbnail(image, width, height);
        } else {
            bitmap =image;
        }
        return bitmap;
    }

    //得到指定大小,压缩后的图片
    public static Bitmap getThumbnails (Context mcContext, Bitmap image, int
widths, int heights) {
        //getResources().getDimensionPixelSize 取出 dimens 中的值
        Bitmap bitmap;
        int width =mcContext.getResources().getDimensionPixelSize(widths);
        int height =mcContext.getResources().getDimensionPixelSize(heights);
        //ThumbnailUtils.extractThumbnail 创建一个指定大小的缩略图
        // source 源文件(Bitmap 类型)
        //width 压缩成的宽度
        //height 压缩成的高度
        int h =image.getHeight();
        int w =image.getWidth();
        if (h >width || w >height) {
```

```
            bitmap = ThumbnailUtils.extractThumbnail(image, width, height);
        } else {
            bitmap = image;
        }
        return bitmap;
    }

    //把 Bitmap 转换成字节数组
    public static byte[] getBitmapByte(Bitmap bitmap) {
        ByteArrayOutputStream out = new ByteArrayOutputStream();
        bitmap.compress(Bitmap.CompressFormat.JPEG, 100, out);
        try {
            out.flush();
            out.close();
        } catch (Exception e) {
            e.printStackTrace();
        }
        return out.toByteArray();
    }

    //把字节数组转换成 Bitmap
    public static Bitmap getBitmapFromByte(byte[] temp) {
        if (temp != null) {
            Bitmap bitmap = BitmapFactory.decodeByteArray(temp, 0, temp.length);
            return bitmap;
        } else {
            return null;
        }
    }

    //把一个 Drawable 转换成 Bitmap
    public static Bitmap drawableToBitmap(Drawable drawable) {
        int width = drawable.getIntrinsicWidth();
        int height = drawable.getIntrinsicHeight();
        Bitmap bitmap = Bitmap.createBitmap(width, height,
                drawable.getOpacity() != PixelFormat.OPAQUE ?
                        Bitmap.Config.ARGB_8888 : Bitmap.Config.RGB_565);
        Canvas canvas = new Canvas(bitmap);
        drawable.setBounds(0, 0, width, height);
        drawable.draw(canvas);
        return bitmap;
    }

    //把 Bitmap 转换成 Drawable
    public static Drawable Bitmap_Drawable(Bitmap bitmap) {
        Drawable drawable = new BitmapDrawable(bitmap);
        return drawable;
    }

    public static boolean isNumeric(String str) {
        for (int i = 0; i < str.length(); i++) {
```

```
            if (!Character.isDigit(str.charAt(i))) {
                return false;
            }
        }
        return true;
    }

    //清理缓存
    public static void cleanTemp(File file) {
        if (file !=null && file.exists() && file.isDirectory()) {
            for (File file2 : file.listFiles()) {
                file2.delete();
            }
        }
    }
}
```

9.2 Web 端后台程序与数据库搭建

9.2.1 后台程序总体说明

"吃都"App 采用的后台服务器是 Tomcat 7.0，开发工具选用 MyEclipse，数据库使用 MySQL。读者可自行安装 JDK、MyEclipse IDE、MySQL 及 Tomcat 服务器。后台代码结构如图 9-24 所示。

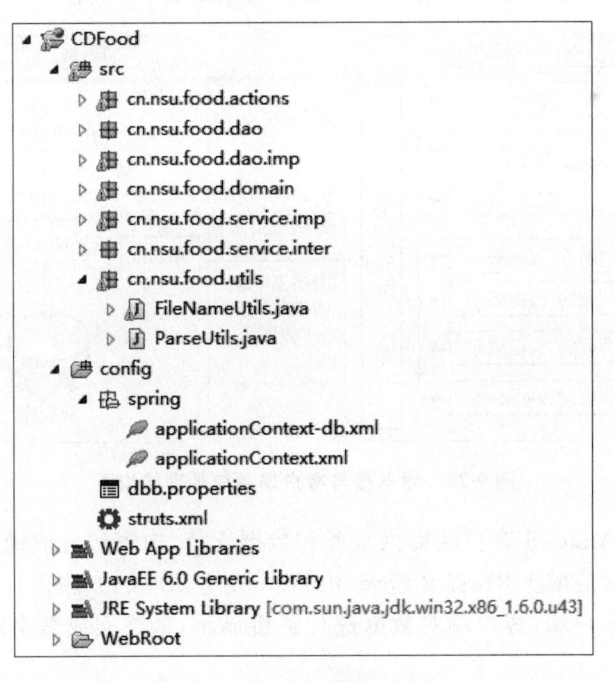

图 9-24 服务器端代码层次图

相关包的说明如表 9-2 所示。

表 9-2　服务器端代码包说明

包　　名	说　　明
CDFood	项目工程名称
src	项目 Java 代码存放目录
cn. nsu. food. actions	Action 层
cn. nsu. food. dao	DAO 层接口定义
cn. nsu. food. dao. imp	DAO 层接口实现
cn. nsu. food. domain	Model 层，存放相关的实体类
cn. nsu. food. service. imp	Service 层接口实现
cn. nsu. food. service. inter	Service 层接口定义
cn. nsu. food. utils	工具包，存放一些公用的静态方法
config	存放 SSH 配置文件
Library、WebRoot 等	项目编译运行相关的配置文件和类库

服务器端采用了最基本的分层方式，结合了 SSH 架构。Modle 层是对应的数据库表的实体类。DAO 层使用了 Hibernate 连接数据库、操作数据库（增删改查）。Service 层引用对应的 DAO 数据库操作，实现对应的逻辑判断。Action 层引用对应的 Service 层，结合 Struts 的配置文件，接收客户端传递的请求数据，并将处理结果返回。以上的 Hibernate、Struts 都需要注入到 Spring 的配置文件中，Spring 把这些联系起来，成为一个整体。"吃都"App 服务器端与客户端的通信处理如图 9-25 所示。

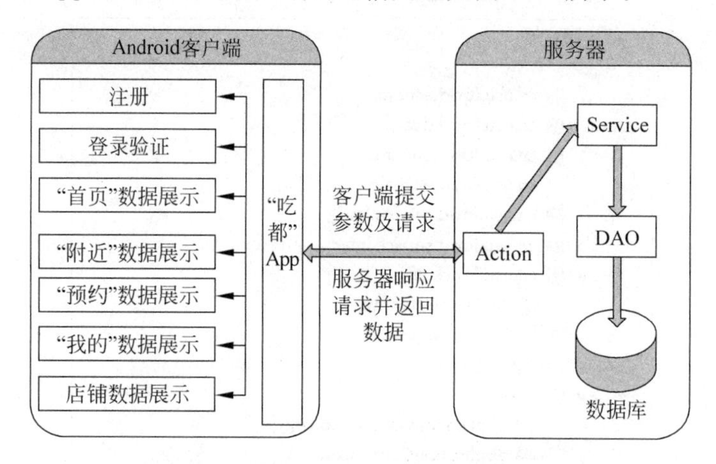

图 9-25　服务器与客户端通信处理结构图

服务器端为 Android 客户端提供服务和数据支持，内容包含商铺、订单和用户等内容。客户端通过对应的 URL 提交请求，由 Action 中相应的方法进行响应，把响应的结果传回 Android 客户端，客户端对数据进行解析显示，最终完成整个服务器和 Android 客户端的交互。

9.2.2 后台数据库表

food 表用于存储所有的美食信息，包括美食的名称、图片、所在店铺等信息，其结构如表 9-3 所示。

表 9-3 food 表

属 性	类 型	主 键	描 述
foodNo	bigint	PK	美食编号
foodName	varchar		美食名称
foodIntroduce	varchar		美食简介
foodIconPath	varchar		美食图片
foodDiscount	double		美食
foodHot	bit		是否为热门美食
marketNo	bigint		美食所在店铺
foodPrice	varchar		美食价格

foodorder 表用于存储用户预约信息，包括预定的编号、时间、人数、总价、预约状态等内容，其结构如表 9-4 所示。

表 9-4 foodorder 表

属 性	类 型	主 键	描 述
orderNo	bigint	PK	预定编号
orderDay	varchar		预定日期
orderTime	varchar		预定时间
orderPeopleCount	int		预定人数
orderState	varchar		预定的状态
orderTotalPrice	double		预定总价
userNo	bigint		用户登录 ID
marketNo	bigint		店铺编号

foodtype 表用于存储店铺类型，其结构如表 9-5 所示。案例中店铺共分火锅、西餐、甜品、饮品等类。

表 9-5 foodtype 表

属 性	类 型	主 键	描 述
typeNo	int	PK	店铺类型编号
typeName	varchar		店铺类型

market 表用于存储所有店铺信息，包括店铺的地址、折扣、简介、图片等内容，其结构如表 9-6 所示。主键 marketNo 作为店铺的唯一标识，可根据 marketNo 查找到相关的店铺信息。

表 9-6 market 表

属　性	类　型	主　键	描　述
marketNo	bigint	PK	店铺编号
marketAdress	varchar		店铺地址
marketDiscount	double		店铺折扣
marketDistance	double		店铺距离
marketHotLevel	int		店铺热门级别
marketIconPath	varchar		店铺图片
marketIntroduce	varchar		店铺简介
marketName	varchar		店铺名
marketPrice	double		店铺价格
marketBigPicture	varchar		店铺大图片
TypeNo	int		店铺类型编号

shareandcollectmarket 表用于存储用户分享及收藏的店铺列表，其结构如表 9-7 所示。type 取“0”时，表示分享；type 取“1”时，表示收藏。

表 9-7 shareandcollectmarket 表

属　性	类　型	主　键	描　述
Id	bigint	PK	序号
type	varchar		类型
marketNo	bigint		店铺编号
userId	bigint		用户登录 ID
date	varchar		日期

user 表用于存储用户的身份信息，包括用户的登录 ID、密码、头像等信息，其结构如表 9-8 所示。userId 具有唯一性，系统会根据 userId 去查询相关的信息。

表 9-8 user 表

属　性	类　型	主　键	描　述
userNo	bigint	PK	用户序号
userId	varchar		用户登录 ID
userPassword	varchar		密码
userAdress	varchar		地址
userIconPath	varchar		用户头像图片
userName	varchar		用户名

本章小结

本章以“吃都”App 为例，讲解了 Android 应用开发的全过程，包括 Android 客户端程序和服务器端程序的开发，便于读者理解案例的执行过程。

在“吃都”App 中涉及的 Android 基础知识在前面的章节中都有讲解，读者可通过实际案例强化对知识的理解。在此基础上，案例也使用了一些 Android 快速开发框架作为提高。读者通过本章案例可初步学会 Android 框架的应用，在以后的 Android 应用开发中能灵活运用。

参 考 文 献

[1] 李刚. 疯狂 Android 讲义[M]. 2 版. 北京：电子工业出版社，2014.

[2] 传智播客高教产品研发部. Android 移动应用基础教程[M]. 北京：中国铁道出版社. 2015.

[3] 李文琴，李翠霞，等. Android 开发与实践[M]. 北京：人民邮电出版社，2014.

[4] 黄宏程，胡敏，陈如松. Android 移动应用设计与开发[M]. 北京：人民邮电出版社. 2014.

[5] 张冬玲，杨宁. 计算机应用：Android 应用开发教程[M]. 北京：清华大学出版社，2013.

[6] 吴志祥，柯鹏，张智，等. Android 应用开发案例教程[M]. 武汉：华中科技大学出版社，2015.

[7] 郭霖. 第一行代码：Android[M]. 2 版. 北京：人民邮电出版社，2016.

[8] Android 中国开发者社区[EB/OL]. https://developer. android. google. cn/index. html.

[9] Android 入门基础教程[EB/OL]. http://www. runoob. com/w3cnote/android-tutorial-intro. html.

[10] Android 开发文档[EB/OL]. http://www. android-doc. com/androiddocs/.

[11] Android 官方培训课程中文版[EB/OL]. http://hukai. me/android-training-course-in-chinese/index. html.

[12] Android Studio 中文社区[EB/OL]. http://android-studio. org/.